LAND AND LAND APPRAISAL

LAND AND LAND APPRAISAL

Robert Orr Whyte

1976

Dr. W. JUNK b.v. — PUBLISHERS — THE HAGUE

ISBN-13: 978-90-6193-546-9   e-ISBN-13: 978-94-010-1577-6
DOI: 10.1007/ 978-94-010-1577-6

Cover design Max Velthuijs

CONTENTS

CONTENTS OF CHAPTERS

Page

PREFACE

This book is designed to present those principles and techniques for land appraisal which are applicable to all developing countries. Examples of specific situations in which these techniques have been or might be adopted are taken primarily from monsoonal and equatorial Asia. It is in this region that the land/food/population problem is most acute. It is also the writer's region of specialization; over the past ten years out of a total of some twenty-five years working in or closely concerned with Asia, an attempt has been made to examine the major problems of land potential in relation to rural economy and nutrition in the whole region, and in particular to show to what extent its different parts resemble or differ from each other.

The geographical scope comprises mainland southern, southeast and east Asia, from Pakistan to the People's Republic of China and Korea, with the insular monsoonal and equatorial lands of Sri Lanka, Indonesia, East Malaysia, Philippines, Taiwan and Japan (part).

International and bilateral agencies and specialists outside Asia repeatedly insist that Asia must learn to feed itself from the produce of its own land, or from imported foods paid for by the exports of primary and secondary commodities and of manufactured products to the developed world.

In assessing the present situation and the actual and potential increases in agricultural production, international agencies base their calculations on the deceptive assumption that the present standards of nutrition may be taken as a base point. However, from rural dietary surveys and other information, it can be seen that apart from a few more fortunate areas, the greater proportion of the people in the rural food-producing areas of monsoonal and equatorial Asia subsists on a diet which provides only some 50 per cent of minimal nutritional requirements. These people represent 85 per cent of the total population of Asia, or about half the population of the world, increasing at the rate of four million a month.

Thus, in more realistic terms of human population and nutrition, the actual demand is for a doubling of Asian production now, to bring standards of nutrition for the rural food-producers to morally and biologically acceptable levels, and for a second doubling of production to allow for increase of population by the end of the century, even assuming early and efficient introduction of methods of population control.

It is impossible to state whether these colossal targets can or cannot be achieved. So little is known of the extent and real capabilities of the land types of Asia; of the small areas of virgin land still awaiting survey to guide optimal use and conservation, of the vast areas of secondary vegetation producing nothing, conserving nothing, of the large areas under traditional (perhaps obsolete) dryland or wetland agriculture, of the grazing resources of semi-arid areas, and of all the lands that have become derelict through unplanned cropping and use of water, or following long sequences of excessive shifting cultivation.

No matter how carefully and accurately the land surveyors assess the true biological potentialities of the land – *The Economist* (22 March, 1975) claims that the world's agricultural resources, used efficiently, could feed a global population fifteen times greater than the present four billion – there is little evidence that the rural Asian, whatever his supposed ideological persuasion, is willing or able to make the greater efforts expected of him in the well-made plans of national and international agencies. Even when he does respond, he becomes exposed to the most efficient methods adopted for the extraction of foods produced in the rural areas for the benefit of urban communities and export. His poverty and low level of education render him unable to resist this lure, while farmers in a totalitarian state are subject to the additional burden of meeting obligatory levies. Thus the standards of nutrition of the rural family fall lower than before, particularly in respect of those types of animal protein that were formerly consumed in the village. Consideration is given in sections 2.7 and 4.3 to the role of rural man as a component of biological ecosystems and to his evolution to the status of a manager of economic ecosystems.

The World Bank (Robert S. McNamara, quoted in *Ceres* 37, 1974) states that the goal should be to increase the production of small farms, so that by 1985 their output will be growing at a rate of 5 per cent per year, thereby doubling their annual output between 1985 and the end of the century. 'But a 5 per cent rate of growth has never been achieved among smallholders in any extensive areas of the developing world.'

An attempt is being made to achieve self-sufficiency in rice in the Philippines, but the International Rice Research Institute has stated that the desirable percentage increases will not be obtained from their new cultivars until there is greater availability of water for irrigation at all seasons, and until these water resources are better managed in the fields. Throughout this book, great stress is laid upon the integrated approach, both in land appraisal itself, and in the application of the recommendations made by the assessors of land potential.

Through the fog and confusion created by constant changes in economic policies and political directives at the national level, the farmer has to make his personal decision regarding the optimal combination of inputs which he can afford and risk for his particular conditions; pending a major effort in extension, the farmer will continue to rely on his own judgement of the variables. Farmers are often said to be conservative and suspicious, 'but it is likely that their judgements are no worse than those of the average banker or stockbroker' (Philip Bowring in *Far Eastern Economic Review*, 28 March,

1975). And in any case, why should the specialist assume that he will be satisfied with the relatively meagre returns from a rice crop, when other crops such as sugarcane, plantation crops, industrial cash crops and even fodder crops when fed through productive bovines may be more profitable in cash return per hectare. Those who support land reform as a social and political goal may consider that it is undesirable to transfer land to various types of plantation crop if this results in the land's moving from tenant farmers to large holdings. The opposing school can refer to more efficient production on communes in the People's Republic of China or on commercial plantations elsewhere in Asia (see 4.3), to the greater national economic benefit.

The role of land appraisal in the overall dynamics of development is to indicate the technical steps which have to be taken to move the level of production from an existing plateau to the next above, and on upwards to optimal production in that specific environment. These steps from one plateau to another become progressively more difficult and more costly to achieve. The techniques to be applied for increased and more efficient production are well known, but what is not known is whether the rural cultivator, in his poor and undernourished condition, can be induced to see any advantage to himself from all the extra expense, risk and effort.

Thus it is all well and good to develop and to apply techniques for land appraisal; they will be of little value without a parallel and sympathetic appraisal of the rural Asian, in studies planned from a rural, not an urban base, on rural sociology, rural economics and rural psychology in relation to acceptance of change, first, for the benefit of the cultivator's own family, second, for his neighbours in the same rural ecosystem, and third, for urban communities and the nation.

Kota Baru, Kelantan, Malaysia
April, 1975.

1 CONCEPTS AND DEFINITIONS OF LAND AND LANDSCAPE

1.1 Expert Consultation on Land Evaluation for Rural Purposes, Wageningen, 1972

This Consultation was convened by FAO in co-operation with the University of Agriculture and the International Institute for Land Reclamation and Improvement, Wageningen. The objective was to encourage the standardization of methods of land evaluation, so that surveyors and users of information from resource surveys might talk the same language.

Discussions were based on proposals put forward in a background document prepared prior to the meeting by two interdisciplinary committees, one within FAO, the other in the Netherlands. The text of this document is published in the report of the Consultation, entitled Land Evaluation for Rural Purposes (Brinkman & Smyth, 1973).

Other documents prepared for the Consultation have been published separately in FAO Soils Bulletin No. 22: Approaches to Land Classification (1974). These papers describe approaches to the classification of agricultural land developed in different parts of the world (but with little reference to Asia). The aspects covered include those discussed in Chapter 6 below, and also in particular the land classification system adopted for irrigation projects by the Bureau of Reclamation of the U.S. Department of the Interior (3.5), land inventory in New York State (6.6) and also a presentation by Riquier of FAO (1974) of the parametric method, designed to express land evaluation in quantitative terms compatible with modern facilities for calculation, and to facilitate communication between the pedologist and the economist (see also 6.1).

Land means different things to different people. Land embraces the atmosphere, the soil and the underlying geology, the hydrology, and the plants on, above and below a specific area of the earth's surface. It also includes the results of past and present human activities as well as the animals within this area, in so far as they exert a significant influence on the present and future uses of the land by man.

Other more limited and overlapping concepts of land use can be identified: land as *space*, three-dimensional, unchangeable and fixed in quantity; land as *nature*, defined in terms of natural or man-made ecosystems influenced by natural processes; land as a *gene resource* (see Cain, 1970); land as a *production factor*, together with labour and capital; land as a *consumer good* or *commodity* as a support for highways, buildings, etc.; land as a source of *pleasure* and *recreation*; land as *location*, in modern economy and politics;

1

land as *property*, exerting so powerful an influence upon man's attitudes and actions; and finally, the related legal and economic connotation of land as *capital*.

A parcel of land may be suited to several uses at a given time. Its value does not depend on physical characteristics alone, but is greatly influenced by social requirements and economic considerations. The influence of the use of one tract of land upon other tracts in the same or adjacent land systems or catchment areas has also to be considered.

The Experts were asked to discuss the following questions: 1. for purposes of land evaluation, is it desirable to define land in terms of a broader range of natural environmental factors than has been customary in the past?
If so: 2. is the definition of land proposed above generally acceptable or in what way should it be improved? 3. within a broad definition, which concepts should be included in all evaluations and which should be considered under specific conditions only? 4. what are the practical consequences of using a broader definition of land in terms of required procedure for land evaluation and in terms of the required organization of institutions engaged in this work?

There was general agreement on the need to distinguish between 'soil' and 'land' in the context of resource evaluation. Soil is regarded as a term in natural science, kinds of soil being areas defined in terms of features relevant to their genesis and physical behaviour. Land is regarded as a term in social science, kinds of land being areas or tracts defined in terms of the features and relations relevant to their use for producing (or conserving?) something and to their value as property. Soil and land may differ not only in their breadth of concept, but also in terms of their identity with reality — mapped areas of soil representing concepts, those of land tending to represent unique tracts of the earth's surface.

A similarity was recognized between the broad concept of 'land' and that of 'ecosystem'. It is difficult to establish standardized terminology because of differences in language (see discussion by Zonneveld, 1972, p. 11), and in allocation of responsibility to national departments and services. It was suggested that vegetation should be excluded from the attributes of land in order to distinguish between the two concepts, land and ecosystem. Tracts of land are always characterized by more or less strict boundaries whereas ecosystems are usually centrally defined concepts. The concept of land may be close to that of agro-ecosystem or cultural ecosystem: ecosystems made or transformed by man. Socio-economic factors are excluded from the proposed definition of land on the understanding that they will be taken into account in the process of interpretative land classification.

To facilitate understanding of land and to ensure that appropriate studies are undertaken for development of irrigation, it may be preferable to describe land in terms of three main conditions: land surface (climate, topography, land cover of vegetation and stones); soil profile; sub-surface drainage (depth and quality of ground water, drainability).

2

1.2 Commonweath Scientific and Industrial Research Organization Division of Land Use Research[1]

In their presentation of the methodology of integrated surveys to the UNESCO Conference on Aerial Surveys and Integrated Studies, Toulouse, 1968, C. S. Christian & G. A. Stewart (1968) define 'the true resource: land'. While climate, topography, soil and vegetation each has its own individual significance in the biological habitat or ecosystem, and therefore in the development of agriculture, animal husbandry and forestry, it is the combined effect of all and the interactions between them which are important. The integrated surveys developed in Australia use the word 'land' to denote this complex (Christian & Stewart, 1947, 1953).

'The word land is used to refer to the land surface and all its characteristics of importance to man's existence and success. It is the integration of all such factors, rather than mere likeness or unlikeness in some of the more obvious observable characteristics, which determines the similarity or dissimilarity of aerial subdivisions in respect to land use potential (Christian, 1958).

'Land must be considered as the whole vertical profile at a site on the land surface from the aerial environment down to the underlying geological horizons, and including the plant and animal populations, and past and present human activity associated with it. There are many features in this total profile, some easily observable like the soil and vegetation, some measurable such as the rainfall and surface slope, but many not so readily observed or measured, such as the internal drainage and aeration features of the soil, the rate of chemical weathering and chemical migration in the soil, the level of microbiological activity, the variations in the micro-climates, and the precise impact of former land use by man. The many features of this total profile vary from site to site and their many combinations and interactions result in a vast array of land types, each with its own potential and limitations for agriculture or forestry, each presenting its own specific barriers to the achievement of

[1] In March, 1973, the CSIRO Executive carried out a re-organization of those Divisions concerned with land. The former Division of Land Research was renamed Division of Land Use Research, to indicate the addition to its previous broad character based on studies of land and water potential of a new socio-economic research group. Those sections in the old Division concerned with experimental field agronomy and crop physiology were transferred to other Divisions. A completely new Division of Land Resource Management, with headquarters in Perth, Western Australia, (Chief, R. A. Perry) was created on the basis of the former Rangelands Research Unit, concerned with the arid zones. These two Divisions, plus the already existing CSIRO Division of Soils, have been grouped into the Land Resources Laboratories, with Dr. E. G. Hallsworth (former Chief, Division of Soils) as Chairman (John McAlpine, personal communicaton, 1974).

However, since the approaches and techniques discussed in this and subsequent sections of the volume were developed by the former Division of Land Research, that title has been retained, except where reference is specifically made to activities of the new Division of Land Use Research.

maximum plant or animal production. Rarely does one feature alone determine productivity. It is the combination of all that is important, and if we are to understand land we must think of it in terms of this complex rather than only of the individual components of it' (Christian, 1963).

It is this complex, which constitutes the habitat at each and every site, which is the true resource, not the individual resource factors of which it is composed. It is the variation in the complex from place to place rather than the variation in individual factors that provides different potentials for development. Thus, attention is focused on the significance of the combination of factors in specific tracts of country, that is, on areas of distinctive 'land types' — the basic unit for the planning of development.

The land surface at one location is regarded as the end product of evolution involving the action of a combination of physical and biological processes, acting on a particular type of geological parent material at certain rates for given periods in certain geological climates (Christian, Jennings & Twidale, 1957). In this process of evolution, particularly in the more recent phases, the shaping of the land form, the development of the soil profile, and of surface and sub-surface drainage features, and the occupation, utilization and modification of habitats by the biological components, have gone on simultaneously. Land is thus a dynamic feature; when one describes the land complex at a site, it is merely a description of the state of the land surface at one moment in a slow evolutionary sequence (Christian & Stewart, 1968).

A site is a part of the land surface which is uniform throughout its extent in land form, soil and vegetation. The small amount of variability which occurs within a site is of such a low order that the variation falls within the units of classification used in each discipline (Christian & Stewart, op. cit.). Each site represents a distinctive type of environment and at each of its occurrences it will provide a similar range of habitats for plants, animals and man. A site is a unit of landscape, a taxonomic land type, which for most practical purposes can be regarded as presenting similar problems for land use wherever it occurs. Although the individuality of sites has to be considered in subsequent technical studies, it is the larger areas of land represented by land units or land systems which will form the basis of development planning (see also discussion of the Australian approach by Zonneveld, 1972, pp. 28-29 and discussion of the approach in the islands of the Pacific, 6.2).

1.3 International Institute for Aerial Survey and Earth Sciences, Netherlands

Concepts of land, landscape, environment and ecosystem are given by Zonneveld (1972). The terms 'land' and 'landscape' occur in various languages belonging to the Germanic group, in which they have in the course of time acquired certain differences in meaning, some slight, others important. The German *Land* implies a neutral indication of a certain (usually large) area,

4

especially in an administrative or political sense. *Landschaft* was originally identical with *Land*, but has developed since the Renaissance (under the influence of art and science) into a more comprehensive and holistic concept, almost identical with 'the environment' (see also 1.7).

Certain soil scientists and certain ecologists claim to have developed the best method of arriving a comprehensive description of the environment (i.e. of the land). The dominating influence of soil survey (in physiographic schools) on land evaluation has affected the concept of *land*. Land classification comes to be considered as a sub-discipline of soil science, almost synonymous with applied soil science. 'The usurpation of pragmatic land classification by soil science is particularly intolerable from the point of view of integrated surveys. Undoubtedly soils play an important role in evaluating the land for agricultural purposes. However, pragmatic land evaluation must take into account all land attributes — rock, relief, vegetation, water, animals and man — including the results of soil survey and soil science, for evaluation and classification of the environment. Within the scope of the integrated survey, it would be best not to associate too wide a range of meanings with the word *land*.'

The term *land* may be used in integrated surveys, and especially in pragmatic land classification, as comprehensively as possible, synonymous with environment and thus with landscape. Any further separation would be artificial. The term *landscape* may also be used in combinations for which the word *land* alone cannot be used, such as soil landscape, geological landscape, geomorphological (landform) landscape, to indicate a certain arrangement of soils, rocks or relief forms in the environment.

1.4 Food and Agriculture Organization of the United Nations, Soil Resources, Development and Conservation Service

The meaning to be given to the term *land* is a fundamental issue which needs to be decided before international standards of land evaluation can be evolved (Smyth, 1971). No precise definition has yet found general acceptance; views differ on the distinction between *land* and *soil*, and on the extent to which the concepts of *land* embrace social and economic attributes. In the CSIRO definition, characteristics of a purely social or economic nature are omitted (but see J. McAlpine, personal communication 1974, in 6.2). However, the actual process of interpretative land classification will surely always include some consideration of socio-economic criteria.

Land is broader in concept than the broadest interpretation of *soil*, since it includes all attributes in addition to soil essential to land evaluation, e.g. macro-topography, vegetation, surface and groundwater hydrology, climate, and size, shape and location of the tract being studied. However, some consider that soil is the one environmental factor that is both stable enough to form a base for land classification and is yet sufficiently flexible for deliberate change by man. One cannot, however, continue to regard interpretative classification of soil and of land as almost synonymous.

See also background document to Land Evaluation for Rural Purposes (1.1).

'The most valuable classifications of land are those which identify needed improvement and which predict potential rather than present suitability and, as Vink (1960) has pointed out, soil is the one environment factor that is stable enough to form a base for land classification while at the same time offering flexibility to man's influence' (Smyth, 1972).

1.5 Land Resources Division, Overseas Development Administration (United Kingdom)

Although the Division operates primarily in Africa, the Americas and the Pacific, the reports of its many surveys for the assessment and development of resources would be of great value to teachers, students and practitioners in Asia (see also below in 3.6.3 and 6.2.3).

The Australian definition of *land* and the CSIRO Division of Land Research division of a surveyed region into land systems have been adopted, with the addition of *land facets*; these are significant in terms of existing and potential land use, and measurement of the areas occupied by them will give accurate estimates of the surface area of the more promising types of land (Baulkwill, 1972).

1.6 The approach of the biogeographer

Dansereau (1957) has defined biogeography as comprising plant and animal ecology and geography, with many overlaps into genetics, human geography, anthropology and the social sciences. The ecological perspective embraces the entire field of the sciences of the environment, sometimes called biogeography, and involving the study of the origin, distribution, adaptation and association of plants and animals. These studies may be conducted at the five levels indicated in Table 1.1 from Dansereau (op. cit.)

Although biogeography has connections with ecology, geography, geology, evolutionary history and economic anthropology (Cox, Healey & Moore, 1973), it is usually studied in conjunction with only one or two of these disciplines. Biologists are rarely satisfied with the mere description of patterns of distribution of organisms in space and time. The biogeographer wishes to discover which environmental factors determine or restrict the distribution of species. It is then necessary to draw upon knowledge from many disciplines, including geology, geophysics, climatology, palaeontology, plant and animal systematics and taxonomy, evolution, physiology and ecology. How do the

6

Table 1.1. A comparison of the criteria and units involved at each integrative level in the study of environmental processes and relations. (Dansereau, 1957).

Level	Sciences Directly Involved	Affinities with Other Sciences	Material Studied	Object of Research	Nature of the Limitations	Methods of Study	Conclusions	Units
I. HISTORICAL	HISTORICAL PLANT AND ANIMAL GEOGRAPHY (AREOGRAPHY)	Geology Evolution Phylogeny Paleoclimatology Taxonomy Geography Paleontology	Phyla to Species	Origin, expansion, and decadence Movement in relation to climatic change Distribution Areal affinities	Geological events, Paleoclimatological fluctuations	Excavation of fossils. Analysis of strata Location of relic areas Plotting of areas Comparison of areas	Evolutionary trends and sequences Occupancy of areas over periods of time Extension Disjunction and former continuity	Fossil floras Fossil faunas Isoflors Floras Faunas Types of area
II. BIOCLIMATOLOGICAL	BIOCLIMATOLOGY	Climatology Meteorology Vegetation science	Species to races Plant formations	Behaviour in relation to climatic area	Climate or climatic factors	Mapping of coincidences of area Study of varvs, pollen profiles, tree rings	Responsibility of individual meteorological factors; cycles	Vegetation zones and formations Life zones Climatic types Isophenes Isobiochores
III. SYNECOLOGICAL	SYNECOLOGY	Autecology Physical geography Pedology Vegetation science	Vegetation Animal populations Communities	Composition, structure and dynamics of communities	Nature of ecosystem, from habitat to biotope	Physiognomic observation Quadratting	Type of association Nature and orientation Static, dynamic, and area description of units	Ecosystems or communities ordained and classified
IV. AUTECOLOGICAL	AUTECOLOGY	Physiology Genetics Anatomy	Species to races	Reaction to habitat factors considered singly or holocenotically	Chemical, physical, and biological factors	Direct measurement of responses Experiment	Nature and gravity of immediate limitations. Extension and depth of possible reaction to individual factors	Ecotypes
V. INDUSTRIAL	HUMAN ECOLOGY	Anthropology Agriculture Forestry Human geography Sociology History	Landscape	Influence of man	Human intervention	Historical recording	Nature, gravity, and duration of disturbance	Land use and resource utilization types

physical environment, the biology of a species and its evolutionary history interact to bring about its present pattern of distribution, and to determine the direction of its future evolution?

The approaches and techniques which may be adopted in research on the interaction of organisms and environment are described in a series of books entitled Studies in Ecology, edited by D. J. Anderson, P. Greig-Smith & F. A. Pitelka, published by Blackwell Scientific Publications (see also Tivy, 1971). These cover (in addition to the text on biogeography noted above, Cox, Healey & Moore, 1973); insect/plant relationships, management of animal and plant communities for conservation, mathematical models in ecology, insect populations, mineral nutrition of plants, measurement of environmental factors in terrestrial ecology, transport processes and plant growth, and the African savanna.

For example, in the first volume in this series, on the quantitative analysis of plant growth (Evans, 1972), the section on experimental techniques deals with the choice of the material for study, principles of experimental design, measurement and control of the aerial and root environments, harvest and measurement of respiration. Under analysis of data are discussed main concepts, first analysis of harvest data, relative growth rate, unit leaf rate, leaf weight ratio, stem/root weight ratios, dry weight/fresh weight ratios (see also Wareing, 1970).

The approach of the biogeographer will obviously underline the human activity and biotic factors as main components of the ecosystems, resulting in various landscapes and land uses (P. Legris, personal communication, 1974). 'But the biogeographer has to take the greatest account of their biological consequences on the equilibrium of other components as the vegetation and soil life and therefore he must base his study on the results of biological researches.'

The central concept in the symposium on the ecology and biogeography of India edited by M. S. Mani (1974) and comprising contributions by specialists in geology, meteorology, botany, zoology, ecology, and anthropology, is that the biogeographical and geomorphological evolution of India constitutes an integral whole, and that the flora and fauna and the distributional patterns that are observed today represent a dynamic phase of this complex evolution. The beginnings of India's biogeographical evolution must be sought, according to Mani (op. cit.), in Madagascar, Indochina and Malaya (see also Whyte, 1975, on the evolution of the Gramineae in particular, of western Monsoon Asia). The ecology of nearly the whole of India, Mani continues, is dominated by the rhythm of the monsoon-rainfall climate supporting a tropical flora and fauna, but containing also numerous remarkable pockets of temperate floras and faunas. The changes in character of the flora and fauna and the distributional patterns and ranges of plants and animals, brought about by the extensive destruction of natural habitats by civilized man are mirrored by the changes in the ecology and status of the primitive communities of man himself in

India. With the influx of civilized races from the northwest and under increasing pressure of their continued advance, aboriginal man in India steadily receded to the small isolated refugial pockets in dense and inaccessible forests, where he is found today as tribal man.

1.7 The approach of the geographer

The writings and teaching of geographers tend to be concerned with the evolution and present status of human societies and cultures at the various stages of development of their economic ecosystems. The land is considered as a background to these presentations, rather than as the essential basis for future evolution and change (Grigg, 1969). Those concerned with estimation of land capability, with the assessment of economic potential and with the definition of successive plateaux of production will use geographical texts as a starting point for their planned progress from a base. This applies more particularly to monsoonal and equatorial Asia, where the land is already used intensively, and where human, social and economic factors govern the extent to which the well-made plans of the land scientist may be achieved.

Brief extracts from the contents of some geographical texts relating to Asia in particular will show the extent to which they relate to the land sciences.
Jules Sion: Asie des Moussons, Part 1, China and Japan (1928).
Climate (mechanism of the monsoons, classification, equatorial, hot and temperate climates, consequences of the monsoons); Vegetation and crops; Relief of China and Japan; Ways of life (nomads and semi-nomads, hunters and fishers, sedentary cultures, sedentary/pastoral cultures).
E. H. G. Dobby: Monsoon Asia (1961).
Monsoon Asia regarded as a whole, and as Eastern Asia, Southeast Asia and Southern Asia; Landforms, topography and physical patterns; Climatic environment; Biogeography; Peoples, races and ways of life; Rural land use.
T. R. Tregear: A Geography of China (1965).
Physical (structure and relief, climate, soils, natural vegetation); Historical; Economic and Social; Regions.
H. Robinson: Monsoon Asia (1966).
Characteristics and differentiation of the monsoon region: Physical background (physiography, climate and climatic regions, biogeography, soils): Resources and economy (water, land use and farming); Human background (peoples and population, cultural features, political geography).
Pierre Gourou: The Tropical World (1966).
Population density, human health in hot, wet regions, soils, types of agriculture, livestock husbandry, human diets, examples from hot, wet Asia.
Tuan Yi-Fu: The World's Landscapes: 1. China (1970).
Landforms; Climate; Vegetation; Man's effect on the land; Landscape and life

in Chinese antiquity; and in Imperial China: Tradition and change in modern China: Communist ideology and landscape.

O. H. K. Spate & A. T. A. Learmonth: India and Pakistan (1967).
The South Asian subcontinent as a unit of geographical study; triple tectonic division, peninsula, Himalayan orogeny, Indo-Gangetic plains; climate; vegetation and soils; people, historical evolution, rural and urban; the economy, agriculture and agrarian problems.

O. H. K. Spate. A. T. A. Learmonth & B. H. Farmer: India, Pakistan and Ceylon (1967).
The regions

Pierre Gourou: Leçons de Géographie tropicale (1971).
Geographical problems in tropical Asia, man and the physical environment, population density, agriculture and animal husbandry (see also study of human geography in the delta of Tonkin, Gourou, 1965; and Gourou Festschrift, 1972, which deals, among other things, with agriculture in the Himalaya, Pakistan, Thailand, Cambodia and China).

J. E. Spencer & W. L. Thomas: Asia East by South (1971).
Cultural geography, peoples, languages and populations, settlements, human health and disease, geomorphology, climatology, marine life, and fauna, soils, plants and agricultural systems, regional expression of cultures.

H. C. Brookfield & D. Hart: Melanesia (1971).
An innovative attempt to apply theoretical techniques to a better understanding of a region; physical landscape, population, culture history, noncommercial Melanesian agriculture, commercial agriculture, mining, the organization of production and distribution in a dualistic economy, organization of society, economic patterns, trading and transport networks, regionalization of economy and society around central places.

1.7.1 Contribution of geographers to the study of land

In his paper on the terminology which has evolved in the discipline of landscape ecology and biogeocenology, Troll (1971) quotes the statement of Sukachev (1953):
'Scientists both of the U.S.S.R. and the West independently came to the conclusion that it was necessary to base the studies of geographical phenomena on the surface of the earth on the related conceptions of the biogeocenosis, ecotope, etc. Unification of opinion in this field on correct methodological bases is quite necessary'.

Troll (1971) hopes for a similar collaboration. For the understanding of the essentials of regional units, relevant to natural regionalization, geography needs the deeper ecological knowledge; on the other hand, ecology should increasingly extend its purview from site analysis to regional differentiation and the

10

mapping of life associations. Out of this, a more rounded earth and life research will develop, to which Troll gives the name 'ecoscience', to distinguish it from 'geoscience', applicable only to the inanimate lithosphere (see also 2.1).

Geography is seen as having progressed in the last two decades from the study of the individual phenomena of nature (climate, landform, waters, soils, plant cover and animal world) to a synoptic consideration of natural phenomena and to a functional investigation of the predominant interrelations between the individual (*Landschaft*) elements in nature. Landforms contribute to the division of the climatic elements: temperature, humidity and rainfall. They are the causes of the modification of macroclimate to topo- or mesoclimate. Nevertheless, morphogenesis is affected not only by variations in lithology and in structure, but also by the character of the climate. Climate, rock and water budget are the basis of plant life. Plant cover affects the water budget, pedogenesis and the differentiation of microclimates. The life communities of plants, animals and micro-organisms are controlled by all the environmental factors.

Thus, Troll (1968) defines *Landschaft* ecology as the study of the main complex causal relations between the life communities and their environment in a given section of a *Landschaft*. These relations are expressed regionally in a definite distribution pattern (*Landschaft* mosaic, *Landschaft* pattern), and in a natural regionalization at various orders of magnitude. An individual *Landschaft* of small entities that recur as types and, like a stone (*Fliese*) in a mosaic, form a characteristic community, is called an ecotope, a development of the old term biotope (Troll, 1971).

Geoecology is defined (Troll, 1972) as the science of the full and complex interrelations between the organisms or biocoenosis and their environmental factors. In such studies, one may concentrate on the regionalization of ecosystems or upon the spatial arrangement of ecotopes (sites, facies, etc.) in a geographical region or landscape pattern. This view leads geographers from the global climatic, vegetation, soil and landscape belts to smaller and smaller landscape units down to the ecotopes at the lower end of the hierarchy. This is preponderantly a geographical or regional approach.

Alternatively, functional and quantitative analyses may be made on single ecosystems regarding the full interrelations between biotic and non-biotic elements (macroclimate, basic rock material, soil profile, ground and soil water, landform, topo- or mesoclimate, biocoenosis with plants, animals and micro-organisms, microclimate, soil climate, etc.). Within each ecosystem. there is exchange of matter and energy between soil, air, water and organisms classified as producers, consumers and reducers. This is the predominantly biological view.

Since 1945, the Bundesanstalt für Landeskunde und Raumforschung has worked with university geographical departments on the mapping of the natural regions of Germany. The functional analyses of individual *Landschaften* and ecotopes from the ecosystem standpoint have not been undertaken, since the necessary measurements could not be made by geographers, further evidence of the need for interdisciplinary research (Troll, op. cit.).

A British counterpart may be seen in papers to a seminar on resource management sponsored by the Social Science Research Council, the Royal Geographical Society and the Institute of British Geographers, summarized by Coppock (1972), and published in full in the *Journal of Environmental Management*. The main objective was to identify topics where research by geographers might be profitably pursued. From the following extracts from Coppock's article (op. cit.) it will be seen that there is an apparent marked difference in the approach of the European and British geographers respectively; the former being more concerned with developing a stronger association in field studies between geographers and the land sciences, the latter considering that the important role of the geographer is in transmitting the technical results of the field workers into terminology, statistics and models which might be understandable to the decision-makers.

In fact, E. M. Yates states in his review of Troll (1972) in *The Geographical Journal* 139, 545-6, 1973, that *Landschaft* science has been generally misinterpreted in Britain and dismissed as sterile description. Although the blending into geoecology has been one of the major advances made in geography, the significance of the advance has been overlooked in Britain, because of the undue attention given to the application of statistical method in the so-called 'new geography'.

According to Coppock (1972) there has been in geography, as in most physical planning, an implicit assumption either that resources manage themselves, or that the decisions of individual resource managers are of little importance. Geographers will now have to take account of increasing public awareness and other changes. The use of natural resources is the meeting place of natural systems and human decisions, and therefore offers an opportunity to unite both branches of the subject.

Five conditions may be identified for ecologically responsible and economically and socially responsive management of natural resources (J. W. Birch to the above seminar), namely, an understanding of the structure and functioning of resource-using systems; an ability to specify and evaluate alternative resource adjustments; an appreciation of the gap between the scientific appraisal of these adjustments and the decision-makers' perception of them; an understanding of the objectives of resource management, as seen by users; and the ability to develop computable models so that solutions could be tested in advance, and subsequently monitored. Geographers might seek to devise composite indices of physical qualities and of land capability. Systems approaches had been usefully applied to river basins, but most systems of interest to geographers were generally less coherently organized. The growing emphasis on systems analysis in ecology and other environmental sciences is likely to have a direct impact on resource management and land use (J. N. R.

Jeffers). Land use is one aspect of resource management where important contributions could be made by the use of mathematical models.

There are five areas where geographers, either because of past contributions or because of the nature of their training, could contribute to resource management (W. R. D. Sewell, to same seminar), by: developing techniques for assessing the magnitude of problems; delineating alternative strategies, especially where human adjustment is involved; examining past performance of particular strategies; determining effective ways of identifying public views; and developing techniques of evaluation, considering multiple goals and multiple means.

The role of the geographer is to provide background data for the decision-makers (P. M. S. Jones), particularly on the climatic consequences of environmental change, the availability of natural resources in the future, and changes in rural communities. Geographers may not have been sufficiently involved in the processes of decision-making (as, for example, the economists have) to be fully aware of the issues involved; 'but unless they are able to demonstrate that they have something distinctive to offer in a field which is so relevant to their discipline, they may well find their role assumed by others' (Coppock, 1972).

BIBLIOGRAPHY

BAULKWILL, W. J., – 1972 – The Land Resources Division of the Overseas Development Administration. *Tropical Science* 14: 305-322.

BEERS, H. W., – 1970 – Indonesia: Resources and their Technological Development. University of Kentucky Press, Lexington. 282 pp.

BRINKMAN, R. & A. J. SMYTH (eds.), – 1973 – Land Evaluation for Rural Purposes: Summary of an Expert Consultation, Wageningen, The Netherlands, 6-12 October, 1972. International Institute for Land Reclamation and Improvement, Wageningen. 116 pp.

BROOKFIELD, H. C. & D. HART, – 1971 – Melanesia: A Geographical Interpretation of an Island World. Methuen, London. 464 pp.

CAIN, S. A., – 1970 – Preservation of natural areas and ecosystems: protection of rare and endangered species. In 'Use and Conservation of the Biosphere,' pp. 143-153. UNESCO, Paris.

CHRISTIAN, C. S., – 1958 – The concept of land units and land systems. In 'Proceedings of Ninth Pacific Science Congress,' volume 20, pp. 74-81.

CHRISTIAN, C. S., 1963 – The use and abuse of land and water. In 'The Population Crisis and the Use of World Resources,' vol. 2, pp. 387–406. World Academy of Art and Science, Junk, The Hague.

CHRISTIAN, C. S., J. N. JENNINGS & C. R. TWIDALE, – 1957 – Geomorphology. In 'Guide Book to Research Data for Arid Zone Development,' pp. 51-65. Arid Zone Research 8, UNESCO, Paris.

CHRISTIAN, C. S. & G. A. STEWART, – 1947 – North Australia Regional Survey, 1946. Katherine-Darwin Region. General Report on Land Classification and Development of Land Industries. CSIRO, Melbourne. mimeo.

CHRISTIAN, C. S. & G. A. STEWART, – 1953 – General Report on Survey of Katherine-Darwin Region, 1946. CSIRO, Australian Land Research Series no. 1.

CHRISTIAN, C. S. & G. A. STEWART, – 1968 – Methodology of integrated surveys. In 'Aerial Surveys and Integrated Studies: Proceedings of the Toulouse Conference,' pp. 233-280. UNESCO, Paris.

COPPOCK, J. T., – 1972 – Research in resource management: a joint SSRC/RGS/IBG Seminar. *Geogr. J.* 138: 466-469.

COX, C. B., I. N. HEALEY & P. D. MOORE, – 1973 – Biogeography: An Ecological and Evolutionary Approach. Blackwell, Oxford. 192 pp.

DANSEREAU, P., – 1957 – Biogeography: An Ecological Perspective. Ronald Press, New York. 394 pp.

DOBBY, E. H. G., – 1961 – Monsoon Asia. University of London Press, London. 381 pp.

ELLENBERG, H., – 1971 – Ecological Studies. Analysis and Synthesis. vol. 2. Integrated Experimental Ecology. Lange and Springer, Berlin, Heidelberg, New York. 214 pp.

EVANS, G. C., – 1972 – The Quantitative Analysis of Plant Growth. Blackwell, Oxford. 734 pp.

FISHER, C. A., – 1968 – Social and economic consequences of western commercial agriculture in south-east Asia. In 'Land Use and Resources: Studies in Applied Geography. A Memorial Volume to Sir Dudley Stamp,' pp. 147-154. Institute of British Geographers, Special Publication 1, London.

FOOD AND AGRICULTURE ORGANIZATION OF THE UNITED NATIONS, – 1974 – Approaches to Land Classification. Soils Bulletin no. 22. FAO, Rome, 120 pp.

GOUROU, P., – 1966 – The Tropical World (4th ed.). Longmans, London.

GOUROU, P., – 1971 – Leçons de Géographie tropicale. Mouton, Paris, The Hague. 323 pp.

GRIGG, D., – 1969 – The agricultural regions of the world: review and reflections. *Econ. Geogr.* 45: 95-132.

JACKSON, J. C., – 1968 – Sarawak: A Geographical Survey of a Developing State. London University Press, London. 218 pp.

KING, R. B., – 1970 – A parametric approach to land system classification. *Geoderma* 4: 37-46.

LO, C. P., – 1971 – Modern use of aerial photographs in geographical research. *Area* 3: 164-169.

MacARTHUR, R. H., – 1972 – Geographical Ecology: Patterns in the Distribution of Species. Harper and Row, New York and London. 270 pp.

McVEY, R. T. (ed.), – 1963 – Indonesia. H.R.A.F., New Haven. 600 pp.

MANI, M. S. (ed.), 1974 – Ecology and Biogeography in India. Junk, The Hague. 775 pp.

OOI, Jin-Bee – 1963 – Land, Peoples and Economy in Malaya. Longmans, London. 426 pp.

PELZER, K. J., – 1970 – Geographical literature on Indonesia. In 'Indonesia: Resources and their Technological Development,' Southeast Asia Studies Reprint Series 42. Yale University, New Haven.

RIQUIER, J., – 1974 – A study of parametric methods of soil and land evaluation. In 'Approaches to Land Classification': Soils Bulletin no. 22, pp. 47-53. FAO, Rome.

RIQUIER, J. & D. C. SCHWAAR, – 1972 – Parametric approach to soil and land capability classifications. In 'Proceedings Second ASEAN Soil Conference, Djakarta, Indonesia.'

ROBINSON, H., – 1966 – Monsoon Asia. MacDonald and Evans, London. 559 pp.

SION, J., – 1928 – Asie des Moussons. I. Généralités, Chine, Japon. Armand Colin, Paris. 272 pp.

SMYTH, A. J., – 1971 – The aims and possibilities of standardizing approach and presentation in the evaluation of land resources. In 'Systematic Land and Water

Resources Appraisal. FAO Latin American Land and Water Development Bulletin no. 1,' pp. 72-80. FAO, Santiago.

SMYTH, A. J., – 1972 – Interpretative classifications of land and soil in land development. In 'International Geography, 1972. Papers of the Twenty-Second International Geographical Congress, Canada,' vol. 1., pp. 279–281. University of Toronto Press.

SPATE, O. H. K. & A. T. A. LEARMONTH, – 1967 – India and Pakistan: Land, People and Economy. (3rd ed.). Methuen, London. 439 pp.

SPATE, O. H. K., A. T. A. LEARMONTH & B. H. FARMER, – 1967 – India, Pakistan and Ceylon: The Regions. (3rd ed.). Methuen, London. 877 pp.

SPENCER, J. E. & W. L. THOMAS, – 1971 – Asia, East by South: A Cultural Geography (2nd ed.). Wiley, New York. London, Sydney, Toronto. 669 pp.

STAMP, L. D., – 1962 – Asia: A Regional and Economic Geography. (11th ed.). Methuen, London. 730 pp.

STEEL, R. W. & R. MANSELL PROTHERO, – 1964 – Geographers and the Tropics. Longmans, London.

SUKACHEV, V. N., – 1953 – On the exploration of the vegetation of the Soviet Union. In 'Proceedings Seventh International Botanical Congress, Stockholm, 1950,' pp. 659-660.

SUKACHEV, V. & N. DYLIS, 1964 – Fundamentals of Forest Biogeocoenology. Oliver and Boyd, Edinburgh. 672 pp.

TIVY, J., – 1971 – Biogeography: A Study of Plants in the Ecosphere, Oliver and Boyd, Edinburgh. 432 pp.

TREGEAR, T. R., – 1965 – A Geography of China. University of London Press. 342 pp.

TROLL, C. (ed.), – 1968 – Geoecology of the Mountainous Regions of the Tropical Americas. Proceedings of the UNESCO Mexico Symposium, 1966. Colloquium geogr. 9.

TROLL, C., – 1971 – Landscape ecology (geoecology) and biogeocenology – a terminological study. *Geoforum* 8: 43–46.

TROLL, C., – 1972 – Geoecology and the world-wide differentiation of high-mountain ecosystems. *Erdwissenschaftliche Forschung* 4: 1-16.

TUAN, Yi-Fu, – 1970 – The World's Landscapes. 1. China. Longmans, London. 225 pp.

VINK, A. P. A., – 1960 – Quantitative aspects of land classification. Paper v.52, Trans. Seventh International Congress of Soil Science IV (Madison): 371–378.

WAREING, P. F., – 1970 – Plant science and food production. *The Advancement of Science* 27: 1-10.

WHITE, G. F., – 1966 – The world's arid areas. In 'Arid Lands,' ed. E. S. Hills, pp. 15-30, Methuen/UNESCO, London and Paris.

WHYTE, R. O., – 1975 – The Gramineae of western monsoon Asia: their antiquity and evolution. *Biotropica* (in press).

WINSTEDT, F. L. & J. E. SPENCER, – 1967 – The Philippine Island World. California University Press. 742 pp.

WINT, G. – 1966 – Asia. Anthony Blond, London. 872 pp.

ZONNEVELD, I. S., – 1972 – ITC Textbook of Photo-Interpretation. Vol. 7. Use of Aerial Photographs in Geography and Geomorphology. Chapter 7. 4: Land Evaluation and Land(scape) Science. ITC, Enschede. 106 pp.

ZONNEVELD, I. S., P. N. de LEEUW & W. G. SOMBROEK, – 1971 – An Ecological Interpretation of Aerial Photographs in a Savanna Region in Northern Nigeria. ITC Publication Series B no. 63. Enschede. 41 pp.

2 THE RURAL ECOSYSTEM AS A BIOLOGICAL ENTITY AND AN ECONOMIC RESOURCE

2.1 Concepts and Definitions

A. G. Tansley (1935) first introduced the term ecosystem: 'The more fundamental conception is . . . the whole system (in the sense of physics) including not only the organism complex, but also the whole complex of physical factors forming what we call the environment . . . We cannot separate
them (the organisms) from their special environment with which they form
one physical system. . . . It is the systems so formed which . . . (are) the
basic units of nature on the face of the earth. . . . These ecosystems, as we
may call them, are of the most various kinds and sizes.'

In a later interpretation, Evans (1956) extends the application of the concept and the term, which Tansley applied specifically to the level of biological
organization represented by units such as the community and the biome, to
include organizational levels other than that of the community. An ecosystem
involves the circulation, transformation and accumulation of energy and
matter through the medium of living things and their activities. The ecologist
is primarily concerned with the quantities of matter and energy that pass
through a given ecosystem, and with the rates at which they do so. Of almost
equal importance are the kinds of organisms that are present in any particular
ecosystem and the roles that they occupy in its structure and organization.

'Ecosystems are further characterized by a multiplicity of regulatory
mechanisms, which, in limiting the numbers of organisms present and in
influencing their physiology and behavior, control the quantities and rates of
movement of both matter and energy. Processes of growth and reproduction,
agencies of mortality (physical as well as biological), patterns of immigration
and emigration, and habits of adaptive significance are among the more important groups of regulatory mechanisms. In the absence of such mechanisms, no
ecosystem could continue to persist and maintain its identity.

'The assemblage of plants and animals visualized by Tansley as an integral
part of the ecosystem usually consists of numerous species, each represented
by a population of individual organisms. However, each population can be
regarded as an entity in its own right, interacting with its environment . . .
to form a system of lower rank that likewise involves the distribution of
matter and energy. . . . The ecosystem thus stands as a basic unit of ecology,
a unit that is as important to this field of natural science as the species is to
taxonomy and systematics. In any given case, the particular level on which the
ecosystem is being studied can be specified with a qualifying adjective — for
example, community ecosystem, population ecosystem, and so forth. All ranks

16

of ecosystems are open systems, not closed ones' (Evans, op. cit.).

The patterns of the earth are expressed as expanses of forests, grasslands and croplands, as rivers and lakes, estuaries and oceans. 'Each is physically and biologically different. Each is occupied by different organisms well adapted to the environment in which they are found. Yet, in spite of the differences, oceans and lakes, forests, grasslands and deserts all function the same. Energy fixed by plants flows through them. Nutrients are deposited in the tissues of plants and animals, cycled from one feeding group to another, released by decomposition to soil and water and recycled again. Rarely are the desert or the forest, the stream, the lake or the sea independent of one another' (Smith, 1972).

Troll (1971, see also 1.7.1) gives a concise account of the origin, development and present meanings and usages of many of the terms in European literature within the scope of the subjects of landscape ecology (geoecology) and biogeocenology, from 'ecology', 'synecology' and 'ecosystem' onwards. Ecology may be defined (Margalef, 1968) as the study of: 'systems at a level in which individuals or whole organisms may be considered elements of interaction, either among themselves, or with a loosely organized environmental matrix. Systems at this level are named ecosystems, and ecology is the biology of ecosystems. . . . Every system is a set of different elements or compartments or units, any one of which can exist in many different states, such that the selection of the state is influenced by the states of the other components of the unit. Elements linked by reciprocal influences constitute a feedback loop.'

Biology distinguishes the following levels in the organization of living substance: the biologically active macro-molecules, the cell, the organism or the species, and the population (UNESCO, 1970). Further definitions are given by McMillan (1960, 1971): a. the terms ecotype, ecotypic differentiation, ecogenetic gradient, refer to genetically based variation that is correlated with habitat, b. community is applied to the sum total of organisms in a given area, c. population is applied to the one or more individuals of a close genetic lineage in a given area, d. ecosystem is applied to the sum total of organisms (the community and its included populations) and to their relations with their environmental surroundings in a given area, and e. ecosystem-type results from the lumping or sorting of certain ecosystems into a particular kind.

Expanding this further, with reference to ecotypes and the functioning of ecosystems composed of grasses and other plant species, McMillan (1971) states (see also Wiens, 1972; Ellenberg, 1971); a. the role of the ecotype in ecosystem function is primarily one of allowing a community of organisms to adjust to its habitat, b. the simultaneous selection of ecotypic variants within different kinds of organisms occupying a given area results in harmonious functions of a particular ecosystem, and c. the selection of ecogenetic gradients results in the continuity of an ecosystem-type over geographic diversity.

17

It is possible to recognize a hierarchical sequence in ecosystems (Ellenberg, 1973), starting with the biosphere as a whole, and descending through the mega-ecosystems (the natural marine, lacustrine, semi-terrestrial and terrestrial, the artificial urban-industrial), macro-ecosystems (forests), meso-ecosystems, micro- and nano- ecosystems. From this hierarchical sequence, Ellenberg proceeds to classify ecosystems in relation to the nature and degree of man's effect on them, from slight to destructive. The types of ecosystems and partial systems named by him, as well as the gradations of man's influences, are independent of floral and faunal geographical data, since they are the outcome of the quality of the abiotic environment and general use of the land. They certainly show a parallel with the vegetation zones in formation of different ranks, to the extent that the prevailing plant life forms are used. These life forms are not represented throughout the world by the same species (or other taxa), since the floras and faunas of the individual regions of the earth have remained more or less separate and have developed independently. The ecosystems of different parts of the earth thus also differ taxonomically from one another, when all abiotic considerations are the same. On the basis of known plant and animal subdivisions of the globe, it is possible to recognize the following biogeographical kingdoms: tropoamerican, tropoatlantic, tropoafrican, tropoasiatic, tropopacific, Australian, the Cape (South Africa), nearctic, and holarctic.

Each of these kingdoms embraces biogeographical territories, provinces etc. that one may best denote by further decimalized detail. These geographical facies are needed only for considering global comparisons; they can be ignored at the regional level, since at that level they are the same for all ecosystems.

Projected analyses of ecosystems in relation to primary productivity must include consideration of the ecotypic status of component populations, and an assessment of the contribution of the ecotype to ecosystem function. 'But the ecosystem level (biogeocenose) is a still higher manifestation of life organization in its stability in geological time. The biosphere as a total combination of ecosystems is the highest level of the organization of living matter on the earth.

'Ecosystems were formed in the process of a long evolution and adaptation of species and populations to the environment and to each other. They are adjusted, stable mechanisms capable of resisting, by self-control, changes in the environment as well as rapid modifications in the number of organisms. But there are limits to self-control. . . . If the changes in the environment . . . are more extreme than the periodic oscillations to which organisms are adapted. the harmony of the ecosystem is irreversibly disturbed. Still more profound consequences occur in landscapes when, under the influence of natural phenomena . . . or erroneous activities of man . . . one or several elements disappear as a whole or in separate trophic chains in the ecosystem. In these cases, the ecosystem suffers catastrophic changes and undergoes a radical reorganization. . . . The knowledge of relationships between populations of organisms of different types in ecosystems . . . makes it possible to manipulate ecosystems by using biological methods' (UNESCO, 1970).

A reviewer of six textbooks on ecology published in America in 1972 and 1973 (G. H. Orians, *Science* 181: 1238–9, 1973) notes that they differ greatly in scope, in their attitudes towards and use of theory, in the degree to which the concept of natural selection (now defined elsewhere as population/environment interactions) pervades the text, and the ways in which higher levels of ecological organization are approached and conceived. They differ, according to the reviewer, in the extent to which adaptation is used as an organizing concept; views differ concerning the mode of operation of natural selection; three develop primarily evolutionary approaches to ecology, three use the concept of natural selection much more sparingly. Colinvaux (1973) for example, states that 'it is a mistake to believe that animals and plants have all evolved primarily as efficient converters of energy. The pressures of natural selection are pressures for survival, and survival may sometimes be concerned more with the efficient use of nutrients . . . than with the efficient use, or even collection of energy . . . ecosystems are built up by the immigrations and adjustments of the species that are their parts. They do not evolve during the process of succession . . . natural selection does not choose between ecosystems, it chooses between individual living things.'

Ricklefs (1973) states that, instead of asking what determines the number of species that can coexist in a community, 'we should ask what determines the outcome of competitive relationships between species and the ability of populations to specialize on different resources.'

And, to return to Colinvaux (1973) 'it is sound ecological thinking to ask 'What are the things of the environment which limit this organism?' It is perhaps not so interesting a question as 'Why has this organism been adapted to this set of limits?' . . . The real question for ecology is: 'Why have animals and plants evolved the particular set of tolerances to which they do own?'

Duvigneaud & Lockie (in UNESCO, 1966) spell out the concepts and definitions of the ecosystem terminology in greater detail, for the benefit of the participants in a Regional Seminar on the primary and secondary productivity of tropical savannas, convened by UNESCO in Nairobi, Kenya, in October, 1966. The following sections A—F are quoted from the preliminary report.

A *Levels of integration of biological material* (populations, communities, biomes).
1. Biological material (principally holo- and heteroproteins and lipoids) are integrated in nature in a number of levels of organization of increasing complexity: cells, individuals, populations, communities. Ecology is particulary interested in populations and communities.
2. A population is a collection of individuals of the same species in a given area at a given moment. Examples: population of *Pennisetum benthamii*: population of *Equus zebra*.
3. A community is an assembly of different species populations (thus of individuals belonging to different species) in a given area at a given moment.
One can distinguish (and study separately if one wishes): plant communities, phytocoe-

noses: animal communities, zoocoenoses: communities of microbes, microbiocoenoses: communities of fungi, mycocoenoses.

4. The collection of phyto-, zoo-, microbio- and mycocoenoses in a given place forms a biocoenose or better a biome. The innumerable living things which make up a biome are linked to one another in various ways of which the principal are food and distribution.

B *Food chains at the heart of the biome. Levels and trophic chains.*

1. *Trophic levels*

a. The producers in the biome are the organisms which, by photosynthesis (rarely chemosynthesis) accumulate potential energy in the form of organic material fashioned from minerals derived from the abiotic environment. These are for the most part macro- or micro- green plants.

b. The consumers are organisms which nourish themselves directly or indirectly from the organic materials elaborated by the producers.

We can distinguish: primary consumers, obtain nourishment directly from the producers, phytophagous organisms, herbivores, plant parasites. secondary consumers, feed upon primary consumers for example, carnivores which eat herbivores. tertiary consumers, feed upon secondary consumers: carnivores which eat carnivores, carrion eaters or scavengers, feed upon the corpses of the preceding consumers or the prey abandoned by carnivores.

c. The decomposers (or bioreducers) assure the progressive mineralization of organic material and its return to the inorganic world. It is a complex group of organisms which should be further sub-divided. They include among others, the insects, fungi, coprophagous bacteria, which feed on corpses not eaten by scavengers. In addition there is the immense population of fungi and saprophytic bacteria which change organic material in the soil to CO_2 and H_2O (respiration of the soil) assuring the continuity of the cycle of nitrogen in mineralization and by the fixing of atmospheric nitrogen; finally restoring to the soil the cations and anions necessary for living things.

2. *Trophic chains*

Food material thus passes from one level to another, the producers, consumers, and decomposers forming a trophic chain.

3. *Trophic web*

The same producer can serve as food for different kinds of herbivore or the same herbivore can feed on many producers. These herbivores can in their turn be eaten by various carnivores. This results, in the biome, in a multiplicity of trophic chains which anastamose in a trophic web.

C *Distributional relations at the heart of the biome.*

These are set up between organisms depending, among other things, on the place occupied in the biome (struggle for food, light, water; search for protection against an unfavourable environment or against an enemy). These border on the determined structure of the biome which in the terrestrial environment more or less coincides with the structure of the phytocoenoses.

D *The ecosystem.*

Each population forming the communities of the biome depends for its existence on factors of the non-living (abiotic) environment. These can be grouped in two large categories: those which depend on climate (climatope) and those which depend on the soil (edaphotope). If one integrates in the same system the collection of populations forming the biome and the various factors of the environment, one obtains the ecosystem; the ecosystem is a functional system which includes any community of living things and their environment. It consists of (Sukachev): biome: phytocoenoses, zoocoenoses, microbiocoenoses, mycocoenoses and all the food chains which unite them; ecotope: (factors of the environment) climatope and edaphotope.

It functions really like a machine with complicated wheel-cog system, utilizing the energy of the sun (light and heat), feeds on water and mineral substances in the soil, elaborates and transforms diverse materials by the action of living beings and their metabolism, making the materials circulate, or sometimes accumulates them all along the trophic chains, and breaks them down by the action of decomposers in an open or closed cycle.

E *Some important definitions*:

Station: geographical site, marked on a map where a population, or a community, or a biome or an ecosystem exists. Habitat: all the conditions of the environment which affect a population, community, biome or ecosystem. Biotope: geographical site, marked on a map where a particular habitat exists. Ecological niche: the total of functional and spatial aspects in a biotope that apply to one population; in an ecosystem in equilibrium, there is only one theoretical ecological niche for each species population as a consequence of adaptations in the struggle for existence. Example: in the *Acacia* savannas of East Africa, the possibility of feeding in a very dry environment on leaves mixed with spines several metres above the ground in trees of a parasol shape constitutes an ecological niche occupied by the giraffe.

F *Biomass, primary and secondary productivity*

1. Biomass: the weight of living organisms or of individuals forming a population or of the populations forming a community. Biomass can be expressed in grams, kg. or tons of material dried at 70^0 centigrade (in order to avoid losses of nitrogen). In terrestrial ecology the biomass is usually expressed as so much per unit area (M^2, Ha, KM^2 depending on the importance of the individual). Biomass per unit of surface particularly in dry savannas or 'standing crop' is more or less the same as production.

2. Productivity (primary and secondary); each trophic level or each population making part of a trophic level has a particular biomass. Productivity of this level (or of this population) is called the biomass formed during a given time (usually expressed as per second, day or year) per unit surface area. Productivity is thus the speed with which new material is produced. Primary productivity is the productivity of the producers, that is, the green plants. Secondary productivity is the productivity of consumers (especially animals or decomposers such as fungi or bacteria). It is also possible to study the total productivity of an ecosystem.

2.2 Diversity, stability and mathematical models

UNESCO (1971) produced a document stressing the merits of applying eco-system models to the rational use of natural resources. It should not be thought that modelling is a substitute for empirical studies but the modelling process can afford a valuable means for guiding and integrating empirical work. The crucial part of ecosystem modelling lies in the sub-models. Each will simulate a particular sub-set of processes within a distinguishable sub-system, e.g. nitrogen transformations in the water; water transfer between soil, plant and air; predation; seed germination; or direct interactions between man and the biota. The sub-models will in general deal with processes on a shorter time-scale than is required for the model as a whole.

The ecosystem model can be used as a guide to rational use and management. The effects of any desired modification of an ecosystem can be tested on the model far more quickly and cheaply than in the field, and without the risk of embarrassing and irreparable repercussions. A simulation model can be incorporated into a computer programme designed to test all possible combinations of management practices before their field application. The unit in land-use management is usually a complex of ecosystems in, for example, a river catchment area. If models are available of the separate ecosystems within such a land unit, together with their spatial interrelations, and if the way in which outputs of one become inputs to another is known, the separate ecosystem models can be combined to give a model of the whole complex.

Experience in ecosystem modelling, which demands good computer facilities and a team of ecologists who have both a broad ecosystem approach and quantitative interests, is at present limited to a few centres in the developed countries. There is, according to UNESCO (op. cit), a need for more such centres, but 'their success depends greatly on the extent to which the modelling team can keep its feet on the ground of biological reality. Otherwise there is a serious risk that the intrinsic interest of the theoretical problems of modelling and computer simulation may lead the team away from the primary objective of ecological prediction.'

Discussion is currently devoted to the relation between diversity and stability in ecological systems. For example, at a symposium on the subject held at Brookhaven in 1969, it appeared to be concluded that each taxon and geographic region should be viewed individually, and generalization should be avoided. However, in his study of stability and complexity in model ecosystems, May (1973, reviewed by D. Simberloff, *Science* 181: 1157/8, 1973) has attempted just such generalizations: 'Natural stability of a system implies that displacement of an equilibrium system (equilibrium meaning that population sizes are either constant or undergoing fixed oscillations) produces another equilibrium system, the parameters of which are determined only by the degree and the direction of the displacement. Continuing cycles caused by neutral stability . . . are dismissed as unrelated to cycles observed in nature, since once the amplitudes are set by environmental disturbance they remain constant indefinitely, no matter what environmental conditions prevail. . . . With neighbourhood stability as the main criterion, the chief conclusion is that, contrary to ecology's central dogma, increased species number and complexity of food web structure usually lead to decreased stability. Hence ecologists ought to focus on those particular types of complexity which produce mathematical stability, since there seems to be a high, but not perfect, correlation in nature between complexity of trophic structure and stability of the community, as manifested by continued existence of all its populations within limited size ranges . . . the plethora of mathematical models oppressing today's ecologists are variations on just a few simply stated themes' (D. Simberloff reviewing May, 1973).

For other works on mathematical models in ecology, see Watt (1968) — predicts effects of different levels of management policies on yield — defines how to find optimal management policy — how to use computers; MacArthur (1972) who considers mathematical ecology in relation to patterns in the distribution of species; also Jeffers (1972) regarding which a reviewer stated that the relation between mathematical models and ecology is still unstable — the modelling of ecosystems is perhaps more difficult than the earlier practitioners had expected. A review of Maynard-Smith's *Models in Ecology* (1974) states that ecology suffers from a surfeit of fascinating but apparently unrelated observations, superimposed upon an acute shortage of general theories.

There are two different ways of using mathematical models, and the type of model constructed varies according to the situation (Charles-Edwards & Thornley, 1974). In order to understand the operation of a system in terms of the mechanisms operating in it, one would construct a mechanistic model. To describe the operation of a system without considering the detail of its mechanisms and components, one would construct an empirical model. The method of mathematical model-building is illustrated in Fig. 2.1. The models developed at the Glasshouse Crops Research Institute in Britain tend to be of the mechanistic type, although they frequently contain empirical elements. Brockington (1971) has described modelling work where the emphasis is somewhat different, being more concerned with extensive systems (see Fig 2.2).

The British Ecological Society discussed diversity in ecosystems at its winter meeting in January, 1974, as follows:
R. M. May (Princeton, U.S.A.) — the use of general mathematical models in studying the relations between species diversity and stability in ecological communities.
J. M. Anderson (Exeter, U.K.) — parameters promoting high species diversity with particular reference to trophic separation, microhabitat selection of species in space and time, and the effect of predators.
M. Webber (Oxford, U.K.) — stability in a multispecies food-web model.
J. L. Harper (Bangor, U.K.) — diversity in grassland ecosystems at the inter- and intraspecific level; factors influencing these types of diversity in plant communities, with special reference to the role of predators.

Under the general title of structure, function and management of ecosystems, the First International Congress of Ecology, held in The Hague, September, 1974, discussed 1. flow of energy and matter between trophic levels. 2. comparative productivity in ecosystems. 3. diversity, stability and maturity in natural ecosystems, and in systems influenced by human activities (G. H. Orians distinguished nine different meanings of stability), and 4. strategies for the management of natural and man-made ecosystems. Eugene Odum (U.S.A.) argued that, historically, the diversity of ecosystems shows two peaks of maximum diversity; the first being energy-subsidised from outside the system or from an earlier resource, while the second, which is self-maintaining as regards resources, is more stable. The implication for man is that, as he comes to dominate more of the world's resources, his strategy must evolve from that of the first to that of the second type of system.

In reply, F. H. Regier (Canada) deplored the ecologist's obsession with

23

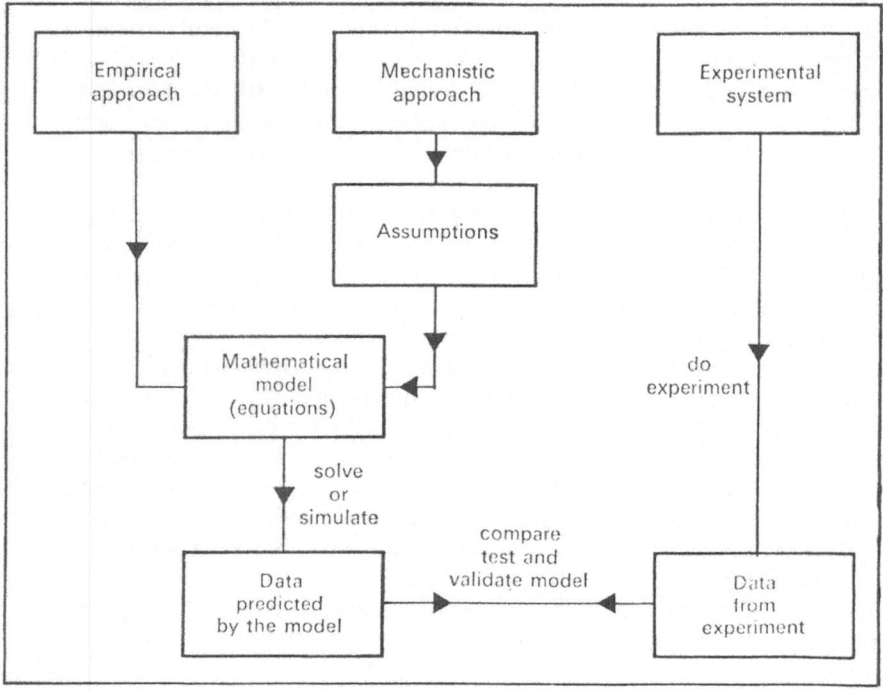

Fig. 2.1. The mathematical modelling method. Diagram illustrates relations of mechanistic and empirical models (Charles-Edwards & Thornley, 1974).

indefinable concepts, and urged the development of falsifiable hypotheses. He challenged a number of widely accepted 'articles of faith', maintaining that there is no necessary relation between diversity and stability, that energy recycles and nutrients flow through ecosystems.

A Mathematical Ecology Group, formed jointly by the British Ecological Society and the Biometrics Society to give ecologists an opportunity to put their sampling, analytical and other problems to statisticians, held its first meeting in London in April, 1974, on the theme: classification in ecology. The following papers were presented (see also discussion under 3.6 on ordination and classification – Whittaker, 1973 – and of vegetation dynamics – Knapp, 1974): P. Greig-Smith (Bangor, U.K.) – analysis of vegetational data; P. Dagnelie (Gembloux, Belgium) – classification; problems more than solutions; G. J. S. Ross (Rothamsted, U.K.) – ordination methods for the presence/absence; species – quadrat matrix; M. V. Brian (Furzebrook, U.K.) – species distribution (ants) using principal component analysis.

24

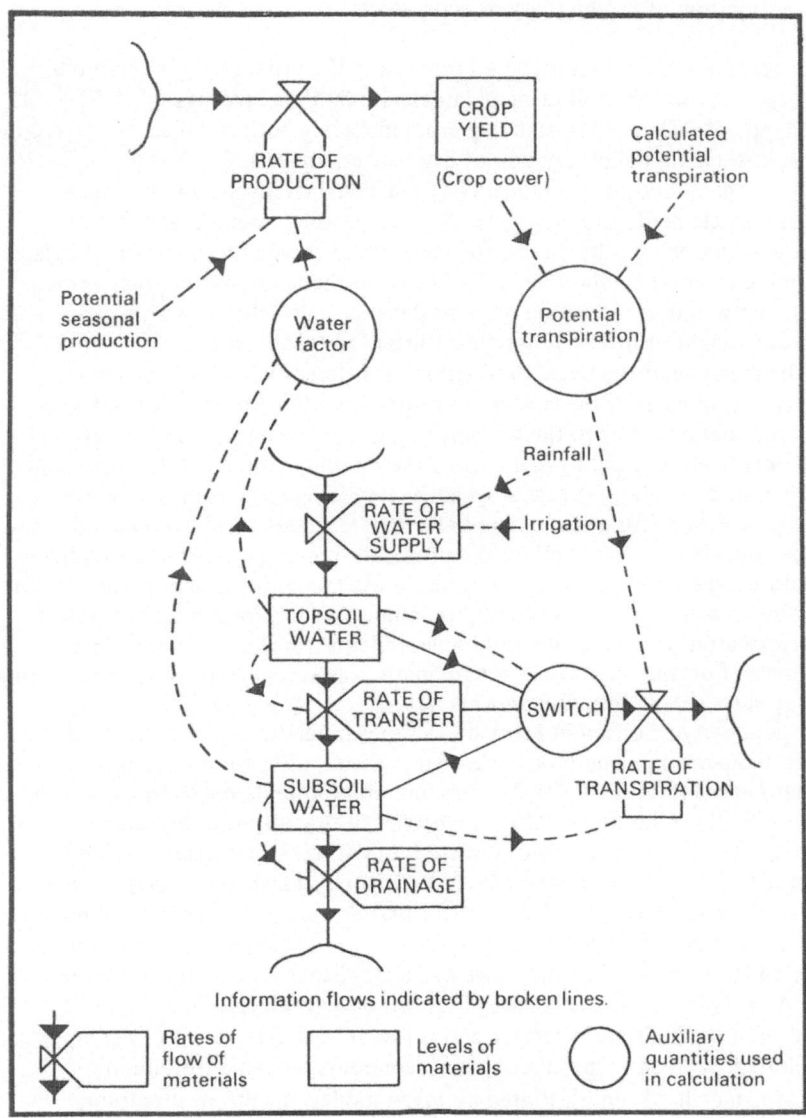

Fig. 2.2. Grass production in relation to water supply: flow diagram of the model shows relations of the components (Brockington, 1971).

2.3 Estimation of productivity of ecosystems

In undertaking the difficult task of estimating the productivity of particular phytogeocenoses, the following parameters should be investigated (UNESCO, 1970, pp. 24-27): a. total standing crop, including both aerial and underground parts, b. total annual production of dry matter, above and below ground, c. chemical composition and mass of food elements involved in biological cycles (ash elements and nitrogen), d. average ash content in annual phytomass production, e. the number of metabolites (based on the study of balance of food elements) involved annually in the synthesis of phytobiomass, and f. intensity of decomposition of plant detritus, calculated from the ratio between weight of litter and weight of litter fall (green parts).

The meaning of the term 'production' is different when used by agriculturists or ecologists of the ecosystem respectively (Smith, 1972). In agriculture, production refers to the amount of grain or the number and weight of cattle per hectare, yield to that part of the standing crop harvested for man's use. To the ecologist, the rate at which energy is fixed by plants is known as gross production (for a school and university textbook on the origin and bases of the concept of energy and on the processes of energy transfer within living structures, see Linford, 1966); since plants use much of the energy they fix for their own respiration, gross production minus this respiration is net production, appearing as plant tissue or biomass. Biomass is usually expressed as grammes of organic matter per square metre, calories per square metre, or other appropriate measure per unit area.

Productivity of different ecosystems varies in relation to nutrient availability, water, temperature, length of the growing season, utilization by animals and similar factors (Smith, 1972). Net production represents the energy available either directly or indirectly to the consumer of organisms, to be passed through the ecosystem in a series of steps of eating or being eaten, a food chain. Feeding on plant tissues are the herbivores, which are, however, able to transform only a portion of the plant matter they consume into animal tissue. Much of the assimilated energy is lost, partly as the heat increment required for fermentation and nutrient metabolism over and above that required for basal metabolism. Part of the net energy which remains after this loss is used for body maintenance and is lost as heat. The remaining energy goes into growth and reproduction and represents secondary production. This sequence has been illustrated by many models in current literature, including that by Gates (1971) as the flow of energy in a natural system, based upon data obtained by Howard T. Odum of the University of Florida (Fig. 2.3).

The highest standing crop of biomass occurs in forests, but a high annual production of phytobiomass is characteristic not only of forests, but also of grass communities such as savannas, prairies, pampas, steppes and landscapes derived from flood plain. The chemical composition of organic matter,

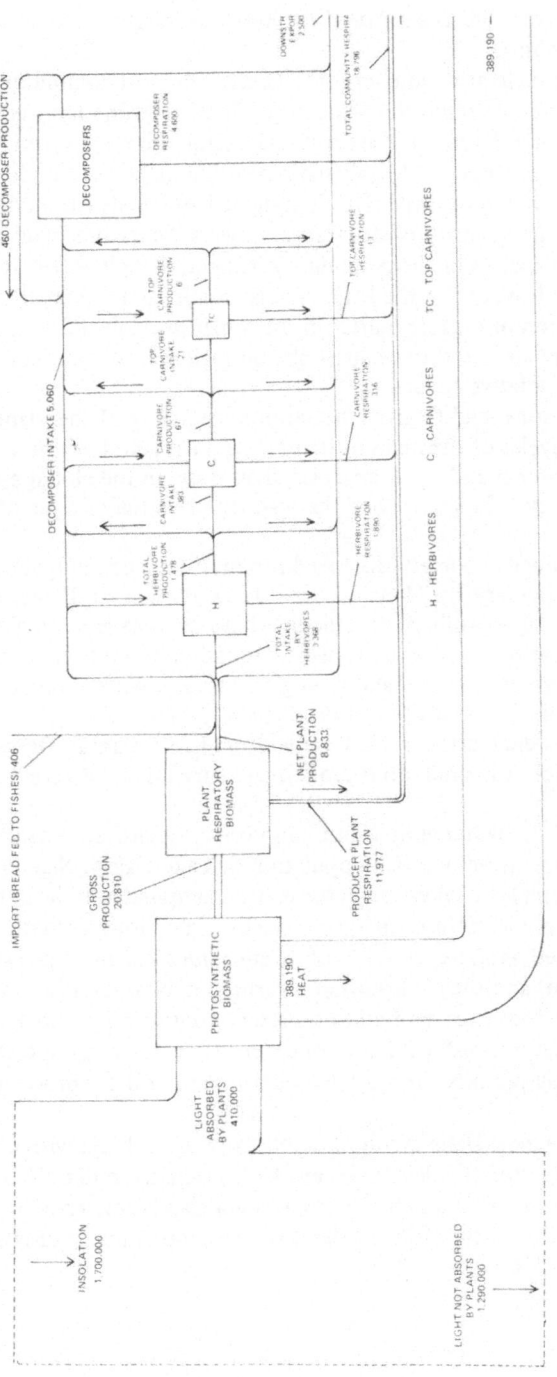

Fig. 2.3. Energy in a natural system. Natural system flows as indicated in a system which consists of a clear, spring-fed stream with vegetation covering the bottom and numerous species of animals living in or near the water. Top carnivores are at top of food chain. Numerals give inputs and outputs in kilocalories per sq. m. per year, with relative energies indicated by the widths of the bands and lines (from Gates, 1971, after Howard T. Odum of the University of Florida).

27

and hence its relative value as a nutrient for livestock, is quite different in different phytocenoses.

The French Institute at Pondicherry (P. Legris, personal communication, 1974) has undertaken work on the primary production of the montane forest in Kodaikanal, Palni hills, and of the scrub-woodland ('dry evergreen forest' of Champion & Seth, 1968), of Marakanam, near Pondicherry. Emphasis is placed on the production of litter. This is collected and analysed monthly, taking account of the quantity of dry leaves, flowers, fruits and twigs produced by the main species of the community. Chemical analyses of fresh leaves and of dried leaves from the litter are also made, and the results related to the decomposition of organic matter in the upper horizons of the soil. Some classical indexes are also used to evaluate the biological activity of soils under various types of vegetative cover.

These investigations tend to give a better understanding of the plant/soil relations and the cycles of the main nutrients, e.g. nitrogen. Carbon, calcium, magnesium, phosphorus and other analyses, like those on the global biological activity indexes, may help in assessing the fertility potential of agricultural soils.

The British Ecological Society discussed productivity and energetics at its winter meeting in January, 1974, as follows: E. D. Ford, J. D. Deans & R. Milne (Edinburgh) — production ecologists have become less satisfied with gross comparisons of production by ecosystems and more inclined to study the complexity of environmental and biological features which control it — illustrated by a project to study forest production.
R. Abel (Oxford) — energetics of a fish population in a stream in Dorset; on two sites, production was similar but growth and survival characteristics different.
D. S. C. Lewis & N. G. Willoughby (Southampton) — quantitative estimation of the trophic composition of a fish population of Lake Kainji, Nigeria.
K. Thompson (Exeter) — *Cyperus papyrus* easily manipulated in both natural swamps and hydroponic culture; an exceptionally high photosynthetic productivity cannot be used as an explanation of its enormous sustained growth rate.
P. J. Edwards (Southampton) — litter fall measured in New Guinea lower montane forest, on four sites, chosen to represent contrasting ecological situations found in primary forest; estimates of extent to which decomposition occurs in the canopy permits a more accurate estimate of litter production to be made.

Reference to the New Delhi symposium on tropical ecology, with emphasis on organic production (Golley & Golley, 1972) is made under 2.6 below. The symposium on the productivity of tropical grasslands convened at Varanasi, under the auspices of the IBP Grassland Biome programme is discussed under 3.6.4: Grass Covers.

2.4 Ecosystems, soil and water

A symposium was organized by UNESCO and IBP in Paris in 1967, to provide for an interchange of information on recent advances in the study of soil ecology, with special emphasis on production and energy flow in soil ecosystems. In addition to the published volume of Proceedings of this symposium (UNESCO, 1967), action was initiated for the production of an IBP handbook on Methods of Study in Quantitative Soil Ecology: Population, Production and Energy Flow.

There are several definitions of 'soil ecosystems' (F. di Castri in UNESCO, 1967). Ecological research on the soil system can be broadly classified into two main groups, each with its individual theory and methods: study of the metabolism, and study of the regulation, corresponding only very roughly to the classic division by function and structure. In many countries, a considerable period is necessary for the establishment of the basic knowledge, especially on the taxonomic and pedologic foundations of soil ecology.

The problems of soil ecology in the different biogeographical regions of the world call for research (J. Balogh in UNESCO, 1967), as also studies on the taxonomy, biomass, and the rate of renewal of the energy of soil organisms, especially in the tropics. In India, the drastic changes which have occurred in the plant canopy due to anthropogenic factors are reflected in changes in the nature of the soil and so in the equilibrium of the ecosystem (K. M. M. Dakshini in UNESCO, 1967). Unfortunately it is particularly in the tropics that studies on the relation between vegetation and its underlying soil as an entity have been meagre. It will be necessary to know, first, the process of soil formation and the interaction between soil and vegetation, and second, the correlation between the distribution of soil types and vegetation types. The application of the catena concept in mapping, if substantiated with the successional trend in vegetation, will give a composite picture of interrelated land forms, soil types and plant communities. 'The generalized world soil maps that have appeared have made little contribution to realistic thinking about tropical soils' (Dakshini, op. cit.).

J. C. I. Dooge (in FAO, 1973) has reviewed the biological significance of the hydrological cycle in problems involving the biosphere. 'Water, whether in the form of surface storage, soil moisture or channel storage supports a variety of living organisms. Under undisturbed conditions, the population of organisms in the soil or in the surface water is the result of an ecological equilibrium. In many cases this equilibrium is a very delicate balance and even small changes in the quantity or quality of water flowing can result in considerable changes in the number or type of organisms present in a given location.

'The thin cover of soil, which is largely due to weathering by water, supports not only vegetal cover but a wide variety of micro-organisms. These act on complex organic material and reduce it to simpler organic forms and ultimately to a mineral form. Their action on the surface layer of the soil is beneficial and results in the development of humus which is the organic detritus of decayed vegetation. Humus has a considerable effect on the hydrological cycle, since the existence of an appreciable layer of humus results in a relatively high rate of infiltration. Since infiltrating water normally percolates downward toward the groundwater table, and the micro-organisms are concentrated on the surface layers where there is an abundant food supply for them, the habitat of the soil micro-organisms is confined to the unsaturated zone.

'A fresh water environment is less harsh than an air environment for plant and animal life. Water, whether in the form of standing water or of running water, is the habitat for a wide variety of organisms showing the same complexity of food relations as the inhabitants of the land surface. Some standing waters tend to fill in as a result of physical and biological development, so that we have an ecological succession from open water to shallow water to emergent vegetation to bogs. At each stage of the development, there will be a particular population complex of organisms which will be in balance in relation to competition for food and form part of a relatively stable food chain. Interference with the natural hydrological cycle in such areas can result in either the destruction of a given type of organism or the explosive growth of another or may tend to accelerate (or to slow down) the development from one type of environment to another.

'Moving water systems have their own types of organism and their own environmental features, some of which are not shared by lake communities. Again any disturbance of the system and in particular any conversion of a running water system to a standing water system will result in a change in the biological community. For example, under primeval conditions, a stream might flow through a wooded valley possibly interspersed by swamps. Thus most of the surface of the water would be relatively shaded and would therefore be populated by plants which would be more at home in shade than where more exposed to sunlight. Such streams would have a relatively low productive ability in terms of plant material. The clearing of the banks of the streams would produce a quite different environment which would soon develop a new population mixture of organisms.

'Many examples can be cited of cases where fish populations disappear from streams for a variety of reasons. Both communities in developing countries who look to good fishing waters as a source of nourishment, and communities in developed countries who also look to such waters as a recreation amenity, suffer damage from such a loss. To maintain a fish population, a stream must have an adequate supply of water at an appropriate temperature, an adequate supply of dissolved oxygen, a source of food supply for fish which may consist of several levels in a food chain, easy access to adequate areas of spawning beds to facilitate reproduction and a number of other environmental factors. Interference with any of these even for a short time can destroy or limit the fish population or may cause them to seek a more favourable habitat elsewhere.'

30

2.5 Weather, climate and the ecosystem

The first part of this section represents a paraphrase of a WMO document seen in draft, A Brief Survey of Meteorology as related to the Biosphere, by C. C. Wallén, together with supplementary new material.

The importance of the atmosphere as an integral and active component of the biosphere depends not only on its passive presence as a mixture of gases, as a source of oxygen and carbon dioxide, but also on its active dynamic role in weather and climate. Unlike the soil, the geology or the biological community, the atmosphere is being constantly modified as the great air masses move and interact. Physical processes beyond the confines of the ecosystem, as well as those within, have their influence.

Weather is the state of the atmosphere at a given time, as defined by various meteorological elements. Climate is a synthesis of the weather of a specified area over a specified interval of time. Weather can change rapidly over very short periods of time, and often show regular rhythms on a diurnal and an annual scale. Climate is relatively stable, although in marginal areas small changes may have large consequences. Certain of its characteristics are determined, within somewhat flexible limits, by fixed parameters of the local environment. Locally, the characteristics of climate may be strongly influenced by other components of the ecosystem, or modified by the activities of man. Although climate may then be influenced by the system itself, it nevertheless always has a decisive influence on other processes occurring in the system.

The basic energy source of an ecosystem is solar radiation. As this is transformed into other forms of energy through the photosynthetic and nutritional chain, higher elements in the system are created, some of which eventually serve as food for man or livestock. Thus any deviation of energy from this purpose may reduce the amount of food ultimately available. An increase in the productivity of a system can be achieved only by a progressively more effective use of solar radiation, or by the introduction into the system of new elements, especially plants, which are able to make more effective use of the existing solar radiation.

It is obviously quite impossible to establish how solar energy is efficiently used in a particular ecosystem without knowing how much incoming energy is available. WMO is co-ordinating a network of radiation stations around the globe and is standardizing radiation measurements (see WMO Guide to Meteorological Instruments and Observation Practices, WMO No. 8, TP3; also WMO Technical Note no. 85, Precision of Pyrheliometric Measurements; WMO Technical note no. 104, Radiation including Satellite Techniques).

Energy flow through an ecosystem is a complex and often inefficient process, rarely more than one per cent of the total solar energy available being intercepted by vegetation and stored as energy for possible use. Apart from its direct significance to the ecosystem, solar radiation has a basic indirect influence by creating the environmental conditions which appear as weather and climate. A particularly important secondary effect of solar radiation is the introduction to the atmosphere of energy for potential use created by evapotranspiration from a vegetative cover.

The meteorological data of great importance in the study of the atmospheric processes of an ecosystem are: radiation (incoming short-wave including spectral composition, outgoing long-wave, and net); air temperature (including low-level lapse rates for special purposes); air pressure; wind direction; wind speed; air humidity (including low-level lapse rates for special purposes); clouds, amount and type; duration of bright sunshine; precipitation, amount and type; evaporation; and, for special purposes within the biosphere: evapotranspiration; soil temperature; soil moisture.

WMO is co-ordinating and standardizing methods of collection, storage and retrieval of meteorological data (much of which has been recorded for many decades) for the purposes of specialized research (see WMO No. 174TB86 — Catalogue of Meteorological Data for Research, Parts I and II).

The methods of studying ecosystems with regard to the implications vary with the scale: large-scale ecosystems — macroclimates; small-scale ecosystems — microclimates; meso-scale ecosystems — local (meso-) climate.

Theoretically one might expect that a macroclimate or local climate could be built up by a mosaic of many microclimates, and that a macroclimate over a larger area could be established by the integration of results from microclimatic studies at a large number of sites. However, since the micro-meteorological studies are concerned with small-scale phenomena, it is unlikely that one could use them to distinguish the large-scale latitudinal characteristics of a macroclimate. Because of the sophisticated instrumentation needed at micro-meteorological stations, it is unlikely that sufficient stations will ever become available to establish the macroclimate of ecosystems.

Some investigators have tried to develop empirical or approximative formulae to determine a key parameter for the microclimate, e.g. evapotranspiration only from measurements at screen level of simple parameters which are available from ordinary climatological stations. Thornthwaite (1948) established the climates of ecosystems on the basis of potential evapotranspiration, the evaporation that should occur from a vegetative surface provided water is available in the soil. His formula may be applied on a monthly and annual basis with data from any station which records temperature and daylength. The expression is very complex, and may in addition break down in areas of temperatures below zero, in very arid regions, on islands, and on coastal zones. The more sophisticated formula to determine potential evapotranspiration developed by Penman (1948) and modified by others, is more applicable to determine large-scale variations in microclimatic conditions, as it appears to be more or less valid in any part of the world. Budyko (1968) has developed a similar formula to establish the heat-balance and evapotranspiration used to delineate an overall picture of world climatic ecosystems.

By international agreement, climatic assessments in macro-meteorological studies are based on a 30-year period; in many parts of the world, data are available for more than three 30-year periods. From these assessments, it is concluded that climate fluctuates considerably, and that these changes in climate are not cyclic in nature but vary in amplitude, phase and period. When dealing with 30-year averages and frequencies, fluctuations of a shorter time span in general become statistically insignificant. Changes of practical significance to an ecosystem may occur over periods of a few decades.

It has been projected that by 1985 the margin between supply and demand for food will be only 3 per cent, whereas adverse weather may cut grain production in the major producing countries by 10 per cent or more (Winstanley, Emmett & Winstanley, 1974). With increasing climatic variability, there is a higher probability of more frequent crop failures induced by weather in the years ahead. But one of the main aspects of climatic change is that the changes vary regionally in nature and magnitude. It is quite feasible that global temperature could change by 0.5 to 1.0 °C. within half a century; it is not the direct effects of this thermal change on agricultural production which are most significant, but the regional changes in rainfall and perhaps the more frequent occurrence of frosts that would accompany the overall thermal change.

It is these changes of climate that are causing concern at the present time, especially in the lands of monsoonal climate in Africa and Asia. Rockefeller Foundation sponsored a conference on the subject in January, 1974. A unit of the (U.S.) National Academy of Sciences is preparing a report on climatic change. The Environmental Data Service of the (U.S.) National Oceanic and Atmospheric Administration maintains a special group to monitor global weather as it affects food production. A workshop convened by the International Federation of Institutes for Advanced Study finds (W. O. Roberts) that a new climatic pattern is emerging, that it will persist for several decades, and that it will adversely affect systems of food production which cannot be readily adjusted in time.

It is necessary to give due consideration also to the contrary views regarding these supposed climatic changes, as expressed by Indian climatologists with reference to their subcontinent ('our rainfall studies from the beginning of this century show that there has been a 30 per cent *increase* in rainfall'); and by American specialists who have investigated the Sahelian situation, as summarized by Wade (1974). In the latter region, it is obvious that as much if not more responsibility for the recent crises should be placed on human and bovine over-population. The effect of overgrazing is dramatically shown by the NASA satellite photo reproduced by Wade — the striking difference between the overgrazed open range and the 'curiously shaped green pentagon' of the Ekrafane ranch, 100,000 hectares, inside a barbed-wire fence, established five years ago when the drought began, divided into five sectors with the cattle allowed to graze one sector each year. It must be concluded that the Sahelian situation has been caused by a major disruption of the traditional social systems

of the cultivators and the herdsmen, the breakdown of the delicately balanced
and well-designed systems of annual transhumance between the high-rainfall
grasslands to the south and the desert fringes to the north, and the greatly
increased demand for cash crops for export (UNESCO, 1974).

Changes with spans of 100 and 1000 years may occur (WMO Technical
Note no. 79; Whyte, in UNESCO, 1963). Meteorologists have not yet been
able to define the natural causes of climatic change — an important field of
research in short- and long-term climatic changes of great importance in forecasting climatic trends and planning the more efficient use of the resources
of the biosphere.

An important step in this direction was the 100-day experiment — the most
extensive ever conducted in weather research — which was started on 15 June,
1974, in the tropical area of the Atlantic Ocean between South America
and Africa. The President of the WMO Experiment Committee, B. J. Mason
of Britain, has stated that the experiment, which was to cost U.S.$ 100 million
and involve between 3000 and 4000 people, would help in making forecasts of
more than one week ahead, of the date and strength of meteorological crises
such as hurricanes and typhoons. It is hoped that a similar massive study may
be possible over Tibet, in co-operation with the People's Republic of China,
since the seasonal behaviour of climates in this area is said to be of fundamental
significance to the monsoonal lands of the northern and even to some extent
of the southern hemisphere.

It is not possible to distinguish between those climatic variations caused by
natural changes in the atmospheric processes from those arising from the activities of man. A local climatic trend in a certain system may well be due to
the combined effect of a slow and gradual change in the general circulation
and the influence of man's introduction or intensification of irrigation,
deforestation, animal grazing, cultivation, etc. A clear picture of what is actually taking place can be obtained only by comparing the natural processes
in neighbouring areas protected from and exposed to anthropogenic influences.
WMO considers it essential that bench-mark ecosystems be established in
various zones, in which natural conditions should remain unchanged over
a long period of time.

The order of magnitude of man's impact on climate and water balance may
be expressed in terms of changes in the Bowen ratio. In very humid areas this
ratio becomes very small, between 0 and 0.5, indicating that the radiation
energy is transformed into energy used for evapotranspiration. In very dry
areas where water is not available for evapotranspiration, the Bowen ratio
becomes quite large, sometimes between 5 and 10. Provision of irrigation
water may reduce the ratio from 5 to around 0, or it may even become negative. Deforestation in tropical regions may cause large and quite dramatic
changes in the Bowen ratio; on the forest/savanna borders in the humid tropics, the Bowen ratio might after deforestation change from an order of
magnitude of 0.3 to around 5, implying a complete change in the water bal-

34

ance. The replacement of large latent heat sources by sensible heat sources may possibly affect the dynamics of the general circulation – WMO recommends that studies of this probability with numerical experiments and mathematical models be started as soon as possible.

The Bowen ratio over an irrigated oasis amounts to –0.3, while over the surrounding semi-desert it is 5. The increasing application of irrigation water in an area on the margin between the arable land and the desert creates an increasing imbalance between the amount of water available through precipitation and the actual evapotranspiration. In the long run, this growing imbalance certainly has an unfavourable effect on the water budget of the whole area, where there is also continued degradation of the natural vegetation through overgrazing.

In formulating mathematical simulation models of natural resource systems, one recognizes static variables, driving variables and output variables. The static variables represent values of measurable properties (biomass, plant conditions, number of animals, water content, etc.). It is the time variation of these variables that is the basic object of the analyses of the system. To create a dynamic model, it is necessary to add the driving variables, of which the most important are the major climatological and meteorological factors (radiation, temperature, precipitation, evapotranspiration and many others). The output variables represent the result when the complex system of algebraic and differential equations is solved by use of the computer (WMO).

2.6 Ecosystems in the equatorial and tropical zones

Many of the studies on productivity in tropical ecosystems are conducted by botanists and plant ecologists. They tend to concentrate primarily on their plants, trees, grasses, crops, as major components of the ecosystems they study. It would be preferable for this type of specialist to forget his basic training and to become an ecosystem scientist who studies the whole biological complex as a unity, with plants as merely one of the component parts. Specialists in faunal distribution seem to be less restricted, saying that vegetation governs or is closely related to the distribution of animals of all sizes, just as the animals are one of the major factors governing the nature of plant communities and the distribution and migration of individual species. Animals can be significant factors in ecosystems without necessarily being consumers.

Tropical forest ecosystems
The total biological rather than purely botanical or zoological approach is characteristic of broad-based studies such as that on the tropical forest ecosystems in Africa and South America (Meggers, Ayensu & Duckworth, 1973). This is evident from the introduction to the volume by J. P. M. Brenan, President of the Association for Tropical Biology in 1971, the year of its prepara-

tion, who states 'Agriculture and forestry no doubt are the most promising activities, but they must be carried out with solicitude and comprehension of the unique characteristics of the natural environment. It is to be hoped that these papers will stimulate efforts to increase our knowledge of the plants and animals inhabiting lowland tropical forest ecosystems. . . . With this knowledge, it may be possible to learn before it is too late, how to utilize and live with these fragile communities without destroying them — perhaps also to appreciate that they possess an innate value not to be measured in terms of money or economic return.'

Fosberg (1973) defines the properties of the humid tropical ecosystem that are basic to an understanding of what is now happening as 1. the rapid leaching out of the soluble bases and other nutrient elements from the soils; 2. the resulting low fertility of the soil proper and the lack of weatherable mineral compounds, e.g. silicates; 3. the concentration of the greater part of the available nutrient supply in the biomass of the tropical forest; 4. the frequent accumulation of nutrients in the lower subsoil zones; 5. the tendency of the sesquioxide mixture dominant in laterized soils to harden to ironstone on exposure to the sun and air; and 6. accelerated erosion that starts immediately when bare soil is exposed to the often torrential tropical rains.

From the point of view of the comparative ecology of rain forest mammals in South America and Africa, Bourlière (1973) notes that the stratification of trees and shrubs in the rain forest is correlated with stratification of animal species, different ecological niches are provided by different food materials (fruits, leaves, etc.), and the availability of food controls the size and distribution of populations of individual species (see 3.6.3 for tropical forest ecosystems).

But, as reviewers of books by geographers and biologists frequently state, man, whether at the primitive (2.7) or economic (4.3) level, is so often forgotten. The reviewer (*Journal of Asian Studies* 33:133) of Bartz (1972) on South Korea notes that it is widely accepted that the geographer investigates the relations which exists between man and his physical environment, and the interaction between man and his natural milieu; yet throughout this book the author has missed the most critical and important element in South Korea that a geographer (who should be the best type of ecosystem scientist) must be aware of — 'the people'.

However, this criticism cannot be directed at the study of tropical forest ecosystems (Meggers, Ayensu & Duckworth, 1973), in which M. P. Miracle considers the Congo Basin as a habitat for man, and F. W. Lowenstein discusses the concept of balance between aboriginal man and his environment in the tropical rain forest.

Grassland ecosystems
The symposium in New Delhi on tropical ecology with particular emphasis on organic production (Golley & Golley, 1972) had as its terms of reference

the production of the primary producers, the only group that receives its major energy input from abiotic sources. The other groups are the consumers, with food chains based on living plants, and the decomposers, with food chains based on dead organic material.

F. B. Golley and H. Leith (in Golley & Golley, 1972) state that the definitions of primary production are confusing, and sometimes lead to the assumption that primary production equals photosynthesis. To include nutrient uptake in the equation, one would have to say that dry matter production equals photosynthesis plus mineral uptake. Since plant dry matter may consist of more than 20 per cent ash, this distinction is considered to be reasonable. Golley & Leith distinguish the following production categories: gross primary production = net primary production plus metabolic (respiration) loss; net primary production = standing crop increase (above and below ground) plus litter; standing crop increase = yield plus waste.

Net primary production is equal to the organic material available for the consumers and decomposers and for export and storage.

Golley & Leith then examine methods of measuring primary production, under the heads of gross and net production, potential and realized production, and the range of production in forests, grasslands (a doubtful distinction is drawn between 'savannas' in Nigeria and 'grasslands' in Bihar) and annual crops.

Any specialist concerned with advising graziers regarding the improved management of their grazing grounds on the basis of ecological succession must feel uneasy with some of the conclusions which arise as a sequel to assessments of primary production. One example will suffice. It is reported (Golley, summary to Golley & Golley, 1972) that *Heteropogon contortus* gives the highest net production at Varanasi, higher than *Dichanthium annulatum,* and higher than 'natural' relatively undisturbed vegetation in Rajasthan. *Dichanthium annulatum* may be out of its true environment at Varanasi, and net production may be expected to be much lower than may be obtained in protected plots at Poona or Palghar in Maharashtra. A direct comparison between herbage cut at Varanasi and in Rajasthan is not possible, because of the higher protein content and therefore higher nutritive value and palatability of herbage in Rajasthan, especially if *Cenchrus ciliaris* is dominant. Further, one notes that, in the contribution of the Indian Grassland and Fodder Research Institute, Jhansi, U. P. (K. A. Shankarnarayan, P. M. Dabadghao, V. S. Upadhyay & P. Rai in Misra, 1974) to the January, 1974, Symposium at Varanasi, that the levels of primary production of species, without and with added nitrogenous fertilizer, in a *Sehima/Dichanthium* cover (Whyte, 1974a, Table 5.2, p. 86), are highest in *Iseilema laxum,* followed closely by *Sehima nervosum* and far behind by *Heteropogon contortus.*

Workers at the Central Arid Zone Research Institute find that it is most difficult to draw specific conclusions regarding the above-ground productivity of grasslands at Jodhpur, Rajasthan. This varies widely because of the charac-

37

teristics of the climate and the livestock management of the arid biome (Gupta, Saxena & Sharma, 1972a). The above-ground productivity and phytomass of three individual grasses, *Cenchrus ciliaris, C. setigerus* and *Lasiurus sindicus* are also highly variable under different rainfall conditions (Gupta, Saxena & Sharma, 1972b).

Sehima/Heteropogon grasslands are major constituents and have high successional status in the *Sehima/Dichanthium* cover, the largest in area of the five grass covers of India, extending over large parts of southern and central India lying south of the Great Indian Plains (Dabadghao & Shankarnarayan, 1973). The *Sehima/Heteropogon* grasslands give a stable expression to the grass cover over large tracts with medium to low rainfall with undulating topography and red-gravelly (murrum) soils. Their production ecology is important (Shankar, Shankarnarayan & Rai, 1973) because of their high potential for the maintenance of free-range livestock. *Sehima nervosum* provides the major contribution towards both photosynthetic and non-photosynthetic community structure. Comparison with grasslands elsewhere in India shows that the standing crop biomass and rate of production are especially high in the *Sehima/Heteropogon* grasslands at Jhansi, U. P. It is not stated whether the experimental area used had formerly been grazed, heavily or otherwise, or protected, and if protected, for how long, before these observations were made.

The ecosystematists also state that mean annual productivities of tropical grassland may equal or exceed that of tropical forests, but that productivity from 'savannas' is less than half that of 'grassland' (again, when is a grass cover a grassland and when a savanna?); also that net production from temperate grassland is half that from tropical grassland; this generalization does not recognize the higher protein content of temperate grasses and the need to compare temperate and tropical on the basis of net production of protein per hectare; nor the higher fibre content for much of the year in tropical grasses in a monsoonal environment. But above all, it is not recognized that the differences in net productivity between grass cover types in tropical environments and between grass cover types in temperate environments are probably greater than those claimed to exist between tropical and temperate grasslands.

The work of the physiologists would appear to be of more direct practical application, which must be the ultimate test; for example, on the comparative light and temperature requirements for the growth of young vegetative material of temperate and tropical grasses, the efficiency of energy conversion through photosynthesis, and hence the potential dry matter production at three levels of increasing complexity (Cooper, 1970, Cooper & Tainton, 1968 in Whyte, 1974a, p. 175).

Ecosystems of arid lands
Physiographically, seven ecosystems are recognized in western Rajasthan (Sax-

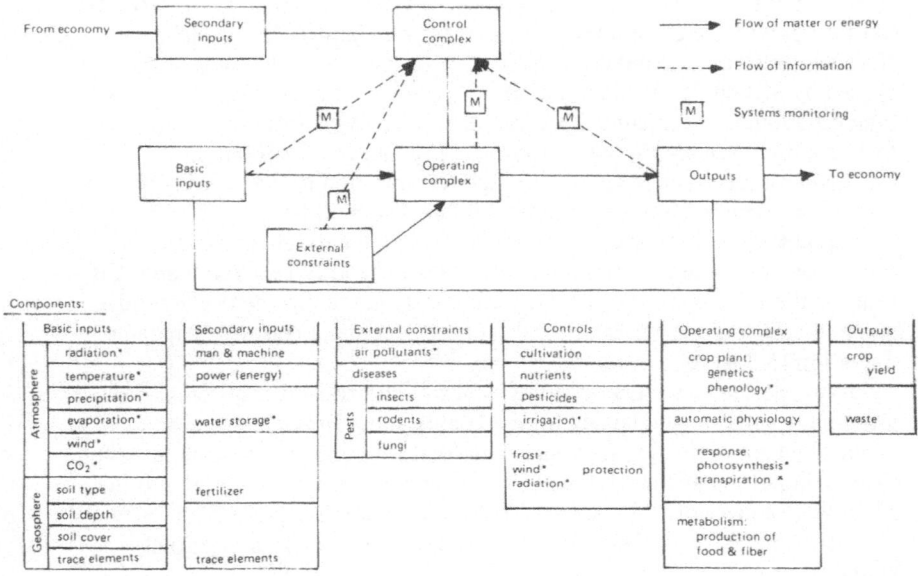

Components:

Basic inputs		Secondary inputs	External constraints		Controls	Operating complex	Outputs
Atmosphere	radiation*	man & machine	air pollutants*	Pests	cultivation	crop plant: genetics phenology*	crop yield
	temperature*	power (energy)	diseases		nutrients		
	precipitation*		insects		pesticides		
	evaporation*	water storage*	rodents		irrigation*	automatic physiology	waste
	wind*		fungi			response: photosynthesis* transpiration *	
	CO₂*				frost* wind* protection radiation*		
Geosphere	soil type	fertilizer					
	soil depth					metabolism: production of food & fiber	
	soil cover						
	trace elements	trace elements					

Flow of matter or energy
Flow of information
M   Systems monitoring

Fig. 2.4. A scheme for a systems analysis of land use in agriculture, the asterisks indicating the components related to weather (Landsberg, 1958).

ena, 1972): 1. the hills and rock outcrops, 2. piedmont slopes, 3. alluvial plains, 4. saline flats and depressions, 5. graded river beds, 6. sand dunes, 7. aquatic areas. The vegetation of these seven land classes is : hills, rock outcrops and piedmont slopes — mixed xeromorphic thorn forest; the extensive alluvial plains — mixed xeromorphic woodland; saline depressions and flats — halophytic succulents; graded river beds — dwarf, semi-scrub; sand dunes — psammophytic scrub, xeromorphic thorn and wooded desert vegetation. Seven types of grass covers are associated with these habitats, and the trend of plant succession in each land class, when moderately grazed, severely grazed or protected, is illustrated in detail (Gupta & Saxena, 1972).

In one of the integrated surveys in Community Development Blocks noted in 5.3.3, the land in the Chhotan Block of Barmer District has been evaluated (Gupta & Saxena, 1971) on the basis of the ecological parameters: topography, soil texture, available nutrients, hazards, depth to available water, quality of water, natural vegetation, percentage slope, etc. (climate is almost uniform within the Block). Beyond the Block limit lies the Rann of Kutch, into which flows the river Luni, ephemeral, flowing for only two months during the year. Gupta & Saxena suggest the possible types of land management which might be adopted for the six agro-ecological units which they recognize (Table 2.1).

Noy-Meir (1973) has reviewed (with 119 literature references) the structure and function of the ecosystems of deserts (or arid lands). The classification of arid ecosystems is generally consistent with the terms and maps of Meigs, as used by McGinnies, Goldman & Paylore (1968): extreme arid (E) — less than 60-100mm. mean annual precipitation; arid (A) — from 60-100mm to 150-250mm; semi-arid (S) — from 150-250mm. to 250-500mm. The higher limits refer to areas with high transpiration in the growing season, e.g. regions with subtropical summer rainfall. The limit between A and E corresponds roughly to the border between diffuse natural vegetation and vegetation contracted to favourable sites only. The limit between S and A is roughly the drier limit of diffuse dryland farming; the limit between semi-arid and non-arid zones is where dryland farming becomes a reasonably reliable operation (5.3.4).

Having defined desert ecosystems as 'water-controlled ecosystems with infrequent, discrete and largely unpredictable water inputs', Noy-Meir (1973) enquires what are the implications of this definition 'for the behaviour of the system, in particular the patterns and dynamics of energy flow in it and the adaptive strategies of its organisms? What are the implications for our attempts to understand this behaviour and to represent it by models, conceptual, graphical or mathematical?' (see Fig. 2.5).

'The distinctions between active and reserve biomass and between the patterns of production and translocation in different types should therefore be important elements in models of arid ecosystems. They are of consequence in modelling not only of primary production but also of herbivory and its effect on long-term production potential. An interesting aspect is the analysis of the different pulse-reserve strategies as optimized strategies in different environments.

'An examination of data from various arid regions shows that the average annual net above-ground primary production varies between 30 and 200 g/m^2 in the arid zone and between 100 and 600 g/m^2 in the semi-arid zone. Estimates of below-ground production are scarce, but are given in some Russian papers. It seems that total production may be 100–400 g/m^2 for arid, 250–1000 g/m^2 for semi-arid communities.

'These data, as well as some which directly relate productivity to precipitation, suggest that a fair proportion of variation in productivity (Y) in arid ecosystems could be accounted for by a linear regression on precipitation:

$$Y = b(P - a) \quad (Y = 0 \text{ if } P < a)$$

where a may be interpreted as the total of "ineffective precipitation" or water losses (evaporation and runoff) and b as the average water use efficiency of the community. The "zero-yield intercept" a is between 25 and 75mm/year.

Table 2.1. India, Rajasthan: Integrated ecological surveys. Characteristics of various agro-ecological units for potential development in Chhotan Block, Barmer District (Gupta & Saxena, 1971).

Agro-ecological zone	Area in sq. km.	Land-use capability class	Habitat	Hazards	Soils	Depth to water table in metres	Water Quality	Natural vegetation
1.	57	VIII	Saline depressions and water-logged areas	Saline – sodic soils	Deep sandy	up to 6m.	$C_5 - S_3$ and above (Highly Saline)	*Sporobolus/Desmostachya* type.
2.	36	VIII	Hillside dunes	Wind and water erosion	Deep sandy	60m. and above		*Acacia senegal/Salvadora oleoides*
3.	10	VII and VIII	Hills and piedmont areas	Water erosion	Shallow mixed with weathered rock fragments	60m. and above		*Euphorbia caducifolia/Acacia senegal*
4.	1165	V	Interdunal hummocky	Wind and water erosion	Deep, calcareous sandy loam	35-60m.	$C_3 - S_5 C_4 S_1$ (Moderately Saline)	*Prosopis cineraria* and *Prosopis/Tecomella* community
5.	1340	VI	Sand dunes	Severe to moderate wind erosion	Deep fine sandy	35 – 60m.	-do-	*Prosopis/Tecomella/Acacia jacquemontii* (low and medium dunes), *Prosopis/Salvadora* (stabilised dunes), *Calligonum/Aerva persica* (unstabilised high dunes) *Calligonum/Panicum* (high dunes with active crest)
6.	553	III and IV	Interdunal plains	Saline water and wind erosion	Deep, sandy to sandy loam slightly calcareous	5 – 35m.	$C_2 - S_1 C_3 S_2$ (Brackish to moderately saline)	*Salvadora/Prosopis/Calotropis/Clerodendrum* community

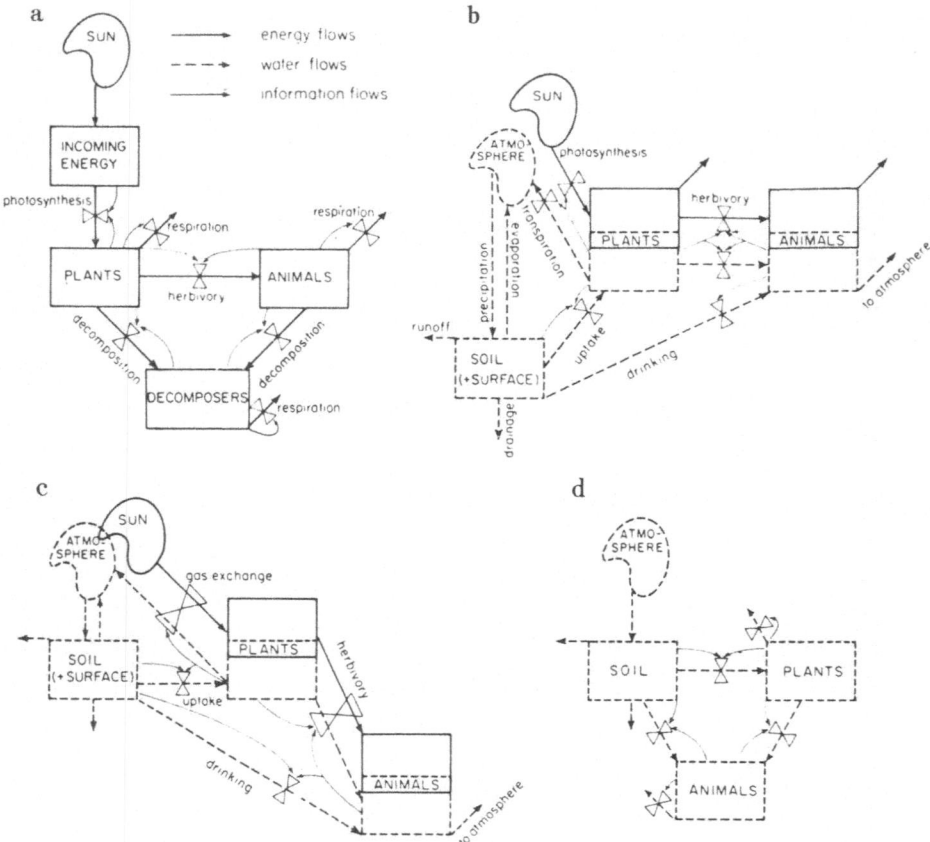

Fig. 2.5. Compartment models of desert ecosystems (Noy-Meir, 1973) a. Energy flow model; b. Energy and water flow models combined (decomposers not shown); c. Same, simplified; d. Water flow model alone.

The efficiency *b* is between 0.5 and 2 (mg. dry matter/g water) (or g/m² mm) for above-ground production, from which one may estimate it to be 1–6 mg/g for total production. This is considerably lower than the A/T values. of 5–50 mg/g measured for individual desert plants over periods of hours by gas exchange measurements, but similar in magnitude to efficiencies reported for irrigated crops in arid climates over a growing period. Apparently adaptations of desert plants for more efficient water use just compensate for energy losses due to the irregularity of the water supply.

'The accumulation of standing live plant biomass and the turnover rate (productivity/biomass) in deserts depend on the dominant type. In ephemeral communities there is 100 per cent turnover of shoots (and of root biomass in annuals) during the growth period of 2–5 months. The numbers for net annual production are equal to peak biomass (mean biomass has little meaning). In fluctuating perennials annual foliage production may be 50–95 per cent of peak foliage biomass, but when stems and roots are included, production is probably only 20–40 per cent of biomass, which in communities where the type is dominant amounts to 150–600 g/m² above ground or 400–2500 g/m² total. These turnover rates are higher than in forest or tundra. In arid and semi-arid communities of stationary drought-persistent trees, shrubs and cacti, annual production may be only 10–20 per cent of a standing biomass of 300–1000 g/m² above ground, i.e. possibly 600–400 g/m² including roots.' (See also Noy-Meir 1974)

2.7 Man as component of tropical rural ecosystems

Introduction of man into the study of the ecosystem
Dr. M. A. Little, Co-ordinator of the Human Adaptability Office of the U.S. National Committee for the International Biological Program, introduced a conference in Annapolis, Maryland, May, 1973, on Man and the Ecosystem by stating that 'truly integrated research incorporating man as a component of ecosystems has never been successfully initiated. . . . One of the principal barriers to truly cooperative multi-disciplinary research is the tradition of individual entrepreneurship among scientists . . . the independent worker mode of investigation . . . is neither useful nor desirable when the complexities of ecosystems are being explored.' The 1973 U.S./I.B.P. conference discussed energy flow in peasant and primitive societies, human settlements and their effect on the environment, and the type of information which was needed in studies of man and ecosystems.

A Workshop sponsored by the U.S./I.B.P. held at University Park, Pennsylvania, August, 1974, discussed the general topic of energy flow in human communities under the following sections: 1. implications of these studies in the fields of ecology, human biology and social science; 2. case studies on methodology; 3. methodological and technical problems associated with energy flow studies; 4. improvement of methodologies employed in measurement of energy flow. The report of this Workshop contains many models which might be used in this type of research.

Table 2.2 quoted from a separate study on rural nutrition in Monsoon Asia (Whyte, 1974b) provides the link between the biological ecosystems discussed above, and the following considerations of man as a biological component of tropical rural ecosystems, before he himself becomes an economic resource (2.7) and after he has acquired the capacity to understand, within his limitations, and to manage the other components of the now economic ecosystems to which he belongs (4.3).

In Table 2.2 are brought together under twelve heads some of the major types of rural economies or ecosystems found throughout the region. These twelve types are further classified in relation to the standards of nutrition generally found among the rural people in each individual type. The productive energies of these peoples to produce food up to the plateau level of their land's potential depends in the final analysis on their correct nutrition, on the creation of adequate incentives to give them the mental and physical energy, so that they may produce more and better plant and animal crops per unit area available to them, to provide food for their own improved subsistence and also for export to urban and industrial centres, and cash crops for export to the developed world. The overall picture which can be seen from Table 2.2 is of a general deterioration and increasing lack of variety and balance in the diet, from the most primitive rural communities down through the swidden and dryland farmers to the wet-rice cultivators, and, potentially at least, rising again in those areas in which some form of alternate husbandry can be evolved and at least some of the produce retained for use on the farm.

These are some of the tropical agroecosystems which are, according to Janzen (1973), misunderstood by specialists in temperate zones and mismanaged by those in the tropics. Since this critical review has rather a socioeconomic than a biological bias, it comes more properly in a separate study.

Anthropology and the total environment
Man is introduced into the text of this study at several stages in human development. First, we consider man as yet another integral, biological component of tropical rural ecosystems, in symbiosis with the flora and the fauna, perhaps cultivating a patch of some specialized food that he cannot obtain from his natural surroundings, Other examples follow of intermediate conditions, through a swidden phase to forms of settled cultivation at a subsistence level, without and then with irrigation. In another section (4.3), man is assumed to have evolved to a higher level of technological competence, at which he can supposedly control and appear to dominate the physical, biological and other economic components of economic ecosystems. The main question then is how does one assess man as an economic resource in himself, and survey his capacity to respond mentally and physically to the demands for increased effort placed on him.

Within Asia, the number of people who may be considered as solely as biological components is now very small, and the area of land which they use for collecting, hunting and primitive cultivation wholly unaffected by neighbouring cultures is minute in relation to the whole land area of the region. In the study of land use and the assessment of potential, it is necessary to consider the extent to which studies of anthropologists concerned primarily with man give a true picture of the whole ecosystem, with man as their primary field of interest, and the approaches and techniques which are adopted in such studies.

44

At this point the marked dichotomy appears between the environmental determinists and those who believe that man, through the mere fact of being human, somehow rises above the influences and hazards of the biological world to which all other living components are subject. This is 'the most salient theoretical controversy among anthropologists' (Salisbury, 1973) at the present time.

In his introduction to a symposium on environment and cultural behaviour, Vayda (1969) states that the main objective at that symposium was to make cultural behaviour intelligible by relating it to the material world in which it occurs. The influence of the material world upon cultural behaviour deserves just as much attention as the traditional subjects of anthropological study; ideologies, human values, former cultural practices, linguistics, motivational patterns, personality structure, etc. Anthropologists are divided in their approach. Some regard items of cultural behaviour as part of fundamental systems that also include environmental phenomena; other show that environmental phenomena are responsible for the origin or development of the cultural behaviour under investigation.

The anthropological dilemma is further discussed by Rappaport (1971a): 'The concept of culture at autonomous, superorganic, characteristic exclusively or almost exclusively of *Homo*, has presented some obstacles to the assimilation of a general ecological perspective for anthropology. If culture is not organic and cannot be understood or explained in terms of the laws governing organic and inorganic phenomena, of what possible relevance to its elucidation can be general ecological theory, bound as it is to biological considerations? Further, how can ecology, which deals with that which is common to all species, be of assistance in understanding culture which presumably occurs almost exclusively among men? . . .

'While it may be that special laws (anthropologists have generally been unsuccessful in discovering them) govern the ways in which culture works, we must look to larger natural systems if we are to understand the functions and effects of cultural phenomena. . . . We can surely say that cultures, or components of cultures, form major parts of the distinctive means employed by human populations in fulfilling their biological needs in the ecosystems in which they participate. . . To so regard culture neither slights what may be its unique characteristics nor demands any sacrifice of anthropology's traditional goals. Indeed, it advances these goals by proposing additional questions to be asked about cultural phenomena. We may ask, for instance, what effects particular social conventions, such as rules of residence and group affiliation, or widespread cultural practices such as warfare have upon the dispersion of human and animal populations over available resources. We may inquire into the effects of religious concepts and rituals upon the birth and death rates and the nutritional status of those who perform them or believe in them, and we may investigate the ways in which men regulate the ecosystems which they dominate.

'In such questions as these we come to the significance of an ecological perspective for anthropology. It leads us to ask whether behaviour undertaken with respect to social, economic, political or religious conventions contributes to or threatens the survival and well-being of the actors, and whether this behaviour maintains or degrades the ecological systems in which it occurs. While the questions are asked about cultural phenomena, they are answered in terms of the effects of culturally informed behaviour on biological systems: organisms, populations and ecosystems. The distinctive characteristic of ecological anthropology is not simply that it takes environmental factors into consideration in its attempt to elucidate cultural phenomena, but that it gives biological meaning to the key terms – adaptation, homeostasis, adequate functioning, survival – of its formulations. . . .

The discrepancy between cultural images of nature and the actual organization of nature is a critical problem for mankind and one of the central problems of ecological anthropology. To cope with this problem the ecological ethnographer must prepare two models of his subject matter. One, let us call it the "cognized model", is a description of the knowledge and beliefs concerning their environment entertained by a people. It is in terms of this model that they act. The second, we may call it the "operational model", is a description of the same ecological system (including all the people) in accordance with the assumptions and methods of the science of ecology. While many components of the physical world are likely to be included in both the cognized and operational models, their membership will seldom, if every, be identical. The operational model includes those organisms, processes and cultural practices which ecological theory and empirical observation suggest to the analyst affect the biological well being of the organisms, populations and ecosystems under consideration. It may include elements of which the actors may be unaware (such as microorganisms and trace elements), but which affect them in important ways. The cognized model, on the other hand, may well include components, such as supernaturals, whose existence cannot be demonstrated by empirical procedures, but whose putative existence moves the actors to behave in particular ways' (Rappaport, 1971a).

The land scientist concerned with the survey of actual or potential use of the habitats occupied by primitive communities is or should be more interested in the optimal conservation of land resources than in the attainment of maximum productivity, because of the nature of the environment so inhabited. At the level of primitive cultures, the field worker cannot underestimate the dominant role of the environment – selection of living sites and plots for cultivation, adaptation to seasonal change by movements, and above all the direct provision of optimal nutrition for the family are the major preoccupations of hunting/collecting and swidden cultures.

In Table 2.2 the standards of nutrition which are characteristic of twelve types of rural economies and land use are indicated. It may be concluded that primitive man in his evolution towards technological mastery is faced with

46

progressively increasing disequilibrium with his habitat. When an advanced stage in this trend towards disequilibrium has been reached, three courses may be taken: a. removal of man as a biological component of the ecosystem (practised with only limited success in forest reserves); b. reduction in human population but maintenance of former economic system, by controlled emigration – for example, from Java to Sumatra, and Luzon to Mindanao; c. replacement of the indigenous land use practices by settled or controlled swidden agriculture in a partly monetized economy.

The techniques associated with the discipline of economic anthropology may be used for the interpretation of the dynamics of economic and cultural change (Nash, 1966), from primitive (small, non-westernized, non-monetized) to peasant (partly monetized) economies. The objective would be to observe the gradual evolution of man from a biological organism with relatively little effective power over his environment, to an economic agent who can manage (and mismanage) and so radically transform the environment . . . Associated with this evolution is the progressive destruction of climax vegetative covers, leading with excessive increase in population to disequilibrium in land use, expressed as loss of water-retaining capacity and erosion at the higher elevations, and disturbance of the water regime and flooding at lower elevations. This has been clearly demonstrated for Java by Made Sandy (1973; see Chapter 7).

Evolution from biological to economic status
There are many cultures in Asia intermediate between total dependence on the environment (e.g. the Punan of Kalimantan), and the limited, seasonal dependence of the monetized subsistence agriculturists (5.3.2). While isolated communities unaffected by contact with the outside world are few, there are a great many which, partly monetized, still devote the greater part of their activities to the subsistence of their families, activities which are still wholly dependent upon the environment. An example may be taken from the socio-ecological study of the Malays of south-west Sarawak (Harrisson, 1970). Over twenty years of investigation were undertaken by a team of workers, some of whom changed, the leader remaining the same. This study is thus perhaps unique in Asia in the length of time involved, permitting a profound knowledge of mental attitudes and the accurate recording of incipient and actual change, and its social, ecological and economic results. The preferred activities of almost all Malays in this area are directly related to water; land activities are an unavoidable supplement, occupying more time where villages have limited access to fishing (Fig. 2.6). Perhaps for this reason, the primary objective of clearing land is not, as throughout much of Asia, the attainment of permanent padis or upland fields, but progression through a preliminary cropping phase, often including padi, to perennial tree crops – rubber, durian or coconut – which do not demand continuous care.

47

Table 2.2. Rural economies, land use, systems of crop and animal husbandry, and standards of nutrition (Whyte 1974b).

Rural economies	Sources of plant foods	Sources of animal foods	Nutritional status
1. Hunters and collectors; low density of population. Few such communities remain.	Wild plants, for roots and tubers, leaves, shoots, seeds, fruits, nuts. Fungi.	Wild fauna, only rarely including larger animals; small animals. rodents, bats, lizards, birds, insects, worms. Some fish trapping. Collection of honey.	Relatively balanced diet, when not crowded and harassed by neighbouring swidden communites. Possible calorie deficiency probably compensated by small physical stature.
2. Hunters and collectors in primary and secondary forest, with swidden. Low population density.	As above, but decreasing use of wild roots and tubers. Cultivated cereals and legumes, roots and tubers, and vegetables adapted to local environment.	Wild fauna, plus scavenging domestic fowl and domesticated wild fowl, ducks, goats. Domestic animals mainly for ceremonial use and consumption. Fishing.	Diversified diet, with more carbohydrate from the cultivated crops.
3. Swidden, with regenerated secondary forest as fallow. Increasing density of population.	Wild plant foods of lower stages in succession, plus the crops as above.	Secondary levels of wild faunal components, following regression of plant cover. Hunting becomes less important. Domestic livestock as above. Cattle and/or buffalo kept by some groups for draft.	Animal protein may be less frequent in diet, depending chiefly on availability of fish.
4. Fishing communities on sea coasts and inland lakes where fishing as the primary occupation is combined with some agricultural and/or horticultural activities	Adapted cereal, root, legume and vegetable crops; collection from wild plants and animals of moderate or little significance.	Small amounts of fish eaten regularly, depending on economic status and length of fishing season.	Diet may be adequate or marginal in calories, and consequently inadequate in protein; calcium requirement met where small fish eaten whole. Vitamin and iron deficiencies.
5. Swidden, with inadequate regeneration of secondary forest; soil fertility tending to fall. Excessive population pressure for this system.	Mainly millets, maize and upland rice, where adapted. Contribution from wild flora less significant. Limited legumes and vegetables on seasonal basis.	Contribution from wild fauna less regular. Scavenging fowl, pigs and goat, dependent upon environment and ethnic traditions. Consumed at festivals. Some fishing.	Diet likely to be inadequate in calories and most nutrients, with very marked seasonal fluctuations.
6. Settled dryland cropping with swidden on infertile soils. Excessive population pressure per unit of cultivable land.	Adapted cereals and grain legumes. Vegetables few and limited to rainy season in absence of water for small vegetable plot. Up to two-years' store of cereal may be retained.	Scavenging or free-ranging fowl, pigs, goats and cattle, dependent upon environment and ethnic tradition.	Diet dependent on access to markets for sale of produce. Diet marginally adequate in calories, with deficiencies of animal and plant protein. and serious deficiency of vitamin A, other vitamins and minerals. Marked seasonality.

Rural economies	Sources of plant foods	Sources of animal foods	Nutritional status
7. Grazing and browsing communities. Migratory, nomadic or settled, with some cultivated land, dry and/or irrigated, at base village or camp.	Mixed cereals, though not usually eaten at same meal, with some grain legumes. Few vegetables.	Sheep, goats, cattle, buffalo, camel, mostly on free-range grazing, with small livestock around base.	Male diet probably adequate in animal protein, therefore in total protein, most vitamins and minerals. Diet of vulnerable groups adequate only while flocks and herds at base village, calories, plant proteins inadequate; consistent deficiency of minerals and vitamins. Major deficit is water.
8. Irrigated plus settled dryland swidden. River valleys, terraced slopes and hill-sides. Excessive population pressure per unit of cultivable land.	Mixture of several dryland cereals, root crops with wet rice. Few legumes. Highly seasonal vegetable supplies. Some collection of wild green leaves.	Scavenging poultry, pigs, draft buffalo. Eaten only at ceremonies. Small wild-life and insects hunted, fish in padis and streams.	Calories marginal; protein, vitamin and mineral deficiencies.
9. Settled dryland agriculture. High to excessive population density.*	As with (6)	As with (6)	As with (6)
10. Paddy monoculture. High to excessive population density.**	Wet rice, with few alternate crops.	Draft buffalo. Ducks. Fish in padi. Scavenging fowl and pigs near homesteads.	Serious protein deficiencies, especially among vulnerable groups, since animal produce sold and usually only small amounts of fish consumed. Deficiencies of thiamine associated with highly milled rice. Deficiencies of calcium, iron, vitamin A, riboflavine predominant; some other vitamins inadequate.
11. Irrigated alternate husbandry. High population density.	Major cereals, grain legumes, fodder crops, vegetables, cash crops grown in crop rotations, with irrigation water available in all seasons, integrated with adapted forms of intensive animal husbandry.	Intensive forms of animal husbandry (dairy cattle, chickens, pigs, and ducks) fed in stalls or special buildings on feeds and fodders produced from cultivated land.	Farmers tend to sell their cash products of high nutritive value to urban markets until economic status permits some retention to family. Thus most diets deficient in protein, vitamins and minerals.
12. Plantation agriculture.	Food grains, some legumes and vegetables may be grown or purchased.	Cows for milk kept by Indian plantation workers. All may keep scavenging chickens, pigs, ducks, goats, depending on location and ethnic preferences.	Amount and quality of food intake dependent upon provision by plantation manager to workers and their families. Great variation in number of meals provided per day by management.

* Greatest cultivated area in Asia.
** Greatest population density in Asia.

'There are certain rhythms inherent in the land itself; or held to be so in fact by the traditions of Malay (and other) peoples in Borneo:' Thus land activities are divided into: 1. those which depend on some terrestrial cycle, not determined directly by man (e.g. orchard fruits); 2. those which depend on some terrestrial cycle, but where man can exercise an appreciable element of control (e.g. rice); 3. those which are not − essentially or rhythmically − cyclical but are affected directly by seasonal or climatic (weather) factors on land (e.g. collecting swamp produce); 4. those which are immediately affected by what happens on the water; mostly those used as 'alternative occupation' by people who would otherwise be on the water − that is, normally men and grown boys only (e.g. rubber). A fifth category includes the effect of remote, external events, such as floods, wars, etc., and the international prices for commodities such as rubber.

These observations were achieved by detailed, continuous observation of daily activities of villagers, of men and women, of different age groups, of interactions with outsiders, throughout the seasons of many years with all their climatic and other fluctuations. The history and pattern of settlement, topography, vegetation, communications, agricultural and fishing methods, the pattern of effort, the marketing of products and the standard of living were all considered.

Another group in Borneo with a totally different culture − the Iban, who subsist by means of shifting cultivation − were studied in relation to their social organization, dwelling patterns, land tenure, agricultural methods, cycle and economics, the organization and division of labour, the amount of time spent on production, yields, consumption, ritual, and − a relatively new factor − the spread of rubber as a cash crop (Freeman, 1970). It is concluded that Iban agriculture and methods of land utilization 'can only be understood in relation to the various groups which characterize Iban society. . . . It is only when the interdependence of economics, social organization and ritual are understood that there can be any genuine appreciation of the nature of Iban agricultural practice.'

The study of the Malays of south-west Sarawak (Harrisson, 1970) was of a relatively limited area but over a very long period of time. A contrasting approach was adopted in a study in Cambodia (Khmer Republic) covering most of the rural population.This was done by means of questionnaires and personal interviews undertaken by trained observers and students with the help of school teachers, religious leaders, monks and peasants throughout the country, but over a more limited time span (Delvert, 1961). The ecosystem picture was built up by reference to specialists − meteorologists, foresters and vegetation scientists, agriculturists, geologists, geomorphologists and soil scientists. Here information of direct interest to land planners was obtained over different Cambodian ecosystems − those who cultivate rice alone, those who cultivate and fish, those who cultivate rice and make a living in forests, those who cultivate rice and horticultural crops, those who grow sugar as well

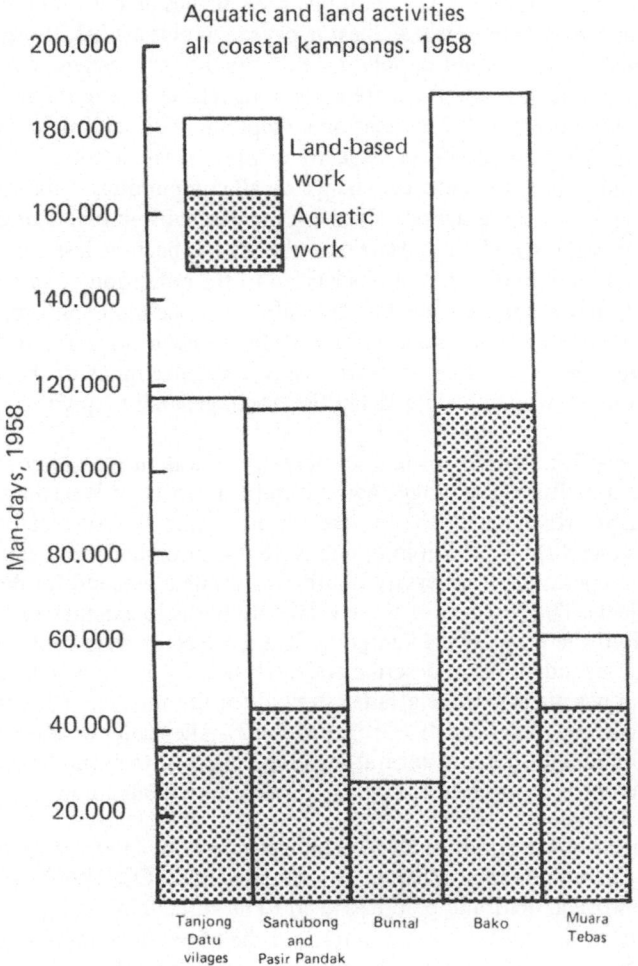

Fig. 2.6. Sarawak: male daily work in Malay villages, 1958 (Harrisson, 1970).

as rice, and those who grow rice and produce craft work; as well as those who raise tobacco, the maize growers and animal husbandmen who also produce soya and haricot on shifting cultivation plots. It was concluded that rice monoculture was the traditional Cambodian activity, and that higher population density has necessitated diversification. Above all there are environmental reasons: 'Le riziculteur est esclave du climat et de traditions très anciennes. Le cultivateur de chamcar, proche encore du jardinier, s'est habilement adapté aux données naturelles et aux conditions particulières de cultures commerciales; il a créé un terroir bien organisé; il est le paysan le plus évolué.' Despite regional differences, the peasant population exhibits the same general characteristics. 'Le rhythme des saisons accusé par la longueur et la sévérité de la saison sèche, par l'importance de l'inondation de septembre et octobre, qui coincide avec les grandes pluies de Vossa, règle avec rigueur la vie de tous.'

Increased population enforces change on all communities whose basic resource — land or fishing rights — does not increase with numbers of people. Among the Malays studied by Harrisson the sea has become less able to meet the increased requirements, and this has led to the extension of activities on the land. Cambodians are forced to diversify from rice monoculture. Among the Ibans rubber as a permanent crop provides a cash income as a buffer against shortage of rice. Iban men are free to tap, since much of the agricultural work is done by women, while Malay fishermen generally tap when they cannot fish.

It appears from studies made elsewhere in Sarawak in six kampongs by the Department of Rural Sociology, Agricultural University of Wageningen, that development, which means change, creates disorganization of social structure, and that some differentiation in opinions, that is disorganization and lack of internal co-operation, is necessary in order to create a demand for development projects. The first aim of these village studies is to examine examples from each of the projects the kampongs had received or were trying to acquire. The second aim is to describe and analyse the ways in which the different kampongs manage those affairs essential for the survival of the village as a unit with its own identity (Grijpstra, 1973). The leader of this survey and four post-graduate students in rural sociology planned to spend between two and four months for each of the village studies, depending upon the complexity of the villages.

Even at his most primitive level, man begins to have an effect on the vegetation of the humid tropics (Papua New Guinea/UNESCO, 1960). At this stage, primitive agricultural practices need to be studied in the light of the complexity of the cultural background, and the effect of proposed changes investigated in order to ensure the continuing welfare of the people concerned (Geddes, 1960). The importance of man, his economy and society, in the acceptance and application of a new technology is often overlooked because: the technology itself is the product of advanced western societies (Dore, 1968). In less developed societies, 'the economy is less fully differentiated from the other spheres of life and economic acts can consequently have deep religious or social significance: "traditional" may be almost synonymous with "sacred".' Attempts at large-scale change should be preceded by a pilot project, the results of which need to be carefully studied not only by agronomists and economists, but also by experienced sociologists. Dore also warns, however, against what he calls the 'Awed Spectator syndrome' — the tendency of anthropologists, economists and ecologists to assume that, because something exists, there is probably a good reason for it, and that it should not be subject 'to the heartless encroachment of colonial authority . . . or the blundering optimism of the more naïve proponents of wholesale change.' One should not 'underestimate the capacity of social systems to adapt, and even to survive the occasional piece of rough surgery.'

The administrative and political structure through which change will be filtered must be carefully examined for possible distortion of the measures; or, in the words of M. N. Srinivas, 'the government proposes, dominant caste disposes.' Those introducing change should be close to the society into which change is to be introduced. There is not yet any established body of sociological precepts enabling reaction to change to be predicted with certainty. 'The closest one can get to prediction, perhaps, is by the intuitive rather than the ratiocinative method — by, that is to say, learning enough about the society and getting a sufficiently intimate acquaintance with the farmers, their language, their jokes, their fears and their sorrows, to be able to think oneself into their skins and guess at their reaction to a proposed change' (Dore, op. cit.).

Table 2.3. India: Evolution of contrasting land use patterns on newly irrigated and non-irrigated holdings within the Rajasthan Canal Project (Bharara, Malhotra & Patwa, 1974).

Type of land	Newly irrigated		Non-irrigated	
	Area (ha)	Percentage	Area (ha)	Percentage
1 Uncultivable:				
Ussar Land	1781	4.71	284	1.38
2 Uncultivable:				
(a) Fallow land	24	0.06	954	4.62
(b) Banjar land	30108	79.69	19298	93.52
Total	30132	79.75	20252	98.14
3 Cultivated:				
(a) Irrigated	7141	18.90	–	–
(b) Unirrigated	32	0.08	99	0.48
(c) Total cropped area	7173	18.98	99	0.48
(d) Double cropped area	1302	0.34	–	–
(e) Net area sown	5871	15.54	99	0.48

There is, however, a growing literature on the results of and reactions to change which can serve as useful parallels (see chapter 8 in Whyte, 1974b). In agricultural development in general, the introduction of irrigation represents a major catalyst of change. In Tehsil Gharsana, district Ganganagar, the introduction of irrigation from the Rajasthan Canal has induced considerable socio-agricultural changes (Bharara, Malhotra & Patwa, 1974). Traditional caste occupations shifted to cultivation and agricultural labour, subsistence farming gradually became commercialized, the mean size of land holding per household decreased, and the landless class gradually became landowners. Agricultural inputs and demand for farm labour increased, and a new set of farming communities sprang up. The area sown, yield, fallow land, agricultural inputs and employment of labour all changed (Table 2.3). In two villages in Mandya District in Karnataka, one with irrigation, one without, growing poverty nas disrupted social and economic relations in the dry village between cultivators on the one hand and the artisans and landless on the other; this has not occurred in the irrigated community, where economic conditions have progressively improved (Epstein, 1962; 1973; Srinivas, 1974). In contrast, however, the introduction of high-yielding varieties of foodgrains in Intensive Agricultural Development Projects in India is not only increasing the disparity between large landowners and small farmers who cannot afford to take advantage of the new technology, but is also breaking down the traditional patron/client relations between increasingly wealthy farmers and tenants, sharecroppers and labourers. Many landowners farm lands they formerly leased, leaving their former tenants destitute; mechanization has reduced opportunities for employment of landless labourers. Labourers, tenants and sharecroppers are worse off in real terms, despite increased wages, following the introduction of payment in cash to replace payment in kind (Frankel, 1971). Here, therefore, a change in the economic ecosystem has led to a deterioration in the condition of many of the human components. It may be asked whether an integrated survey, including a rural sociologist, before the introduction of these fundamental changes, might not have devised appropriate checks in the overall project to prevent the serious increase in rural insecurity and tension which has occurred.

Study of man in arid ecosystems
The Central Arid Zone Research Institute, Rajasthan, although primarily concerned with the analysis and improved management of the land resources, gives full consideration to the place of man in the desert ecosystems throughout the arid zones of India. The crucial problem of the Rajasthan Desert is one of human ecology; man has over-exploited the resources of water, plants and soils, has disturbed the ecological balance and therefore caused progressive degradation of the resources. Socio-economic surveys have been made of sedentary and nomadic groups. Historical, political and cultural factors, combined with climatic and geographical factors, have given rise to nomadism as an adaptation to the environment. The nomads of the arid zone may be grouped into four categories: a. the pastoral nomads (Raikas, Sindhis, Parihars, Billochs, etc.); b. the trading nomads (Banjaras, Ghatti-wala Jogis and Gowarias); c. artisan nomads (Gadoliya Lohars, Sansis and Sattias); and d. various other nomads (Nats, Kalbeliya Jogis). (Cent.al Arid Zone Research Institute, 1974.)

Until recently, these nomads had an important complementary role in the economy of the region, but today, modernization has rendered obsolete many of their specialist services and the mutual interdependence between nomads and the sedentary peoples has broken down. The nomads and their livestock have become a menace to the local land and society, and it has become essential to settle them. The objective is now to rehabilitate the trading nomads, the artisans and nomadic cattle breeders of western Rajasthan, while retaining as much as is practicable of their social and economic way of life.

It has, for example, been found that the Gadoliya Lohars (artisan nomads) must be settled in small, agnatic, scattered groups within their own chokla, to avoid feuding between clans. The symbiotic relation with the sedentary populations may be continued through marketing facilities for the products of the artisan nomads. The Banjaras (trading nomads) may be settled in kinship groups (tandas) under the Panchayat Samitis, with their economy built upon agriculture, animal husbandry and trading in cattle and salt (CAZRI, op. cit.).

Socio-economic surveys in Rajasthan show that population growth is higher in the arid zones than elsewhere in India, and that within the arid zone, the percentage increase is higher in the areas with less rainfall. Density of population has increased from 36 persons per sq. km. in 1961 to 45 persons per sq. km. in 1971. The traditional large joint households are disintegrating into small, nuclear households. Caste still governs most economic and social relations. Socio-economic surveys, for example, one of Challakeri Taluka of Karnataka State, when combined with data from Basic Resource Surveys (see 5.3.3) provide the basis for land transformation projects in Community Development Blocks (CAZRI, op. cit.).

Study of man in rice ecosystems
It is necessary to consider the anthropological emphasis on the significance of environment on cultural phenomena in relation to the predominant economic land ecosystems of Asia whose ultimate object is the production of the major foodgrain in Asian nutrition, rice. In general, it may be said that the environment governs the type and intensity of the rice-producing ecosystem, and that the characteristics of these ecosystems are reflected in the agrarian and socio-economic structures of the communities. So far, classification of Asian rice ecosystems has been based on topography, seasonal climate, availability of supplementary water and methods of cultivation and harvesting — see Table 2.4 for a general classification (Delvert, 1973), and the specific examples of systems in Malaysia, Table 2.5, from Hill (1970); see also Löffler (1973) for systems adopted in the high altitude regions of southern Asia, and Geertz on Bali (1972).

See also Fig. 5.2, showing how rainfall and irrigation govern crop growing seasons, and Fig. 5.3 showing cropping patterns practised in Asian rice-producing regions (from Kung, 1971).

Table 2.4. Classification of rice agriculture (Delvert, 1973).

I   DRYLAND RICE CULTIVATION
    a. ladang = swidden (plains or mountain areas)
    b. upland rice, grown in cultivated fields
II   PADI RICE GROWN UNDER NON-INTENSIVE CONDITIONS
    a. 1. rainfed padis (most common)
         riz léger ou hâtif (short growing season – 3-4 months)
         riz moyen (average growing season – 6 months)
         riz lourd   (long growing season – 8-9 months)
       2. padis flooded by river or lake water (always riz lourd)
         – normal rice (generally not transplanted)
         – floating rice
    b 1. padis cultivated only for family subsistence
       2. padis cultivated chiefly for marketing
       3. special case of dry season rice ('boro' in Bengal – always early ripening)
III INTENSIVE PADI RICE
    Intensively cultivated padis are always irrigated
    a. one crop (Upper Burma)
    b. two crops, only one being rice (Chinese type, Japanese type)
    c. double-crop rice
      one crop in rainy season, one crop in dry season
      North Vietnam and south China type
      Java-Bali
      central Tamilnad
    d. Padis in which rice is associated with commercial crop
      central and east Java
      e.g. rice/tobacco or three rice crops and one of sugar-cane in a three-year cycle
    e. terraced padis

Table 2.5. Malaysia: Part of a figure showing system characteristics adopted in a scheme of classification of systems of rice cultivation (Hill, 1970).

(a) Topography	1a	Naturally flat plains or terraces
	1b	Artificial terraces
	2	Slopes
(b) Season	1	Wet
	2	Dry
	3	Wet & dry
(c) Cycle	1	Permanent
	2	Temporary (leading to perennial)
	3a	Shifting, short cycle
	3b	Shifting, medium cycle
	3c	Shifting, long cycle
	3d	Cycle within cycle

56

(d) Annual rotation	1a	One rice crop, monoculture with fallow
	1b	One rice crop, rotation, other crops
	1c	Promiscuous cultivation
	2a	Two crops, rice/rice
	2b	Two crops, promiscuous
	3a	Continuous, rice/catch/rice
	3b	Continuous, rice & other/rice
	3c	Continuous, rice/rice/rice
(e) Water supply & control	1a	Rain/flood, bunded flat fields, drained
	1b	Rain/flood, bunded flat fields, undrained
	1c	Rain/flood, unbunded flat fields
	1d	Rain/flood, unbunded sloping fields
	2a	Irrigation, river & gravity canals
	2b	Irrigation, river & current–driven pumps
	2c	Irrigation, river & animal–powered pumps
	2d	Irrigation, river & mechanical pumps
	2e	Irrigation, tanks & gravity canals
	2f	Irrigation, tanks & animal–powered pumps
	2g	Irrigation, tanks & mechanical pumps
	2h	Irrigation, wells & animal–powered pumps
	2i	Irrigation, wells & mechanical pumps
(f) Preparation for planting	1	Slash only
	2	Slash and burn
	3	Animal trampling
	4	Hoe
	5a	Plough, animal drawn
	5b	Plough, mechanical
(g) Planting method	1a	Direct sowing, broadcast
	1b	Direct sowing, dibbled
	1c	Direct sowing, drilled
	2a	Transplanting, dry nurseries
	2b	Transplanting, wet nurseries
	2c	Transplanting, other nurseries
(h) Harvesting–cutting	1a	Panicles plucked
	1b	Panicles cut singly
	2a	Sickle reaped, base of culms
	2b	Sickle reaped, near panicle
(h) Harvesting–drying	1a	Sundrying on racks
	1b	Sundrying on ground
	1c	Sundrying in shooks
	1d	Immediate threshing
	2	Artificial drying
(h) Harvesting–threshing	1	Trampled by man
	2	Trampled by animals
	3	Threshing boxes
	4	Flail threshing
	5	Mechanical threshing

Different methods of rice cultivation in southeast Asia have marked implications in human ecology (Hanks, 1972). 'Seen in an ecosystem, man is no longer the measure of all things nor the master of nature. He is bound intimately to the grain he would grow.' The energy requirements of rice gathering, and of cultivation by swidden, broadcasting and transplanting may be examined in relation to 1. the number of people in the work force, 2. the duration of work through the crop season, and 3. the area of land to be worked. The input of energy also depends on the cultural definition of workers and work days, level of industry, work quality and perseverance. Each mode of cultivation is appropriate to different densities of population, different economic resources and varying availability of land. 'While shifting cultivation best fits among scattered people, transplanting presumes many people working small holdings, while broadcasting forms an intermediate step between the two' (Table 2.6).

'One value of this ecological approach lies in the new perspectives offered on certain old problems. For instance, the dyked fields are customarily regarded as important stages in the development of riziculture. Our data, however, point not only to mounding dykes in level fields but also to leveling dykes to resume another mode of cultivation. Rather than a stage in development, these dyked fields become adaptations to an environment with a moderately dense population, limited area for cultivation, and the market incentives for increasing production. These modes of cultivation may well have originated in sequential fashion, yet their contribution to human society bears no necessary relation to this sequence, for the technically final and most complex may be discarded for the simplest, when conditions are right. Thus technological evolution may well proceed in different directions and at different rates than social evolution.' A concept of ecological holding and optimal output coming from optimal input is studied, with the corollary that social problems 'are to be understood basically as imbalance caused by insufficient input in relation to output or excessive input in relation to output. In both types of social problem we may expect to find more or less chronic shortages, perhaps of food, labour, land or tools' (Hanks, op. cit.).

The dichotomy discussed at the beginning of this section, which exists among anthropologists – and even more between anthropologists and ecologists – is illustrated in many studies of Asian communities. The minority – Rappaport, Harrisson, Hanks, Leach (1964, 1968) are examples – are aware that a culture – whether it be expressed in cropping pattern, social structure or ritual – evolves because of the environment in which it is located. The majority believes – or appears to believe – that culture is pre-eminent, using from the environment that which fits into its concepts and discarding the rest. One may again turn to Rappaport to redress the balance: 'Men are gifted learners and may continually enlarge and correct their knowledge of their environments. But their images of nature are always simpler than nature herself, and often incorrect, for the ecological systems in which men live are complex and subtle beyond

Table 2.6.   The human factor in relation to methods of rice cultivation
(from Hanks, 1972).

A.   Requirements of human labour per unit of land per crop season

| Method of Cultivation | Labour (man-days/acre) | |
	Average	Range
Shifting	67	42–103
Broadcasting	17	14–24
Transplanting	53	24–133

B.   Size of cultivated fields worked by single household

| | Field size (acres) | |
	Average	Range
Shifting	3.8	1.0–6.8
Broadcasting	10.5	8.0–12.3
Transplanting	5.5	2.0–12.3

C.   Direct manpower requirements per crop season

	Manpower rate (man-days/acre)	Field size (acres)	Manpower/ Crop season
Shifting	67	3.8	255
Broadcasting	17	10.5	179
Transplanting	53	5.5	292

D.   Total human labour input for growing a single rice crop (in man-days)

| Type of manpower requirement | Method of cultivation | | |
	Shifting	Broadcasting	Transplanting
Direct	241	179	292
Indirect	4	122	138 + X
Total	245	301	430 + X

(X = unknown long-term labour input needed for irrigation systems)

E.   Relation of input to output during a single crop season

Method of cultivation	Input man-days/ crop season	Output (short tons of rice)	Output/ Input x 1000
Shifting	245	2.451	9.08
Broadcasting	301	6.237	20.39
Transplanting	430	5.363	12.47

their comprehension.'

Not unnaturally, the land use planner will derive more helpful, practical instruction from the school which considers the ecosystem as dominant than from those for whom the ecosystem is merely the incidental background against which their cultural structures are formulated.

BIBLIOGRAPHY

ALDRICH, S. R., – 1972 – Some effects of crop-production technology on environmental quality. *Bio-Science* 22: 90-95.

AMIRAN, D. H. K., – 1966 – Man in arid lands – 1. Endemic cultures. In 'Arid Lands,' ed. E. S. Hills, pp. 219–237. Methuen/UNESCO, London and Paris.

AMIRAN, D. H. K., – 1966 – Man in arid lands – 2. Patterns of occupance. In 'Arid Lands,' ed. E. S. Hills, pp. 239–254. Methuen/UNESCO, London and Paris.

BALOGH, J., – 1967 – Biogeographical aspects of soil ecology. In 'Methods of Study in Soil Ecology. Proceedings of the Paris Symposium,' Ecology and Conservation 2. UNESCO, Paris.

BARTZ, P. M., – 1972 – South Korea. Clarendon Press, Oxford. 203 pp.

BHARARA, L. P., S. P. MALHOTRA & F. C. PATWA, – 1974 – Some socio-agricultural changes as a result of introduction of irrigation in a desert region. *Annals Arid Zone* 13: 1–10.

BOURLIÈRE, F., – 1973 – The comparative ecology of rain forest mammals in Africa and tropical America: some introductory remarks. In 'Tropical Forest Ecosystems in Africa and South America,' ed. B. J. Meggers, E. S. Ayensu & W. D. Duckworth, pp. 279–292. Smithsonian Institution, Washington D.C.

BOURLIÈRE, F. & M. HADLEY, – 1970 – The ecology of tropical savannas. *Ann. Rev. Ecol. Syst.* 1: 125–152.

BROCKINGTON, N. R., – 1971 – Using models in agricultural research. *SPAN* 14: 26–29.

BUDYKO. M. I.. – 1968 – Solar radiation and the use of it by plants. In 'Agroclimatological Methods: Proceedings of the Reading Symposium,' pp. 39–53. Natural Resources Research 7. UNESCO, Paris.

di CASTRI, F., – 1967 – Les grands problèmes qui se posent aux écologistes pour l'étude des écosystèmes du sol. In 'Methods of Study in Soil Ecology. Proceedings of the Paris Symposium,' pp. 15–31. Ecology and Conservation 2. UNESCO, Paris.

CENTRAL ARID ZONE RESEARCH INSTITUTE, – 1974 – Fourteen Years of Arid Zone Research (1959–1973). CAZRI. Jodhpur. 56 pp.

CHAMPION, H. G. & S. K. SETH, – 1968 – A Revised Survey of the Forest Types of India. Government of India, Manager of Publications, Delhi. 404 pp.

CHARLES-EDWARDS, D. A. & J. H. M. THORNLEY, – 1974 – Mathematical models and crop science. *SPAN* 17: 57–59.

CLAPHAM, W. B., – 1973 – Natural Ecosystems. Collier–Macmillan, New York/Macmillan, London. 248 pp.

COLINVAUX, P. A., – 1973 – Introduction to Ecology. Wiley, New York. 622 pp.

COLLIER, B. D., G. W. COX, A. W. JOHNSON & P. C. MILLER, – 1973 – Dynamic Ecology. Prentice-Hall, Englewood Cliffs. 564 pp.

CONKLIN, H. C., – 1969 – An ethnoecological approach to shifting agriculture. In 'Environment and Cultural Behaviour,' ed. A. P. Vayda, pp. 221–233. Natural History Press, New York.

COOPER, J. P., – 1970 – Potential production and energy conversion in temperate and tropical grasses (106 references). *Herbage Abstracts* 40: 1–15.

COOPER, J. P. & N. M. TAINTON, – 1968 – Light and temperature requirements for the growth of tropical and temperate grasses. (71 references). *Herbage Abstracts* 38: 167–176.

DABADGHAO, P. M. & K. A. SHANKARNARAYAN, – 1973 – The Grass Cover of India. Indian Council of Agricultural Research, Scientific Monograph. New Delhi. 713 pp.

DAKSHINI, K. M. M., – 1967 – Ecological studies on soils of the tropics with special reference to wet and semi-arid lands of north India. In 'Methods of Study in Soil Ecology. Proceedings of the Paris Symposium.' pp. 39–42. Ecology and Conservation 2. UNESCO, Paris.

DELVERT, J., – 1961 – Le Paysan Cambodgien. Mouton, Paris, The Hague. 740 pp.

DELVERT, J., – 1973 – Un projet de classement des types de riziculture. In 'Vergleichende Kulturgeographie der Hochgebirge des Südlichen Asien,' ed. C. Rathjens, C. Troll & H. Uhlig, p. 106. Franz Steiner Verlag, Wiesbaden.

DENT, J. B. & J. R. ANDERSON, (eds.), – 1971 – Systems Analysis in Agricultural Management. Wiley, Sydney. 394 pp.

DOOGE, J. C. I., – 1973 – The nature and components of the hydrological cycle. In 'Man's Influence on the Hydrological Cycle,' pp. 2–18. FAO, Rome.

DORE, R. P., – 1968 – Climate and agriculture: the intervening social variables. In 'Agroclimatological Methods: Proceedings of the Reading Symposium,' pp. 201–207. Natural Resources Research no. 7. UNESCO, Paris.

DUBOS, R., – 1970 – Man and his ecosystems; the aim of achieving a dynamic balance with the environment, satisfying physical, economic, social and spiritual needs. In 'Use and Conservation of the Biosphere,' pp. 177–189. UNESCO, Paris.

DUCKHAM, A. M., – 1971 – Human food chains. In 'Systems Analysis in Agricultural Management,' ed. J. B. Dent & J. R. Anderson, pp. 348–379. Wiley, Sydney.

DUVIGNEAUD, P., – 1967 – Ecosystèmes et Biosphère. Minist. Educ. Nation. et Cult., Documentation 23:2, Bruxelles. 137 pp.

ELLENBERG, H., – 1971 – Ecological Studies. Analysis and Synthesis. Vol. 2. Integrated Experimental Ecology. Lange and Springer, Berlin, Heidelberg, New York. 214 pp.

ELLENBERG, H., – 1973 – Die Okosysteme der Erde. Versuch einer Klassifikation der Okosysteme nach funktionalen Gesichtspunkten. In 'Okosystem–Forschung,' ed. H. Ellenberg, pp. 235–265. Springer–Verlag, Berlin, Heidelberg, New York.

EMLEN, J. M., – 1973 – Ecology: An Evolutionary Approach. Addison–Wesley, Reading, Mass. 494 pp.

EPSTEIN, T. S., – 1962 – Economic Development and Social Change in South India. Manchester Uni. Press. 353 pp.

EPSTEIN, T. S., – 1973 – South India, Yesterday, Today and Tomorrow. Macmillan, London. 273 pp.

EVANS, F.C., – 1956 – Ecosystem as the basic unit in ecology. *Science* 123: 1127–1128.

FOSBERG, F. R., – 1973 – Temperate zone influence on tropical forest land use: a plea for sanity. In 'Tropical Forest Ecosystems in Africa and South America,' ed. B. J. Meggers, E. S. Ayensu & W. D. Duckworth, pp. 345–350. Smithsonian Institution, Washington D.C.

FRANKEL, F., – 1971 – India's Green Revolution. Princeton University Press. 232 pp.

FREEMAN, D., – 1970 – Report on the Iban. Athlone Press, London. 317 pp.

GATES, D. M., – 1971 – The flow of energy in the biosphere. *Scientific American* 225(3): 89–100.

GEDDES, W. R., – 1960 – The human background. In 'Symposium on the Impact of Man on Humid Tropics Vegetation,' pp. 42–56. Administration of Territory of

Papua and New Guinea/UNESCO Science Co-operation Office for South East Asia.

GEERTZ, C., – 1972 – The wet and the dry: traditional irrigation in Bali and Morocco. *Human Ecology* 1:23–39.

GOLLEY, F. B., – 1972 – Summary; In 'Tropical Ecology,' pp. 407–413. Athens, Georgia.

GOLLEY, P. M. & F. B. GOLLEY (compilers), – 1972 – Tropical Ecology. Athens, Georgia. 413 pp.

GOLLEY, F. B. & H. LEITH, – 1972 – Bases of organic production in the tropics. In 'Tropical Ecology,' pp. 1–26. Athens, Georgia.

GRIJPSTRA, B. G., – 1973 – Sociological research in rural development. *Borneo Res. Bull.* 5: 12–16.

GUPTA, R. K. & S. K. SAXENA, – 1971 – Integrated ecological surveys for agricultural development in the arid zones of India. 1. Chhotan Community Development Block in Barmer District of Rajasthan. *Ann. Arid Zone* 10: 85–98.

GUPTA, R. K. & S. K. SAXENA, – 1972 – Potential grassland types and their ecological succession in Rajasthan Desert. *Ann. Arid Zone* 11: 198–218.

GUPTA, R. K., S. K. SAXENA & S. K. SHARMA, – 1972a – Aboveground productivity of grasslands at Jodhpur, India. In 'Tropical Ecology,' ed. P. M. Golley & F. B. Golley, pp. 75–93. Athens, Georgia.

GUPTA, R. K., S. K. SAXENA & S. K. SHARMA, – 1972b– Aboveground productivity of three promising desert grasses at Jodhpur under different rainfall conditions. In 'Eco-physiological Foundation of Ecosystems Productivity in Arid Zone,' 'Soviet National Committee for IBP, International Symposium, pp. 134–137. Nauka, Leningrad.

HANKS, L. M., – 1972 – Rice and Man: Agricultural Ecology in Southeast Asia. Aldine, Chicago, New York. 174 pp.

HARDESTY, D. L., – 1972 – The human ecological niche. *American Anthropologist* 74: 458–465.

HARRISSON, T., – 1970 – The Malays of South-West Sarawak before Malaysia: A Socio-Ecological Survey. Macmillan, London. 671 pp.

HILL, R. D., – 1970 – Peasant rice cultivation systems with some Malaysian examples. *Geographia Polonica* 19: 91–98.

HILLS, E. S., 1966 – Arid lands and human problems. In 'Arid Lands,' ed. E. S. Hills, pp. 1–13. Methuen/UNESCO, London and Paris.

JANZEN, D. H., – 1973 – Tropical agroecosystems. *Science* 182: 1212–1219.

JEFFERS, J. N. R. (ed.), – 1972 – Mathematical Models in Ecology, Twelfth Symposium of the British Ecological Society. Blackwell, Oxford, London, Edinburgh. 406 pp.

JOHNSON, A., – 1974 – Ethnoecology and planting practices in a swidden agricultural system. *Amer. Ethnologist* 1: 87–101.

KAUSHIK, S. D., – 1959 – Types of human settlement in Jaunsar Himalaya, *Geogr. Rev. India,* December: 1 – 17.

KNAPP, R. (ed.), – 1974 – Vegetation Dynamics: Handbook of Vegetation Science Part 8. Junk, The Hague. 368 pp.

KREBS, D. J., – 1972 – Ecology: The Experimental Analysis of Distribution and Abundance. Harper and Row, New York. 694 pp.

KUNG, P., – 1971 – Irrigation Agronomy in Monsoon Asia. FAO, Rome. 106 pp.

LANDSBERG, H., – 1958 – Physical Climatology (2nd ed.). Gray Printing Co., Dubois, Pa. 446 pp.

LEACH, E. R., – 1964 – Political Systems of Highland Burma. University of London. 324 pp.

LEACH, E. R., – 1968 – Pul Eliya: A Village in Ceylon. Cambridge at the University Press. 344 pp.

LEWIS, H. T., – 1971 – Illocano Rice Farmers. University of Hawaii Press, Honolulu. 209 pp.

LINFORD, J. H., – 1966 – An Introduction to Energetics with Applications to Biology. Butterworth, Oxford, Edinburgh, London. 232 pp.

LÖFFLER, L. G., – 1973 – Zur Klassifikation des Reisanbaues. In 'Vergleichende Kulturgeographie der Hochgebirge des Südlichen Asien, ' ed. C. Rathjens, C. Troll & H. Uhlig, pp. 107–108. Franz Steiner Verlag, Wiesbaden.

MacARTHUR, R. H., – 1972 – Geographical Ecology: Patterns in the Distribution of Species. Harper and Row, New York. 270 pp.

McGINNIES, W. G., B. H. GOLDMAN & P. PAYLORE (eds.) – 1968 – Deserts of the World. University of Arizona, Tucson. 188 pp.

McMILLAN, C., – 1960 – Ecotypes and community function. *Amer. Naturalist* 94: 245–255.

McMILLAN, C., – 1971 – Ecotypes and ecosystem function. In 'Ecology. Foundations for Today,' eds. R. S. Leisner & E. J. Kormondy, Vol. 3, pp. 38–41. American Institute of Biological Sciences. Brown and Co., Dubuque, Iowa.

McNAUGHTON, S. J. & L. L. WOLF – 1973 – General Ecology. Holt, Rinehart and Winston, New York. 710 pp.

MARGALEF, R., – 1968 – Perspectives in Ecological Theory. University of Chicago Press.

MARTIN, J. F., – 1973 – On the estimation of the sizes of local groups in a hunting-gathering environment. *American Anthropologist* 75: 1148–1168.

MAY, R. M., – 1973 – Stability and Complexity in Model Ecosystems. Princeton University Press, Princeton, N.J. 236 pp.

MAYNARD-SMITH, J. (ed.), – 1974 – Models in Ecology. Cambridge University Press, London. 146 pp.

MEGGERS, B. J., E. S. AYENSU & W. D. DUCKWORTH (eds.), – 1973 – Tropical Forest Ecosystems in Africa and South America: A Comparative Review. Smithsonian Institution, Washington D.C. 350 pp.

MISRA, R. (ed.), – 1974 – Proceedings of IBP Symposium and Synthesis Meetings on Tropical Grassland Biome. Banaras Hindu University, Varanasi. 174 pp.

NAROLL, R. & R. COHEN (eds.), – 1970 – A Handbook of Method in Cultural Anthropology. Natural History Press, New York. 1017 pp.

NASH, M., – 1966 – Primitive and Peasant Economic Systems. Chandler, San Francisco. 166 pp.

NETTING, R. McC. – 1974 – Agrarian ecology. *Ann. Rev. Anthropology* 3: 21–56.

NIX, H. A., – 1968 – The assessment of biological productivity. In 'Land Evaluation: CSIRO Symposium,' ed. G. A. Stewart, pp. 77–87. Macmillan, London.

NOY-MEIR, I., – 1973 – Desert ecosystems: environment and producers. *Ann. Rev. Ecology and Systematics* 4: 25–51.

NOY-MEIR. – 1974 – Desert ecosystems: higher trophic levels. *Ann. Rev. Ecol. Syst.* 5.

OVINGTON, J. D., – 1962 – Quantitative ecology and the woodland ecosystem concept. In 'Advances in Ecological Research,' ed. J. B. Cragg, pp. 103–198. Academic Press, London.

PAPUA AND NEW GUINEA, ADMINISTRATION OF TERRITORY OF/UNESCO, – 1960 – Symposium on the Impact of Man on Humid Tropics Vegetation. UNESCO Science Co-operation Office for South East Asia. 402 pp.

PENMAN, H. L., – 1948 – Natural evaporation from open water, bare soil and grass. *Proc. Roy. Soc., London,* ser. A., 193: 120–145.

RAPPAPORT, R. A., – 1971a – Nature, culture and ecological anthropology. In 'Man, Culture and Society,' ed. H. Shapiro, pp. 237–267. (Revised Edition.) Oxford University Press.

RAPPAPORT, R. A., – 1971b – The flow of energy in an agricultural society. *Scientific American* 225(3): 117–132.

RICKLEFS, R. E., – 1973 – Ecology. Chiron, Newton, Mass. 862 pp.

SALISBURY, R. F., – 1973 – Economic anthropology: *Ann. Rev. Anthropology* 2: 85–94.

SAXENA, S. K., – 1972 – The concept of ecosystem as exemplified by the vegetation of western Rajasthan. *Vegetatio* 24: 215–227.

SCHMITHÜSEN, J. (ed.), – 1972 – Okologie der Biosphäre. Vorträge einer Arbeitssitzung des 38. Deutschen Geographentages Erlangen–Nürnberg, 1971. Junk, The Hague. 200 pp.

SESTAK, Z., J. CATSKY & P. G. JARVIS (eds.), – 1971 – Plant Photosynthetic Production. Manual of Methods. Junk, The Hague. 818 pp.

SHANKAR, V., K. A. SHANKARNARAYAN & P. RAI, – 1973 – Primary productivity, energetics and nutrient cycling in *Sehima-Heteropogon* grassland. 1. Seasonal variations in composition, standing crop and net production. *Tropical Ecol.* 14: 238–251.

SHEALS, J. G., – 1969 – The Soil Ecosystem. Systematics Association Publ. no. 8. Academic Press, London.

SMITH, R. L., – 1972 – The Ecology of Man: An Ecosystem Approach. Harper and Row, New York, London. 436 pp.

SRINIVAS, M. N., – 1974 – Winners and losers: two villages revisited. *South Asian Rev.* 7: 231–236.

TANSLEY, A. G., – 1935 – The use and abuse of vegetational concepts and terms. *Ecology* 16: 284–307.

THORNTHWAITE, C. W., – 1948 – An approach toward a rational classification of climate. *Geogr. Rev.* 38: 85–94.

TIVY, J., – 1971 – Biogeography: A Study of Plants in the Ecosphere. Oliver and Boyd, Edinburgh. 432 pp.

TROLL, C., – 1971 – Landscape ecology (geoecology) and biogeocenology – a terminological study. *Geoforum* 8: 43–46.

TROLL, C., – 1973 – Die Höhenstaffelung des Bauern – und Wanderhirtentums in Nanga Parbat-Gebiet (Indus-Himalaya). In 'Vergleichende Kulturgeographie der Hochgebirge des Südlichen Asien,' ed. C. Rathjens, C. Troll & H. Uhlig, pp. 43–48. Franz Steiner Verlag, Wiesbaden.

UNESCO, – 1963 – Changes of Climate. Proceedings of the Rome Symposium. Arid Zone Research 20. UNESCO, Paris. 488 pp.

UNESCO, – 1966– Regional Seminar of Tropical Ecology: Primary and Secondary Productivity of Tropical Savannas. Preliminary Report. UNESCO Regional Centre for Science and Technology for Africa, Nairobi. mimeo. 39 pp. + bibliography.

UNESCO, – 1967 – Methods of Study in Soil Ecology. Proceedings of the Paris Symposium. Ecology and Conservation 2. UNESCO, Paris. 303 pp.

UNESCO, – 1970 – Use and Conservation of the Biosphere. Natural Resources Research 10. UNESCO, Paris. 272 pp.

UNESCO, – 1971 – The place of ecosystem modelling in the Man and Biosphere Programme. Paper to International Co-ordinating Council of the MAB Programme. UNESCO, Paris, mimeo. 7 pp.

UNESCO, – 1974 – Le Sahel: Bases ecologiques de l'Aménagement. MAB Technical Notes, UNESCO, Paris. 99 pp.

VAYDA, A. P. (ed.), – 1969 – Environment and Cultural Behaviour: Ecological Studies in Cultural Anthropology. Natural History Press, New York. 485 pp.

WADE, N., – 1974 – Sahelian drought: no victory for Western aid. *Science* 185: 234-37.

WADSWORTH, R. M. (ed.), – 1968 – The Measurement of Environmental Factors in Terrestrial Ecology. Eighth Symposium of the British Ecological Society. Blackwell, Oxford, London, Edinburgh. 324 pp.

WATT, K. E. F., – 1968 – Ecology and Resource Management: A Quantitative Approach. McGraw-Hill, New York. 450 pp.

WATTS, D., – 1971 – Principles of Biogeography. McGraw-Hill, London. 402 pp.

WAX, R. H., – 1971 – Doing Fieldwork: Warnings and Advice. University of Chicago Press, Chicago and London. 395 pp.

WHITTAKER, R. H. (ed.), – 1973 – Ordination and Classification of Communities. Part 5. Handbook of Vegetation Science. Junk, The Hague. 738 pp.

WHYTE, R. O., – 1974a – Tropical Grazing Lands: Communities and Constituent Species. Junk, The Hague. 222 pp.

WHYTE R. O., – 1974b– Rural Nutrition in Monsoon Asia. Oxford University Press, Kuala Lumpur, London, New York, Melbourne. 296 pp.

WIENS, J. A., – 1972 – Ecosystem Structure and Function. Proceedings 31st Annual Biology Colloquium, Corvallis. Oregon State University Press, Corvallis. 176 pp.

WINSTANLEY, D., B. EMMETT & G. WINSTANLEY, – 1974 – Climatic changes and the world food supply. (Paper to) Conference of American Society of Civil Engineers, Biloxi, Miss. (mimeogr.). 38 pp.

WITKAMP, M., – 1971 – Soils as components of ecosystems. *Ann. Rev. Ecology and Systematics* 2: 85 –110.

3 EVALUATION OF PHYSICAL AND BIOLOGICAL COMPONENTS

Man's use of natural resources is the indispensable means to maintaining and improving the human condition. In his use of natural resources, however, man has, until recently, been unable to foresee – far less to assess and take account of – the impact of his actions on his wellbeing, or even on the resources themselves. Historically, in many regions of the world soils have been depleted, forests cut down and oceans and lakes have served as sinks for man's wastes. Why then, has concern over the environmental aspects of natural resource use and management recently reached major national and international proportions?

(Extract from Chapter 1 of Document A/CONF. 47/7, summarizing an approach to integrated management of resources, presented to the U.N. Conference on the Human Environment, Stockholm. June, 1972, paragraphs 1, 2, 11, 12, 13, 14).

What is new is the magnitude of the claim on the world's resources and, hence, on the environment as a whole, together with a new awareness of cause-and-effect relationships. That claim flows from unprecedented rates of population growth, swiftly rising incomes and per capita demand which – magnified by scientific and technological advances – imposes demands on natural systems which may exceed their capacity to respond. As a result: a. the impact on the environment is such that, for the first time, the question could legitimately be raised whether the life-support system of the planet might be damaged beyond repair; b. in some parts of the world, especially where consumers are better informed through improved communications and better supplies by rising levels of affluence, there is mounting concern that demand may – in the case of some resources – overwhelm the supply.

Integrated management requires expanding the planning process, where the economy is suitably organized. It also requires a high degree of co-ordination in the management of sectoral activities. In some cases development will be guided by articulated national goals – economic, social and environmental; in other circumstances, the experience which can be drawn from the development process itself will yield guiding principles over time.

Land use planning, in accordance with intrinsic land capabilities and limitations, is a major tool of integrated resource management. Its use can reduce, if not avert, adverse environmental effects. Land inventories should be undertaken, where appropriate, to provide comprehensive surveys of the potential of land. Such surveys – including aerial surveys and remote sensing – can form one basis for land use and resource planning for agriculture, forestry, recreation or wildlife, and may comprise a classification system which delineates the potential and constraints for each sector analysed.

Central to the integrated approach is the participation of specialists who can, between them, understand the full range of environmental elements and their dynamics while being equally sensitive to economic, social and cultural considerations. The process involved ranges from initial surveys, in the case of development projects, through feasibility studies, investment, management of the completed project and, very importantly, its subsequent evaluation in terms of the environmental impact it has had.

Integrated resource management need not require a massive reorganization of existing governmental institutional structures. Mechanisms providing for co-ordination and for promoting co-operation between existing institutions offer the most practical immediate courses of action. Such mechanisms could include inter-agency or intra-agency working groups, task forces drawn from several agencies to work together on a specific project, and liaison and ex-change of information between governmental agencies concerned about on-going and planned activities.

3.1 Geology, Geomorphology

The following paragraphs have been adapted from the ITC Textbook of Photo-interpretation, Vol. III: Use of Aerial Photographs in Geography and Geomor-phology, Chapter VIII 4: Land evaluation and land(scape) science (Zonneveld, 1972).

Geological information is required for a thorough understanding of land units as a whole, and for more specialized study of the landform, soils, hydro-logy and general ecology of an area. For specialists in soils, vegetation and land use, mineralogical, lithological and stratigraphical data may help in selection of sample sites, delineation of boundaries, etc.; they are thus of value as survey tools and as indicators for the other disciplines, or in the general classification of land units.

However, the procedure adopted in a normal geological survey is difficult to combine with, for example, observations on soil and vegetation, in such a way that the specialists in these different disciplines can move together in the field. The deep incisions of the geologist are of little interest to the soil scientist or vegetation ecologist, who are more concerned with the surface, and who probably know more about the upper layers than the geologist. Much more time and energy are required for a good geological survey than for a soil or vegetation survey. Most land survey teams do not, therefore, have a geo-logist as a regular member.

The results of a geological survey preceding a land survey are of great help to the land surveyors, and particularly to the pedologist in respect of data on lithology and subsoil hydrology. Conversely, a land survey preceding a geological survey may be of great help to the geologist. The type of landscape will determine which attributes of the land are most useful for the geologist. Landform is usually a valuable indicator for geological survey; in flat terrain, e.g. the vast flat savannas in Africa and elsewhere, vegetation can be useful in this respect.

In turning to geomorphology, one comes to consider the features of the land at two contrasting levels, the macro- or regional scale, and the micro- or local scale. According to Dobby (1961), two features dominate Monsoon Asia: it·lies south and east of the Tibetan Highlands, which in breadth, altitude and continuity create within the Asian continent a major impediment to human activity; it is broken by bays and gulfs of the Indian and Pacific Oceans in such a way that most parts are within 500 miles (800 km.) of a coast. In human affairs, in trade and in land development for economic return, the consequence of landforms and dispositions largely turns upon accessibility and routes of communication.

The landforms may be classified (Dobby, op.cit.): a. the Himalayan/Island Arc of folds, everywhere young, still unstable, modelled chiefly by water erosion continuing at a great pace upon the parallel ranges; the Arc divided into two: the Kirthar/ Himalaya/Arakan section, and the Andaman/Java/ Sulawesi/Japan section. b. continental Monsoon Asia with its peninsular combinations of highlands and plains; the southern subcontinent, the highlands of eastern China, and the highlands of southeast Asia.

Spencer & Thomas (1971) have produced 'a somewhat imaginative set of maps' (their own words — their Figs. 6.5 to 6.14) to indicate the general patterns into which fall the structure and landscapes of Monsoon Asia. They are intended to serve as a skeleton upon which to build their detailed discussion of the geomorphology and surface landscapes of the several broad regions (repeated by Robinson, 1966, from the 1954 edition of Spencer & Thomas); on the south, the Indian Triangle, including the territory from southern Tibet to the island of Sri Lanka, formed by the junction during the Tertiary between the South Asian Plate, which moved from its original position as part of Gondwana in the southern hemisphere, with the mainland of Asia (Hallam, 1973), 'on the southeast, the Borneo—Malayan Fan which comprises the mainland countries of Burma, Thailand, Laos, Cambodia, North Vietnam, South Vietnam and West Malaysia (the last, with parts of western Indonesia, considered by some to have been also part of the South Asian Plate), on the east the Chinese Checkerboard ranging from south China to northern Manchuria, with its roots deeply buried in central Asia, off the coast the southern and northern Island Arcs, reaching from Sumatra and New Guinea on the south, to the Kuriles and Sakhalin on the north' (Spencer & Thomas, 1971).

At the micro- or local scale, there are, according to Zonneveld (1972), profound misunderstandings about the relation between holistic or integrated surveys and geomorphology, both in the evaluation of land survey for geo-

68

morphological studies and in the determination of the value of geomorphology for land surveys. It is necessary for geomorphologists trained in and familiar with temperate landscapes to recognize the marked differences which are found in the tropics, in terms of what Tricart & Cailleux (1972) call climatic geomorphology. Different reliefs are produced in periglacial, temperate and tropical landscapes under the influence of physical, chemical and biological agencies; the actions of man may transform the terrain in several ways, and in fact (according to J. A. Steer, reviewing the above – *Geographical Journal* 139: 544–45, 1973) there is abundant scope for research on morphogenic processes induced by man (Tricart, 1972a). So much that is taken for granted in the evolution of temperate lands does not apply in the tropics.

Geomorphology may be divided into various subsciences, including:– morphology *sensu stricto* (geomorphology, geomorphometry, the qualitative and quantitative description of features of relief):– classification (the typology of landforms); – dynamics or genesis (study of formation of various forms of relief by the factors influencing the earth's crust).
It is the Dutch view that classification of landform must make use of characteristics derived from morphometrical description. Information from dynamic studies can be used as guiding principles for the classification.

A land survey team should contain a specialist with an interest in 'static morphometrical typological geomorphology' (Zonneveld, 1972), as one whose main interest is in dynamic situations (landslides, avalanches, earthquakes, etc.) may have little interest in this task. Although on occasion, a specialist in another discipline may perform the description and recognition of a landform, such an analysis then tends to be more general and physiognomic. It is stated on occasion that static geomorphology should play a dominant role in the choice of characteristics as well as of the guiding principles for land(scape) classification. In such a case, the words landform and land unit are used confusedly, almost as synonyms. A land unit of any rank order is a landform that is also described in terms of vegetation, soils, etc.

The Dutch specialists at ITC do not agree with this. It depends entirely on the situation, the type of area and the aim of the survey, as to which properties of which land attributes should be used on any particular site and at which level in the classification. There are examples in which it is useful to employ relief properties at the highest level, but the reverse may also be true. The situation differs when one is dealing with general classification or map legend; in the latter case, the mapping scale in relation to the general character of the land is most important.

Extreme emphasis on the use of geomorphology in the classification of land is found in certain schools of so-called applied geomorphology. The landform is central, and soils, vegetation, lithology, land use, etc. are studied to evaluate the general ecological properties of the land, already primarily described in terms of landform. Pragmatic land classification may be carried out on this basis. Such schools (see Tricart, below) resemble the schools of physiographic soil survey or of vegetation science on a physiographic basis; these latter have soils or vegetation rather than landform as their main focus of interest. Vegetation scientists in particular sometimes feel that the applied geomorphologists claim too much for their discipline, where latitude, altitude, vegetational history and the influences of the anthropogenic factor have played a significant if not a dominant role on any particular site, irrespective of landform.

It may be questioned whether shortages of time, money and trained specialists can justify the inclusion of a geologist/geomorphologist in teams responsible for assessing land potential for production of food and agricultural commodities under Asian conditions of land use and demography. A geomorphologist may make an important contribution in a study of land for its engineering properties, as in Malaysia. Here it has been possible to classify part of the country into 16 land regions with 42 recognizable land systems, each with characteristic soils and rocks (see 6.2.4 and R. S. Millard, contribution to discussion, *Geographical Journal* 138: p. 450, 1972). In that part of West Malaysia comprising Malacca, Negri Sembilan, parts of Johore, and Selangor, there are three broad groups of regions corresponding with the granitic rock, the sedimentary sandstones and shales, and the alluvial plain around the coast. One may apply this classification in locating new quarries where there is only a thin mantle of weathered rock, with good quality rock easy of access below it.

Another geomorphological study of Malaysia is concerned with the Quaternary history of north Sarawak, particularly with the Subis Karst and adjacent areas when the Niah limestone caves were occupied by *Homo sapiens* in the late Quaternary (Wall, 1967). Limestone outcrops cover about 200 sq. miles of Sarawak; they range in age from Upper Carboniferous to Miocene, and are composed predominantly of pure, massive to poorly bedded rock with well-defined but widely spaced joints and low primary porosity and permeability; the limestone areas have probably been exposed to erosion in an equatorial climatic regime since the early Pliocene (Wilford & Wall, 1965). Evidence for the proposed five phases in the Subis Karst, which resulted from glacio-eustatic changes and regional uplifts, is based mainly on interpretation of the present morphology, such as erosion surfaces, terrace distribution and old river courses, conclusions which are not fully accepted by local specialists.

In contrast to the many studies and mappings of the geology and geomorphology *per se* of the Asian region, we must consider those geomorphological studies which have been made on land already intensively used, with a view to assessing its potential for greater and/or more conservative production of food and other commodities from the land. We may take two examples, one from the arid lands of Rajasthan (Pandey, 1969, Ahmed, 1969) and one from the UNESCO volume on deltas.

The Middle Luni Basin in Western Rajasthan has been the subject of integrated survey by the Basic Resource Studies Division of the Central Arid Zone Research Institute, Jodhpur, for a number of years (see Annual Reports of

Institute, and series of papers in *Annals of Arid Zone*, of which particulars appear in the most recent, Singh, Pandey & Ghose, 1971). The principal land systems designated are: 1. Precambrian granite and rhyolite hills; 2. Quaternary older alluvial plains; 3. Quaternary younger alluvial plains; 4. Pleistocene recent sand dunes.

The techniques adopted were as follows. A composite geomorphological study of landforms of 15,222 sq. km. has been compiled from the detailed studies of landforms made with the help of photo-interpretation and field traverses of various Panchayat Samiti Blocks. First, aerial photographs on the scales of 1:25,000, 1:31,000 and 1:40,000 representing different parts of the region were laid down in respective runs to give a composite view of the entire region and to make the photo-patterns representing different landform features. Each stereo-pair of photographs was studied under a mirror stereoscope and the annotations marked on the back of the photographs. The query points were marked in different photo-patterns for field traverses and investigation. The survey party made detailed reconnaissance surveys and noted relevant points and information. The boundaries of geomorphological units already drawn on photos in the laboratory were checked and new boundaries drawn for features of micro-relief. In the laboratory the boundaries of different landform features were transferred on the base map of 1:126,720 scale by Zeiss Sketchmaster. The final geomorphological map of the region was prepared on 1:253,440 scale. After completion of mapping and analysis of sediments, the written description of each geomorphic unit was prepared.

The region has been affected by both water and wind erosion; water erosion is confined to the hilly and piedmont zones. Slight to very severe wind erosion has turned the fertile land in the western and northwestern parts into wasteland (or alternatively, excessive cultivation and inadequate agronomic practices on land which should never have been cleared of vegetation has made that land susceptible to wind erosion which would have not occurred under a climax vegetation cover). The recommendations for conservation measures for control of water and wind erosion are those that would be recommended by a mere agronomist, who might thus be justified in asking why a costly geomorphological survey is essential on land which has long since regressed far beyond any approach to a man/land equilibrium. An integrated survey in such an area can show only an ideal state of management on land units which can never be achieved because of the excessive and irremovable pressure of human beings and livestock on land at its nadir of productivity before final abandonment.

From these lands of low actual and potential productivity in the absence of new and reliable sources of irrigation water, we turn to the great river valleys and deltas of Asia, the areas of maximum concentration of rural and urban peoples and still having a large potential if not actual productivity. Believing that an integrated approach to the scientific problems of deltas, their geography, geology, hydrology, pedology and biology, would be a desirable objective, UNESCO convened a symposium in Dacca (now Bangladesh) in 1964, under seven sections: geomorphology, sedimentation and pedology, hydrography and hydrology, vegetation biology, human influence, descriptive studies and classification of deltas (UNESCO, 1966a).

The introductory paper by Tricart (1966; see also 1963) on the place of geomorphology in the study of the development of tropical deltas deals more particularly (according to H. T. Verstappen's contribution to the discussion) with the relation to pedology, and may be said to stress the static aspect of geomorphology (see above), the mapping of geomorphological phenomena, rather than the dynamics, the changes in the phenomena in which engineers are primarily interested. See also report of a geomorphological reconnaissance of Sumatra by Verstappen (1973). The following paragraphs represent an adaptation of Tricart's paper to the Dacca Symposium.

. The geomorphology of tropical deltas is determined by a complex combina tion of factors: the glacio-eustatic fluctuations of the sea-level — a world-wide phenomenon; subsidences, often disparate even within the confines of the delta; the Quaternary fluctuations of the palaeoclimate in the river basin, which are to some extent on a zonal scale but also vary locally; the peculiarities of the interplay of mutually antagonistic shoreline and fluviatile processes which vary from point to point of a given delta and over a period of time.

Modern geomorphology has satisfactory methods for the analysis of this combination of factors, which are presented, notably, in the form of detailed geomorphological maps. These maps, using the Tricart-CGA system, show the following: a. the nature of the materials outcropping or lying near the surface; b. the genetic nature of the land forms, treating each as a morphogenetic unit (alluvial levees, slack water settling areas, secondary deltas at breaches in the levee, mangrove swamp shorelines, etc.); c. certain topographic characteristics of these land forms not clearly shown on the topographical base (height of undercut banks, for instance); d. the ages of land forms and, in the case of those of ancient origin, the age of the various changes they have undergone since their original formation (e.g. degradation of ancient alluvial levees by subsequent wind deflation).

Thus a picture is presented from which a landscape's dynamic morphology can be grasped. Its bases are field-work (with the intensive use, on the ground, of air photographs) and systematic laboratory study of samples (genetic identification of land forms, determination of processes, dating of accumulations, alterations, etc.). Possible uses are: Soil maps: the detailed geomorphological map is the base for soil surveys. The overall economies obtainable by this approach may amount to 50 per cent, and considerable time is saved. Agronomy: the map gives the agronomist directly usable data, notably on inundation pattern, erosion, etc.; it also provides valuable information for the ecologist. Engineering: the map supplies information on the nature of land formations and deposits of raw materials (e.g. for the building of earth dikes or dams), and on the process currently at work (undercutting, sand movements on beaches or under the action of the wind, etc.); it allows defences against dangerous processes to be planned, permits sites for engineering works and communication routes to be chosen and irrigation and drainage works to be adapted to the natural conditions.

Ohya (1966) contributed a paper to the same symposium on a comparative study of the geomorphology and flooding in the plains of the Cho-Shui-Chi in Taiwan, the Chao-Phya in Thailand, the Irrawaddy and the Ganges (being the report of the expert session on delta areas organized by ECAFE, with members from Japan, the Netherlands and Taiwan).

3.2 Topography

This is not a component of land systems, rather an amalgam or association of the major components discussed in this chapter, under specific geographical conditions which have arisen in geological history as a result of the rafting of tectonic plates, the uplift of mountains, fluctuations in seabed, volcanic action, etc. Yet, because the nature of the macro- and micro-topography is an important factor which has governed the present vegetative cover, land use and social structure, it should be taken into consideration as a major geographical factor in any studies of land potential for changed use and development.

Troll (1972a) and his associates have studied the geoecology of the high-mountain regions of Eurasia from the approach noted in 1.7, with particular reference to definition and differentiation, the upper limit of aridity and the arid core of High Asia, geoecology in the Mt. Everest region of the Khumba Himalaya (W. Haffner), the three-dimensional zonation of the Himalayan system – south to north, southeast to northwest, and the vertical gradation (C. Troll), and the climates of High Asia to the west, south, north and east of the Himalayan massif (U. Schweinfurth).

This work was extended in the next volume in the series (Rathjens, Troll & Uhlig, 1973) to comprise the comparative cultural geography of the high-mountain regions of southern Asia, giving the geographic approach to the study of field systems, common fields and settlement systems (9 authors), rice cultivation (8 authors) and systems of nomadic and migratory grazing (7 authors). See also the Report of the UNESCO Regional Seminar on the Ecology of Tropical Highlands, held in Kathmandu (UNESCO, 1968a).

The innumerable books and scientific papers published on this most significant topographical unit in the whole Asian region since the beginnings of British India contain a wealth of knowledge and experience on climate, vegetation, forestry, grazing, cultivation and the human factor, of great value in any consideration of the land potential. Naturally, any such review will consider not only the highland areas as such, but also how their use and development affect the actual and potential productivity of the much more valuable lands in the plains. The Siwaliks and the front ridges of the Himalaya have been so deforested that it is not a matter of development but of reclamation, if their former role of conservation and control of water flows in a balanced highland/lowland land use symbiosis is to be revived.

Much of the literature on the mountains of South India has been reviewed by Blasco (1971) in his study, according to the Toulouse school, on their forests, savannas and general ecology. On the basis of floristic inventory, vegetation ecology, meteorological condition (especially its marked variability and frequency of dry periods promoting fire – Legris & Blasco, 1969), and pedology, Blasco (1971) defines a tropical mountain stage of vegetation above 1700 to 1900 m., rich in plant families differing from those at lower altitudes and in herbaceous endemic species.

Topography has played a major role in the migrations of peoples, crops and domestic animals along the routes governed by the nature and direction of the mountain systems and their associated river valleys (Fig. 5.1 in Whyte, 1974b). The location and design of the Asian village and its adjacent cultivated land have evolved in relation to factors of topography. Trade in commodities carried on men's backs or transported by draft animals has followed natural contours through types of vegetation and terrain which favoured such movement.

In due course, shortage of level or gently sloping land led to the introduction of terraced agriculture on the steeper slopes which had been cleared of forest, probably through an intervening period of shifting cultivation (5.3.1). Where ample water at high elevations was available, these terraces could be irrigated, as in parts of Indonesia and the Philippines; otherwise they would be used for seasonal rainfed upland crops, as in the foothills of the Himalaya in Uttar Pradesh and other north Indian States. If terraces are well designed and constructed and well-maintained, they represent an excellent land-use technique for the conservation of soil and water. If, however, they are allowed to deteriorate even slightly, the erosion hazard is very great, as it is also with the modern construction of highways in mountainous country where the layout has not been designed in correct conformity with the topography.

Clarkson (1968) has recognized an analytical dichotomy when dealing with resource problems: 'reality here has two faces – that recognized by the group being studied and that recognized by the investigator.' This analytical discontinuity has direct implications for both current and future resource utilization. Where natural resources are being used by a group with one set of values, and the control of that utilization is the responsibility of a second group with another set of values, conflict is inevitable. That conflict may reflect a series of social problems out of all proportion to the actual social or economic elements which produced the conflict in the first place. This 'dichotomy' between those who have assessed the potential of any tract of Asian country and who say that it should be managed according to modern scientifically-based methods, and the users of that land for decades or centuries past, is one of the socio-economic-political aspects to be discussed in the sequel to the present volume.

Under the head of topography as a synthesis of major components of

74

'land', one may refer to the situation in the Cameron Highlands, Pahang State, Malaysia, described by Clarkson (op. cit.). The conflict here is between, first, the specialists who prepared the 1962 soil map of Malaya on the basis of the classification of the U.S. Department of Agriculture (Panton, 1964), and who stated that all land above 225 metres elevation (with more than 25° slope) on the Malayan peninsula (40 per cent of total land area) should be classed as 'unsuitable for extensive agricultural development' apparently unaware, according to Clarkson, of the large tracts of land at these elevations of less than 25° slope where, for example, the multi-million dollar vegetable growing industry has developed in the Cameron Highlands; second, the Central Electricity Board which blamed the siltation of its reservoir (to be silted to the point of disuse within eight to twelve years from 1965) on the clearing of the primary vegetation by Chinese vegetable gardeners, whereas Clarkson would place the main responsibility on road cuts, municipal and private building sites; and third, the Chinese vegetable farmer, cultivating 400 hectares of land (240 legally, 160 illegally) in the Cameron Highlands.

Continuing his critical analysis, Clarkson (1968) compares the prescriptive and pragmatic norms for the assessment of potential for this type of land, a 'dichotomy' which may be seen in differing degrees of intensity and complexity throughout the hill lands of Asia, where one is considering what is good for the land, and traditional for its economic occupants. The prescriptive rational view analyses the Cameron Highlands 'as a part of an abstract class which might be called "tropical uplands with excessive slope and heavy rainfall." As a member of this class the Cameron Highlands area is assumed to possess specific characteristics of soil, vegetation, hydrology and erosion rates. Observations in the Highlands and in other, presumably similar, tropical uplands have led to certain generalized conclusions based, eventually, on rational scientific knowledge of physics and chemistry' (see basis of rules for cultivation of slopes below).

'The Chinese farmer sees only practical problems which involve the retention of the growing mixture he has created and thus the husbanding of his investments of labour and capital. His goals are set by the pragmatic problem of gaining a livelihood from his bit of land. He attempts to reach this goal by a traditional, time-tested, set of farming practices. His technological bag of tricks may be added to or modified in response to particular problems encountered in specific situations. Change, when required, is based on trial-and-error practice, on the experiences of his neighbours or forebears, or on a memory of early stories or advice usually orally received. Because of the integration of the technologic, social and economic systems, change which is introduced in this manner is usually assimilated because it takes account of this traditional integration. Change introduced by agencies external to the Chinese community is frequently a prescriptive, rationalized abstraction based on that external group's normative judgment, e.g. what would be "good" for the Chinese or society as a whole. Such introduced change frequently fails because it cannot

adapt itself to the need for integration of diverse parts of the overall system'
(Clarkson, op. cit.).

The official view of the Highlands is based on the rules of the Natural Resources Board of Pahang State, which the Central Electricity Board was instru-
mental in drafting and which it carefully tries to follow (rules in Table 3.1
quoted from Clarkson, op. cit. pp. 143–4).

3.3 Soils

3.3.1 Soil surveys or land surveys

The soils of the region have been surveyed, classified and mapped on a broad
macro-scale that is not readily adapted to local land planning, and also on a
relatively small scale, the latter sometimes carried out according to a system
which is in fact a land survey. Even some data belonging strictly to geomorphology (relief) or hydrology (groundwater resources) are used in the evaluation of the so-called soil map. Some definitions of soils, including relief, hydrology and soil temperature are so broad that they are almost identical with
'land'. Many reconnaissance soil surveys made with the help of aerial photography may be readily converted into maps of land units with only a slight
extension of the legend; for some, only the title need be changed as the legend
in table form already includes attributes such as soil, geology, landform and
vegetation. Why then, have they been called soil surveys and not land surveys
(Zonneveld, 1972)?

Nevertheless, soil survey and sampling are essential facets of the field part of
an overall land survey. Pragmatic classification of land for agricultural use
requires various types of information which can be provided only by study of
soil (structure, waterlogging capacity, permeability, nutrient status and chemistry, profile). The type(s) of soils occurring in a land unit can be used to
characterize land units of various hierarchical ranks. Soil is usually a basis for
evaluation or means of classification, rather than a survey tool or indicator
(as with geomorphology and vegetation). The smaller the scale of the survey,
the more the soil surveyor depends upon other indirect observation tools.
Most soils of the earth are covered by vegetation, natural or planted. Aerial
photographs and remote sensing even in bare areas show only the topmost
layer; even there, the characteristics of the subsoil must be derived indirectly
by inter- and extrapolation (Zonneveld, op. cit.).

The taxonomic systems of soil classification do not provide units suitable
for mapping purposes at the small scales which are necessary in reconnaissance
surveys (Christian & Stewart, 1968). The complicated and irregular distribution of soils geographically has led soil surveyors to seek mapping units (e.g.
catenas in East Africa) which are actually composite units. The most common
composite mapping unit used in reconnaissance surveys, the 'soil association',
is defined by the U.S. Department of Agriculture as a group of named taxon-

Table 3.1. Malaysia: Basis of rules for cultivation of slopes evolved by the National Resources Board of Pahang State (from Clarkson, 1968).

The theoretical bases of the rules are:-

(a) That above a certain angle of slope the single act of removal of natural vegetation will lead to erosion and/or land slides on such a scale as to (i) make the re-establishment of economic crops a doubtful venture and (ii) be likely to cause damage to other cultivation etc. by the transport and subsequent deposition of solids from the area.

(b) That below a certain angle of slope no specific rules, beyond those normally assoc-ciated with good husbandry, are necessary to conserve the fertility of the soil and to prevent damage elsewhere by soil wash.

(c) That between the two angles of slopes quoted in (a) and (b) above, it may be necessary to take into account one or more of the following: – (i) the method of clearing the natural vegetation, (ii) the method of re-establishing a vegetative cover, (iii) the type of crop to be grown, (iv) the subsequent cultivation of the crop, (v) mechanical methods to prevent the loss of soil, (vi) soil type.

The following list gives a broad classification of the points at issue: –

(i) Method of clearing natural vegetation.

 (a) Selective felling.
 (b) Felling (no burn).
 (c) Felling and burning.
 (d) Clean clearing (c + removal of stumps).

(ii) Type of crop to be grown. Certain crops listed below should only be grown on level or gently undulating land by reason of their cultural requirements. Crops listed (a), (b), (c) below can be grown on slopes in certain cases as *catch crops* during the period of establishment of the permanent crop.

 (a) annual crops – tobacco, vegetables, cereals, pulses, etc.
 (b) biennial and ratoon crops such as tapioca and pineapples.
 (c) perennial crops that require clean weeding, such as coffee.
 (d) miscellaneous crops such as coconuts and pepper.

(iii) Crops likely to be grown on sloping land are, therefore, restricted to: –

 (a) Rubber
 (b) Tea
 (c) Oil Palms
 (d) Fruit Trees
 (e) Cacao
 (f) Bananas (usually a perennial catch crop) and Manila Hemp

(iv) The subsequent cultivation of the crop: –

 (a) No weeding
 (b) Selective weeding
 (c) Clean weeding

77

(v)   Mechanical methods to prevent the loss of soil.

      (a) Minor streams dammed and run-off regulated.
      (b) Gullies dammed and piped to reduce speed of run-off.
      (c) Contour planting.
      (d) Silt pits dug on contour.
      (e) Terracing.

(vi)   Soil type and relationship to crop:–

      (a) Granite   Recommended for Oil Palms and Tea.
                         Better soils suitable for Cacao.
                         Should be alienated only sparingly for Rubber.
                         Suitable for Bananas and Manila Hemp.

      (b) Pahang Volcanic   Reserved for Cacao, possibly Oil Palms.

      (c) Sandstone and Shale   Mainly suitable for Rubber but not for other crops.

omic soil units regularly and geographically associated in certain proportionate patterns. The taxonomic soil units may range from soil phase through soil types, soil series, soil family up to great soil group.

Both soil mapping and landscape mapping have the problem of mapping composites of units, i.e. patterns, and both recognize the importance of recurring patterns in the landscape. In soil mapping, landform and especially vegetation are usually given secondary consideration to soil itself. In landscape mapping landform and vegetation are of first importance because of their influence on the photographic image, although in sparsely vegetated areas the reflective qualities of soils are also important. The morphological characteristics of soils, which are those mainly used in their agricultural classification, cannot be interpreted directly from an aerial photograph – soil units defined as landscape units are preferable (Christian & Stewart, op. cit.).

Now increasingly, the interrelations between land features have come to be recognized. It is rare not to find at least some associated fields combined in studies of specific areas, particularly where the surveys have a practical objective. Soil surveys are rarely conducted without consideration of topography, geology, hydrology and to a lesser extent climatology. Studies in plant and animal ecology are obviously always related to the environment. The trend towards the true integrated survey was demonstrated by the work in Africa of Trapnell and his associates (Trapnell, 1953; Trapnell & Clothier, 1957; Trapnell, Martin & Allan, 1950); they recognized and classified major vegetation types and regional soil groups, first independently and then in combination, thus obtaining vegetation/soil units to be used as a basis on which to define land classification units. This led in Africa to the many surveys conducted by the Land Resources Division of the (British) Overseas Development Administration. The history, organization, functions, work systems and projects of

78

the Division are described in Baulkwill (1972) (see Table 6.1), in relation to the whole process of resource assessment and development. The soil survey procedures applicable especially to tropical and subtropical environments are described by Stobbs (1970), in a contribution to an international symposium (World Land Use Survey, 1970). This starts with the premise that, despite the large number of soil surveys carried out since 1945, these have not influenced agricultural development, nor contributed to increased productivity to the degree which might be expected; agricultural development has occurred over the 25-year interval regardless of, rather than as a result of soil surveys. Stobbs quotes from Smith(1965) who divides his soil surveys respectively into exploratory, reconnaissance and detailed (the last subdivided into low, medium and high-intensity surveys); each type is characterized by the type of soil unit used on the accompanying maps with consequent implied distinctions of scale of publication (Table 3.2). Although agreeing in general with the principle of integration, Stobbs (1970) believes that in the more intensive and detailed surveys, it is likely that field work should be concentrated upon one aspect of the environment   soil where cultivation is the objective, grassland ecology on semi-arid land (see also 6.2.3.).

Students wishing to proceed further in the subject of soil survey in Asia and the Far East may refer to World Soil Resources Report 41 (FAO, 1971a). The papers presented are grouped under four main heads: Problem soils, their occurrence and management (8 papers); Soil fertility and crop production (9 papers); Land resources and potentialities (16 papers); Aerial application of urea in India (2 papers).

Those in the third section relevant to the present discussion are:
Y. P. Bali: Use of soil surveys for planning agriculture
F. W. Hilwig & R. L. Karale: Physiographic systems and elements of photo-interpretation as applied to soil survey in the Gangetic Plain (U.P.)
M. Soepraptohardjo: Soil survey for agricultural development in south Sulawesi, Indonesia
M. L. Dewan: Land resource inventories and their utilization in meeting the food needs of the world (6.6.)
Sarot Montrakul, Thamo Kamkaen & Pongpit Piyapongse: Survey of physiographic features which limit the rice cultivation area in the Central Plain of Thailand
S. V. Govinda Rajan: Soil survey organization and service for land use planning
A. W. Moore: Regional soil data bank for future evaluation and projection (6.6.)
N. D. Rege & N. Patnaik: Country Report   India, soil surveys
C. R. Panabokke: Country Report – Sri Lanka soil surveys
Chun Soo Shin: Land suitability classification in Korea (6.3.2.)

In his concluding paper to the Paris symposium on methods of studying soil ecology (UNESCO, 1967a), J. van der Drift defines the problems in soil ecology requiring most urgent attention, under the heads of taxonomic and

Table 3.2. Types of soil survey (from Stobbs, 1970, after Smith, 1965, with details of scale derived principally from U.S. Department of Agriculture Handbook 18: Soil Survey Manual).

Type	Map units	Scale	Purpose
Exploratory	Associations of phases of great soil groups	1:1,000,000 and smaller (schematic maps)	1 To locate areas of substantial soil difference (inventory) 2 Locate more detailed work 3 Test legend
Reconnaissance	Associations of phases of soil series or higher categories (great soil groups or families)	1:62,500 to 1:500,000	1 To survey areas suited only to extensive use 2 Pre-detailed survey to locate and define such work
Detailed Low intensity	Phases of associations of series	1:30,000 and smaller	For forestry and grazing development areas
Medium intensity	Phases of soil series	1:10,000 to 1:30,000	For arable development areas
High intensity	Phases of soil series specified on a denser sampling pattern	1:7,920 and larger	For very intensive development areas e.g. irrigation, urban expansion

physiological problems, physical and chemical characterization of the soil as an environment for plant and animal life, environmental influences and inter-relations between soil organisms, effect of soil organisms on the soil and the soil community. 'The synecological approach starts from the soil community as such, and tries to describe it and to analyse its function. It is quite certain that a complete description of a soil community in all its aspects has never been made, and probably never will be made: the enormous variety of organisms living in the soil, their specific demands for quantitative studies, and their fluctuations, yearly, seasonal or daily, dependent on the character of the species, will prevent this.'

3.3.2 Soil survey and classification

Humid tropical Asia
Within the geographical scope of their contribution on soils to the Natural Resources of Humid Tropical Asia (UNESCO, 1974) — India, Sri Lanka, Bangladesh, Burma, Thailand, Laos, the Republic of Viet-Nam, the Democratic Republic of Viet-Nam, southern China, Malaysia, Indonesia, the Philippines and New Guinea — Dudal, Moormann & Riquier (1974) note that

a number of classifications have been used. The best known are the '7th approximation' of the U.S. Department of Agriculture, and the classification worked out by Dudal & Moormann (1964). Several pedological maps have been published, often with physiographical or textural local legends, or with a list of series of unclassified soils; comparison with other regions and other countries becomes very difficult. In connection with the Soil Map of the World, a new FAO/UNESCO classification has been drawn up to serve international purposes. Under each head, Dudal, Moormann & Riquier (1974) give particulars regarding nomenclature (summarized here), morphology, chemical and physicochemical characteristics, environmental conditions (brief references to environment and vegetation here), geographical distribution, and land use and agricultural potential. Their paper also has an appendix being a list of some 180 maps of soils and related conditions in the countries coming within their terms of reference.

Fluvisols – These soils are derived from recent fluvial deposits, without any diagnostic horizon other than a slightly humiferous one, a gley horizon or a thionic horizon. Synonyms: alluvial soils, entisols (7th approximation). Well-drained virgin soils may carry forest or bamboo, poorly drained soils often grass, highly acid alluvials marsh vegetation such as Cyperaceae and *Melaleuca leucodendron*, saline alluvials mangrove.

Regosols – This group has little or no profile differentiation apart from an ochric A horizon and a gley horizon at a depth of over 50 cm. from the surface. The electric conductivity of these soils is below 2 m mho and they do not possess the features of cambic or oxic horizons. They correspond to the entisol order of the 7th approximation. Group has wide climatic range, and natural vegetation determined more by climate than by intrinsic soil properties.

Vertisols – includes the dark clay soils of warm regions variously called black cotton soils, regurs and tirs; the term regur has been applied to such soils in Indonesia, the Republic of Viet-Nam and especially in India; margalitic soils and black earths in Indonesia; compact dark savannah soils in Burma; terres noires basaltiques in Khmer Republic and Viet-Nam; Guadalupe clay in the Philippines; grumusols in the Dudal/ Moormann classification. Climate strongly seasonal, dry season of four to seven months, natural vegetation mixed grass/forest savanna or an open forest.

Andosols – name first used in Japan by the U.S.A. Natural Resources Section, and later adopted in Indonesia; also called high-mountain soils, mountain black earth, humic mountain soils, and black latosols; fall within the andept sub-order of the inceptisols in the 7th approximation. Found on volcanic ashes from sea level to 2,500 m., rainfall between 1800 and 7000 mm., annual mean temperatures 14 to 20°C.; at high altitudes, natural montane vegetation may be replaced by plantations of *Pinus merkusii*, *Agathis alba* and *Eucalyptus*; tea up to 1500 m.; cinchona between 1500 and 2000 m; arabica coffee in Java.

Cambisols – many characteristics in common with the group of similar name in temperate climates; called immature brown loams in Sri Lanka, brown forest soils in the Dudal/ Moormann classification; in the FAO/UNESCO classification, most of these soils fall among the dystric, eutric, chromic, calcic and humic cambisols. Climatic range ill-defined; lowlands or cool mountains, but not in everwet zones; natural vegetation open forest to tropical rain forest.

Podzols – term used in Indonesia, West Malaysia, Sarawak; known locally as padang soils or kerangas soils; also ground-water podzol and humus podzol; in FAO/UNESCO classification they are mainly gleyic podzols, humic podzols and orthic podzols. Normally occur on level or slightly undulating terrain in everwet regions with annual rainfall exceeding 2000 mm., natural vegetation on lowland podzols is so-called kerangas forest (Sarawak and Borneo), with orchids and mosses abundant; degraded kerangas is replaced by padang vegetation, scattered groups of stunted trees over patches of ground mosses and resembling a heath forest; vegetation changes with altitude.

Chromic Luvisols – called non-lateritic red loams, non-lateritic grey-brown sandy loams, terra rossa, non-calcic brown soils and reddish brown earths in Sri Lanka; red-brown loams and brown loams, and red-brown earths in Australia; non-calcic brown soils in Indonesia and Republic of Viet-Nam; red-brown savanna soils in Burma; red-yellow Mediterranean soils in Khmer Republic, Laos and Thailand. Found in regions with long and marked dry season, rainfall less than 1500 mm., and mean annual temperatures of over 20°C.; dominant natural vegetation open forest, scrub often with spiny shrubs, and anthropic savanna.

Gleysols and *Planosols* – all hydromorphic soils except those too rich in organic matter (histosols) and those with heterogeneous textural horizons (fluvisols); majority are low humic gley or grey hydromorphic soils in former American terminology. Climate does not appear to play leading role in their distribution on poorly drained lowlying ground; natural vegetation wet grassland, sometimes with scattered shrubs and trees.

Acrisols – the term red-yellow podzolic soil used in southeastern United States applied in Indonesia, Republic of Viet-Nam and Sri Lanka; also red, yellow or yellowish-brown lateritic soils in Indonesia and Sri Lanka, because their clay possesses low silica: alumina ratios; grey podzolic soil in the lower Mekong; grey earths in Viet-Nam; some resemble the ground-water laterite and podzolic laterite of Australia. Form especially on acid to moderately basic parent materials, annual rainfall above 1500 mm., no marked dry season, average annual temperatures generally above 20°C.; natural vegetation lowland tropical forest, open forest dominated by dipterocarps, sometimes pine forest in highlands (Viet-Nam, Laos, Khmer Republic) or short grass savanna.

Nitosols – formerly called lateritic soils, a term now too broad and confusing; red earths, Rotlehme, terres rouges; dark red and reddish-brown latosols by Dudal/Moormann; in latest FAO/UNESCO legend, nitosols, on account of their argillic horizon which is weakly expressed, bringing them close to latosols, but also because of the existence of shiny (*nitidus* = shining) surfaces on the aggregates. Undulating country or low hills, rainfall 1000 to 3000 mm., dry season less than four months, annual temperature above 22°C.; natural vegetation primary rain forest or forest savanna, or secondary forest.

Ferralsols – once called lateritic soils or latosols; equivalent of the oxisols of the 7th approximation. Located on undulating land, often peneplains, and sometimes in mountainous areas; mostly fossil formations, hence found between rainfalls ranging from 600 to 3000 mm., and where dry season is slightly marked (Kalimantan) or strongly pronounced (Viet-Nam, northeast Thailand, northeast Sri Lanka); natural vegetation primary rain forest or forest savanna; forests on ferralsols have been less cleared for shifting cultivation than have those on nitosols.

Histosols – organic soils variously termed bog soils, half-bog soils, peaty soils, marsh soils or muck soils – a wide variety of soils formed under environments ranging from lagoonal marshes to poorly drained mountain depressions (dystric histosols); annual rainfall 1500 to 2500 mm., sometimes short dry season, mean annual temperature 24 to 26˘C; natural vegetation peat swamp forest which varies in composition according to local drainage conditions and thickness of the peat layer; *Melaleuca leucodendron* sometimes dominant.

India

The soil groups of India are given in the Handbook of Agriculture (ICAR, 1961), which it is understood is in the process of revision (A. B. Joshi, personal communication, 1973): Alluvial soils, including deltaic, coastal and inland alluvium; Black soils of varying types, including the typical black soil or regur of the Deccan plateau; Red soils, including red loams, yellow earths, etc.; Laterite and lateritic soils (Patnaik, in FAO, 1971a); Mountain and hill soils; Arid and desert soils; Saline and alkali soils (Abrol & Bhumbla in FAO, 1971a; Abrol, Dargan & Bhumbla, 1973; see also U.S. Salinity Laboratory, 1954); Peaty and other organic soils. Also, acid sulphate soils of Kerala (Murthy in FAO, 1971a).

The techniques adopted in soil fertility research are described by Kanwar & Mahapatra (in FAO, 1971a), soil testing and tissue testing techniques by Ramamoorthy, Pathak & Bajaj (in FAO, 1971a), the screening of crop varieties for their susceptibility to micronutrient deficiency and toxicity, and the evaluation of methods for determining available micronutrients in soil by Randhawa (in FAO, 1971a), and the work on the Soil Test Crop Response Correlation Scheme conducted by the Indian Agricultural Research Institute, New Delhi, the Agricultural Universities at Ludhiana, Hissar, Jabalpur and Hyderabad, and the Departments of Agriculture of Bihar, West Bengal, Maharashtra and Tamil Nadu (by Ramamoorthy & Velayutham (in FAO, 1971a).

Rege & Patnaik (in FAO, 1971a) trace the origin of scientific interest in the nature and characteristics of Indian soils, especially their geological and mineralogical features, to the establishment of the Geological Survey of India in 1846, and of a specific agricultural interest following the Voelcker report in 1893. These authors list the important soil surveys conducted from the 1950's to time of writing, including broad reconnaissance surveys, pre-irrigation soil surveys including the FAO/UNDP survey in the Rajasthan Canal Project (Day in FAO, 1971a), and surveys in river valley projects. Govinda Rajan (in FAO, 1971a) outlines the programme for the All-India Soil Survey with its headquarters in IARI, New Delhi and four Regional Centres located at Delhi, Nagpur, Bangalore and Calcutta, representing the four major soil regions of the country. Bali (in FAO, 1971a) of the Resources Inventory Centre in the Ministry of Food and Agriculture, New Delhi, discusses types of soil surveys in general terms, and stresses the value of integrated surveys especially for large undeveloped areas.

Sarawak, East Malaysia
The procedures adopted in a survey of the soils of the Central Sarawak Lowlands are described by I. M. Scott, in Sarawak Department of Agriculture Soil Survey Memoir no. 2 (in preparation at time of writing); they are quoted at length here to indicate some of the practical difficulties to be faced in soil surveys on terrain of this type. For Sabah, see Soil Map of Sabah, Malaysia (1965), Tuaran, Cartographic Division, 1966 – scale, 1:500,000.

In compiling the soil map of part of Sarawak, information from a number of sources has been used.

Geological maps on a scale of 1:250,000 are available but are of little assistance in localities underlain solely by Tertiary sedimentary rocks, as no attempt is made (or could be made at that scale) to map lithology, In the northeast of the Area, however, the geological pattern is complex and these maps have proved a useful guide to many soil association boundaries.

Air photographs are available for the entire Area but vary greatly in date and scale. A coverage of recent photography on a scale of 1:25,000-1:30,000 is available for most parts of the Area, together with coverage on 1:10,000 scale for some localities. Many of the semi-detailed survey projects have had the benefit of air photo coverage flown in the year of the survey and at a scale requested by the soil surveyor.

Contour maps were not generally available during the course of the project, but an extensive part has now been included in published contour coverage on 1:50,000 scale. For some localities, however, contour mapping from photogrammetric plots on 1:10,000 scale had been completed by Lands and Survey Department and were used as base maps for semi-detailed surveys in upland areas. The reliability of these plots depends partly on the degree of forest cover in the photography used for interpretation, but they have, in general, proved very reliable.

Other base maps available for the entire Area on 1:50,000 scale mainly comprised compilations of drainage and coastal details, settlement, roads and footpaths. As these details were derived from air photo coverage dating from 1949–1956 the information is of variable accuracy. Many longhouses have moved; footpaths which showed well on the air photographs included many which were new tracks to hill rice farms cleared in the year prior to the photography and which have since been abandoned; even the coastline has altered significantly in some localities within a decade. An important phase of the pre-field work was to up-date and correct these maps on the bases of recent air photographs.

Levels of investigation, three levels were employed: reconnaissance, semi-detailed and detailed. Reconnaissance surveys were further sub-divided to indicate the level of detail achieved.

As the project was geared to an assessment of the agricultural potential of the Area, some tracts of both deep basin peat and mountainous terrain were mapped solely by airphoto-interpretation or were briefly investigated on the ground by broad reconnaissance. On such surveys, ground checks were confined to broadly spaced traverses or rapid path inspections. The soil map was thus largely derived from airphoto-interpretation and was circulated on a small scale unless part of a larger project.

The bulk of the Area – comprising those localities with at least some land suitable for development for agriculture – was surveyed at reconnaissance or detailed reconnaissance level, the distinction being made on the density of ground observations. The sampling density may relate to soil complexity or the degree of agricultural interest in the locality but may also be a reflection of the standard of communications.

On reconnaissance surveys no formal traverse grid was attempted. Access was mainly by river, using longboats or speedboats on the larger streams and paddle canoes on smaller side-streams. Generally the most significant soil cross-section for sampling purposes was that on a line at right-angles to the general direction of the stream, extending from the river-bank alluvium into the hills flanking the bottomland (or to the swamp plain basin peats in downriver areas). In the dissected lowlands the navigable drainage net is relatively dense. Rather than continue a traverse beyond one or two days work, it was found preferable to move to a new point on the river and traverse a fresh line through an adjacent locality. This method involved less positioning time for the field parties than a series of parallel traverses, and the field camp (generally occupied by two parties) was commonly moved only once a week. It was also advantageous to locate traverses in relation to the river pattern as, for many areas, the drainage was the only information reliably plotted on available base maps. The traverse coverage largely reflects the drainage network. Traverses are generally 1.6 to 2.4 km long and fan out from the main waterways. Footpaths were traversed where they would be confidently located on the photographs but were roughly compassed by the field party as a check. As a result of information gained during the reconnaissance of large areas, localities with potential for development were identified and many of these were later surveyed at a semi-detailed level.

Methods of survey for semi-detailed projects varied with terrain, soil conditions, and

84

availability of base maps. In upland areas photogrammetric contour plots on 1:10,000 scale were used as base maps if available. Soil boundaries interpolated on such plots were more accurate than those based on an airphoto-mosaic and, wherever possible, an advance request was made to Lands and Survey Department for such contour data (usually at 7.5 metre interval) prior to the field survey. Where time did not allow preparation of such maps, or no ground control was available for plotting, an airphoto-mosaic base was employed.

Airphoto-mosaics were invariably used as base maps in coastal and swamp surveys where there was no significant relief, and the mosaic faced from photostat negative prints. A number of photostat positive copies of the laydown could then be made for use in field and drawing office. Provided the original bromide photographs were available for comparison, the loss of definition on such mosaics was not a serious drawback. The lack of control was unimportant provided the survey area was reasonably small (800 to 2000 hectares was an average project) or unless the mapped boundaries were later married to a contour plot. In the latter case, the field information usually had to be replotted on the new base and boundaries redrawn.

Mosaic base maps were prepared at 1:10,000 scale if photography on that scale was available. Where the coverage was at 1:25,000 or 1:30,000 scale, enlargements were prepared to give a field mapping scale of 1:15,000 or 1:12,500.

Field sampling on semi-detailed surveys generally followed a formal traverse grid. An irregular grid was employed in some localities where many footpaths were available for traversing provided they showed clearly on the photograph coverage. On swamp and coastal surveys, however, a formal sampling grid was essential, as the soil pattern is generally complex and little guidance to interpolation of boundaries is given by the vegetation or terrain data. The soil map was therefore entirely a rationalization of the field sampling records and its accuracy a function of the sampling density. This was best controlled by parallel traverses.

These methods were also used for detailed surveys, although on such work the sampling density was close and little use was made of air photographs. Survey at this level was confined to two special project areas (a forest experimental nursery and a proposed agricultural station) and to a number of sample strips. Detailed contour mapping was available for the project areas. For the sample strips a contour map was prepared, by levelling along each sampling traverse.

In addition to report maps prepared for each survey project, the individual project maps were combined on 1:50,000 scale master sheets and semi-detailed (and detailed) project maps were generalized as necessary for plotting at this scale. The soil map is based on this compilation. Further generalization was necessitated by the reduction to 1:125,000 scale and any of the original mapping units, which were preserved in the 1:50,000 scale compilation, were grouped together at this stage.

Sampling densities. The sampling density achieved by each survey method and implied by the local survey type name has varied. Densities have been calculated for representative survey projects within the Area, and shown in Fig. 3.1, where sampling density, publishing scale for the individual project map, and nomenclature for the survey type have been compared with those adopted by the United States Department of Agriculture and FAO. There is little agreement with either scheme, but as these assume conditions in which a regular sampling grid over the surveyed area gives optimal reliability to the soil boundaries, it is doubtful if comparisons with methods used in the present Area are valid. In much of the upland zone, soils vary rapidly along each traverse, and are unrelated to landform or vegetation. Even in semi-detailed investigation of some upland tracts, the complexity of the pattern is too great for differentiation of family or series mapping units. On reconnaissance surveys, close sampling along each traverse indicates clearly the range of soils occurring in complex within one upland locality, and adjacent traverses, separated by 3 to 4 km., may indicate that this range varies little over a broader area.

Field methods. Each field party normally comprised five locally recruited labourers together with the team leader (either the surveyor or an assistant). Traverses were cut on a bearing controlled by a 50 mm. prismatic compass and were sampled as the line was cut. Traverse measurement was in units of 30 metres on early surveys and 25 metres for later projects. The most convenient measure in the thick vegetation was cut lengths of plastic-coated three-ply telephone cable, with copper or equivalent binding marking the measuring points near either end. Numbered pegs were placed at each 25 m. and soil sampling on the traverse confined to pegged points. On both reconnaissance and semi-detailed surveys, a standard sampling distance of three pegs (75 m.) was adopted.

The soils were sampled to a depth of 1.2 m., using an Edelman barrel auger. Sarawak soils are almost invariably moist and rarely so stony that it is difficult to operate this type. In swamp areas, the auger was discarded when the peat mantle exceeded 1.2m. in favour of a pole cut on the spot and used as a probe. The end of the pole was pointed and small notches cut immediately above the point and at 25 cm. intervals at higher levels to a length of 3 m. Underlying clay is caught satisfactorily in the notches. On traverses directed into the basin swamps, the sampling interval was extended to 125 m., once the zone had been reached where the peat mantle was consistently deeper than 3 m.

Soils were described at each sampling point by standard methods. Samples were discarded after field description unless required for pH or conductivity determinations. In addition to soil descriptions, records were made at each sampling point of vegetation and land use, landform type and slope. A landform profile of the traverse was also drawn and details of settlements, streams and paths were recorded to check the location of the traverse on the air photograph or base map.

Traverse description forms of various designs were discarded in favour of lined hard-covered pocket-sized notebooks without an itemized format. An index to traverses, retained samples, levelling records, etc. was added to each notebook before the field party returned it to the office; the book number was included with the traverse number on transparent overlays of the traverse pattern, which were prepared on 1:50,000 scale for use in conjunction with the master sheets of soil mapping units. The adequacy of the record kept by assistants in notebooks with no itemized format depends greatly on their experience, training and interest.

Approximately 170 profile pits or new road cut exposures within the area were sampled for laboratory analysis. Horizon samples of about half a kg. were retained for analysis in plastic bags with an outer cloth bag cover. Two bags of material from each horizon were retained for the more important profiles and for topsoils or other samples in which the organic fraction was very high. For certain profiles, oriented samples of selected subsoil horizons were collected in 1 in. tins with a view to preparation of thin sections.

Due to the difficulty of communications in many parts of the area, it was common practice for reconnaissance projects to be undertaken by up to five field parties, operating from two or three field camps. In areas where geological and other data suggested that new soils were likely to be encountered, all field work was undertaken by the leader and parties under assistants were directed to other parts of the project.

Brief checks on the assistant's records were made during the course of the survey, but little attempt was made to plot the soil data and build up the soil map until each project was completed. Field camp conditions usually made this impracticable. It was important, however, to keep a running check on the traverse progress, as traverse lines proposed prior to the survey might not prove possible in the field. Starting points established from the air photographs were sometimes difficult to locate on the ground. A number of factors could halt the traverse half-way, ranging from flooded swamp tracts and difficult vegetation conditions to hornets' nests or ripening hill rice (which cannot be entered under customary law). Alternative traverses had therefore to be chosen during the course of the survey.

Map compilation. On completion of the field survey the soil records were classified by the surveyor and plotted on the traverse grid (on 1:50,000 scale maps for reconnaissance surveys and at various field mapping scales for semi-detailed projects). The airphoto-

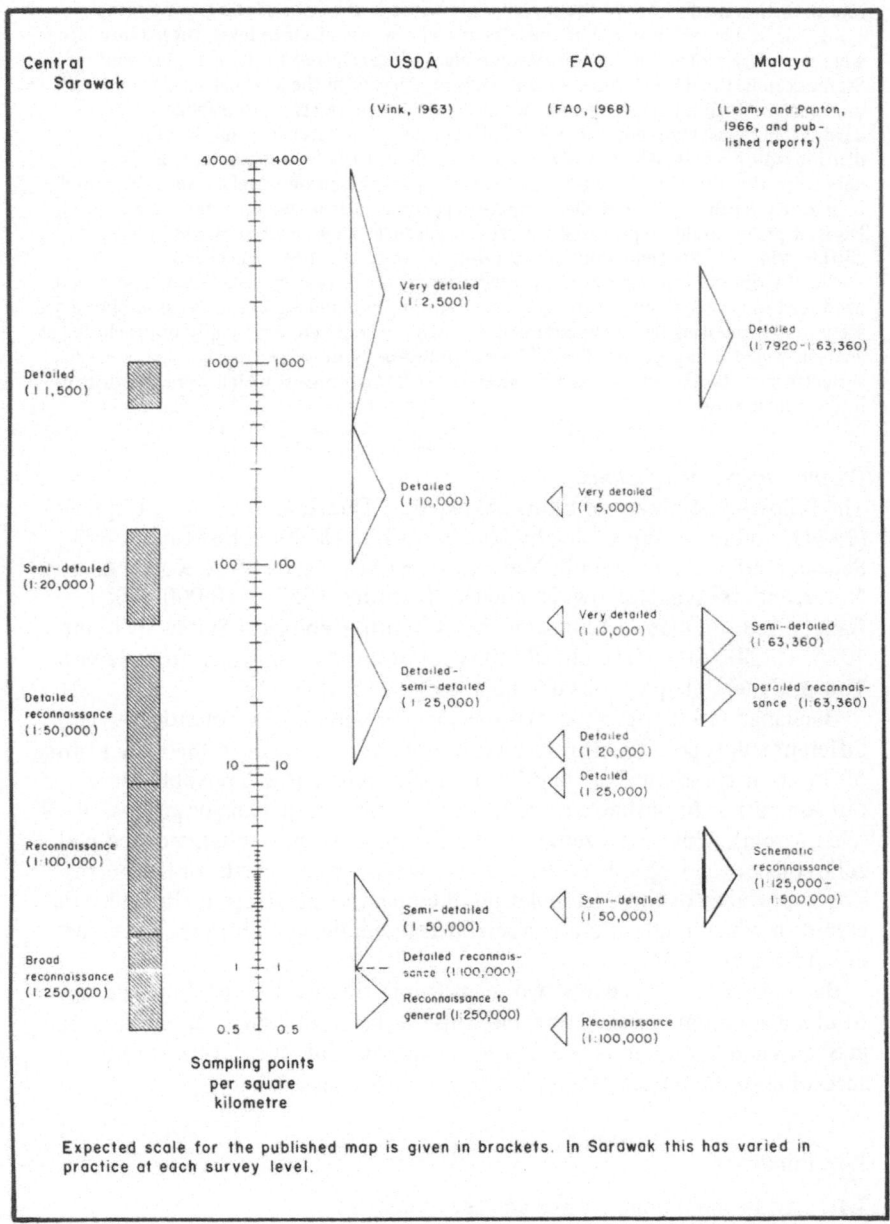

Central Sarawak

USDA (Vink, 1963)

FAO (FAO, 1968)

Malaya (Leamy and Panton, 1966, and published reports)

4000 — 4000

Very detailed (1:2,500)

Detailed (1:7920-1:63,360)

Detailed (1:1,500)

1000 — 1000

Detailed (1:10,000)

Very detailed (1:5,000)

Semi-detailed (1:20,000)

100 — 100

Very detailed (1:10,000)

Semi-detailed (1:63,360)

Detailed reconnaissance (1:63,360)

Detailed reconnaissance (1:50,000)

Detailed-semi-detailed (1:25,000)

Detailed (1:20,000)

Detailed (1:25,000)

10 — 10

Reconnaissance (1:100,000)

Schematic reconnaissance (1:125,000 - 1:500,000)

Semi-detailed (1:50,000)

Semi-detailed (1:50,000)

Broad reconnaissance (1:250,000)

1 — 1

Detailed reconnaissance (1:100,000)

Reconnaissance to general (1:250,000)

Reconnaissance (1:100,000)

0.5 — 0.5

Sampling points per square kilometre

Expected scale for the published map is given in brackets. In Sarawak this has varied in practice at each survey level.

Fig. 3.1. Central Sarawak Lowlands. Sampling densities and survey types used in the area and by other soil survey organizations (Scott, 1974).

87

interpretation made prior to the survey was revised in the light of the ground investigations, and soil association boundaries drawn. At a reconnaissance level, these boundaries were normally plotted on the air photographs and transferred to the mapping base by Stereosketch. Final boundaries were traced, together with the key and base map details, on transparent film (Carbelon or Ethelon) at field mapping scale to publication standard. Final maps were reduced as Kodalith positive transparencies, and Dyeline prints distributed on a scale which was generally half that of the field mapping scale. The soil data were also added to 1:50,000-scale master sheets based on standard sheet lines and a contact Kodalith negative of the compilation produced from which a limited number of Duostat prints could be prepared for office reference. These were replaced by a new edition when further field work added information to the sheet concerned.

The distributed soil maps accompanied a report on the survey concerned. These were produced in mimeographed form for local consumption and were usually issued immediately map processing had been completed. Analytical data were generally not included as analysis was still in progress. The soils were only briefly described and discussion of the agricultural potential emphasized those aspects of development which were of priority interest at the time.

People's Republic of China
The following soil maps of China are listed by Dudal, Moormann & Riquier (1974): Soil regions of China, by Ma Yung Chin. Nanking, Institute of Soil Science, 1956. 1:20,000,000; New soil map of China, by V. A. Kovda & N. I. Kondorskaya, Moscow. Dukuchaev Institute, 1957. 1:10,000,000; Generalized soil map of China, by James Thorp. Geological Survey of China, 1936. 1:7,500,000; Generalized soil types of north-west China, from James Thorp, General map of soils of China, 1936. 1:2,500,000.

Buchanan (1970, pp. 81—82) notes that the agricultural potential of the different soil types of China were known to the cultivators in the early history of the country, so much so that by the 14th century it was possible for a Chinese author to outline a classification of soils into five major groups: black (chernozem), white (sierozem or arid soils), blue (swamp soils), red (red podzolic soils of the tropical south) and yellow (the yellow earths of the north). The importance of this accumulation of empirical knowledge is shown by the extent to which modern Chinese scientists are drawing on it in their schemes of land classification.

Buchanan (op. cit.) adapts two maps from other workers to show the altitudinal zonation of vegetation and soils on the Liaoning-Hopeh borders, and in Shensi and Szechuan (Fig 3.2) and the distribution of soil types in China according to the Soviet pedologists (Fig 3.4). See also 5.2.

3.4 Climate

3.4.1 Arid zone, humid tropics and the monsoons

Ecologists would propose that climate or aerial environment is the most, or one of the most important components of biological and economic ecosystems in monsoonal and equatorial Asia, since it determines the distribution of the

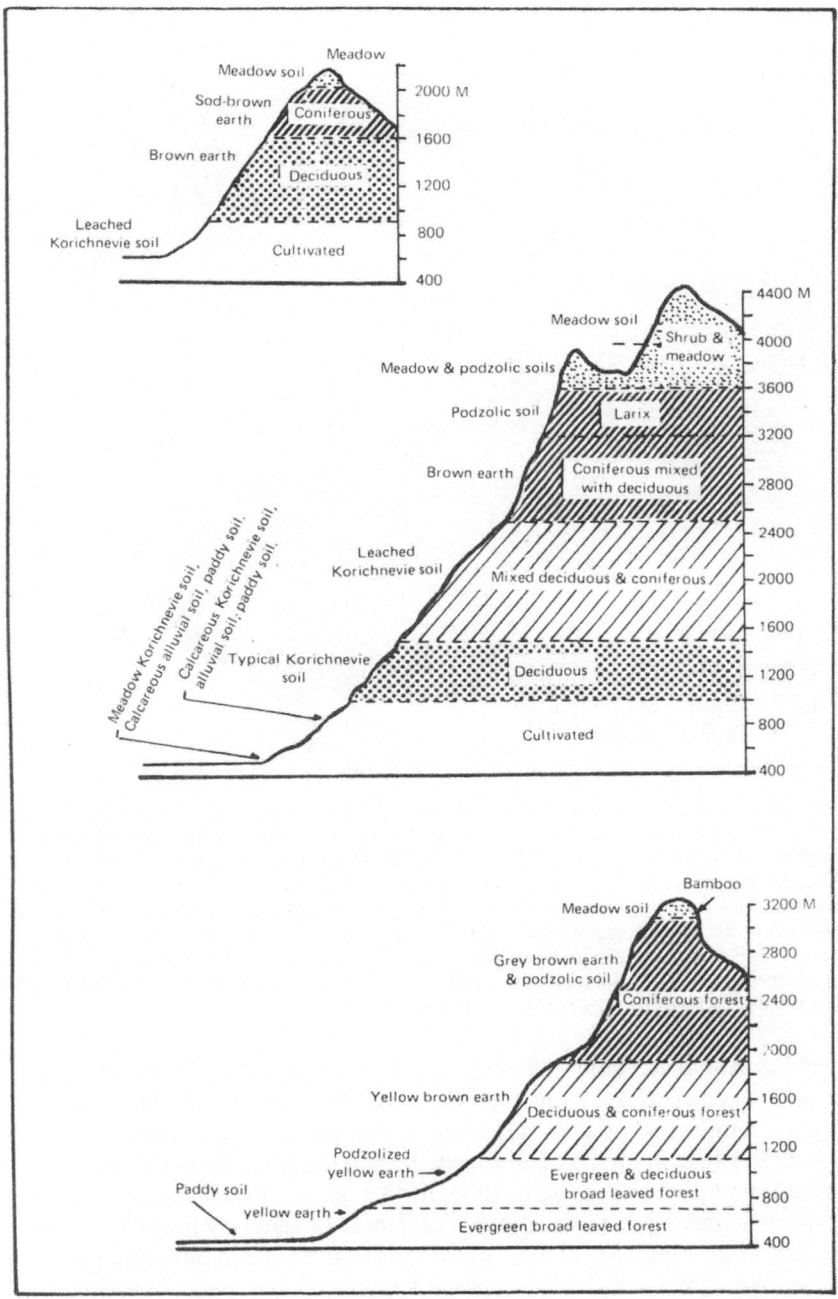

Fig. 3.2. People's Republic of China: Altitudinal zonation of vegetation and soils in north and central China (from Buchanan, 1970, after Wang Chi-wu, 1961).

89

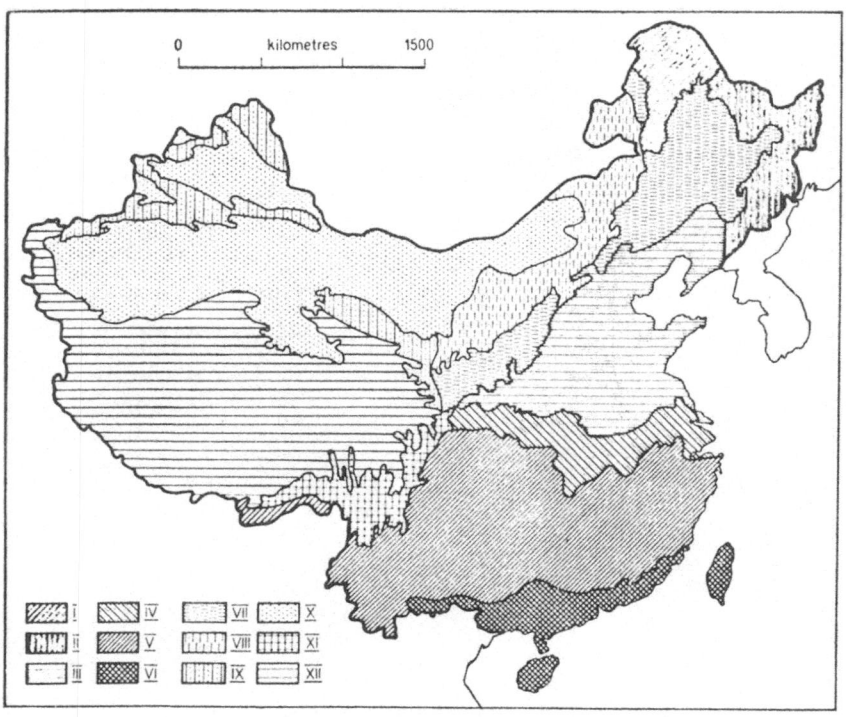

Fig. 3.3. People's Republic of China: The vegetation–soil regions of China (from Buchanan, 1970). I. Coniferous forest-podzol region; II. Mixed coniferous and broadleaved forests – podzol and brown forest soils; III. Deciduous broadleaved forest – brown forest and Korichnevie soils; IV. Mixed deciduous and evergreen broadleaved forests – yellow podzolic and yellow Korichnevie soils; V. Evergreen broadleaved forest – yellow and red podzolic soils; VI. Tropical monsoonal rainforest–yellow lateritic soils; VII. Forest steppe – chernozem and siero-Korichnevie soils; VIII. Steppe – chestnut soils; IX. Mountains or northwest China; X. Semi-desert and desert – sierozem and desert soils; XI. Mountains and plateaux of eastern Tibet; XII. Tibetan plateau.

types of natural vegetation, the main potential crops and the land use; it also plays an important part in the formation and evolution of soils, and in the intensity and nature of erosion (P. Legris, personal communication, 1974). Not all the meteorological factors have the same significance in relation to natural vegetation or local agriculture, neither are all of them measured even at the main meteorological stations. The data most generally available are on amount of precipitation, number of rainy days and temperature; synoptic stations also give values for atmospheric humidity in the form of relative humidity, number of days of fog and mist (see 3.4.4 for the ombrothermic diagram adopted at the French Institute, Pondicherry, and for studies on bioclimates made at the same centre).

90

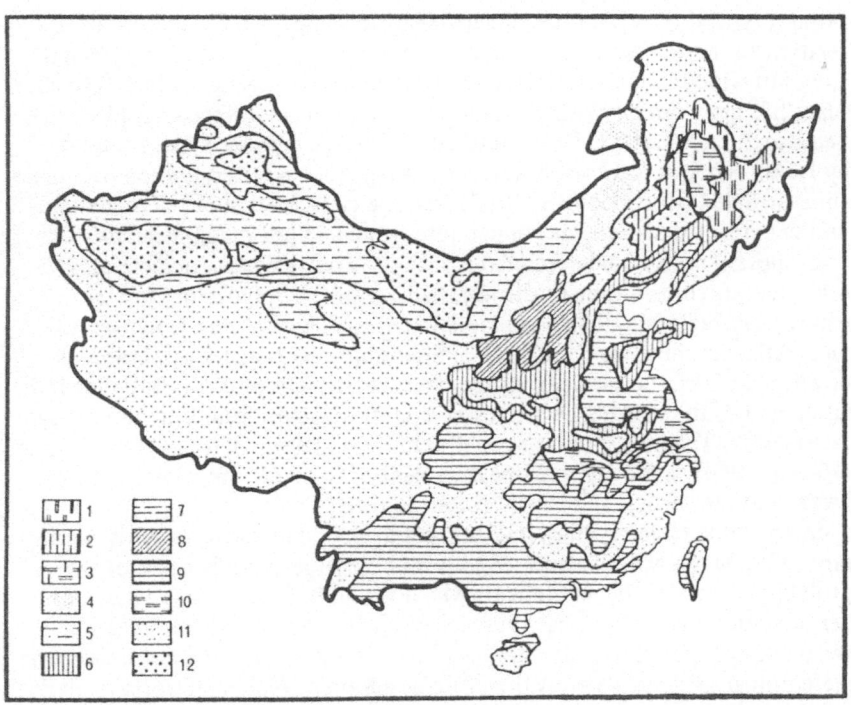

Fig. 3.4. People's Republic of China: The pattern of soils (generalized by Buchanan, 1970, from map by Guerassimov, 1957). 1. Grey forest soils; 2. Leached chernozems; 3. Chernozems; 4. Chestnut soils; 5. Semi-desert and desert soils; 6. Brown forest soils; 7. Calcareous alluvial soils; 8. Grey-brown loessial soils; 9. Yellow and red lateritic soils; 10. Non-calcareous alluvial soils; 11. Mountain soils; 12. Desert sands.

Note the wide extent of mountain and semi-desert and desert soils (which together occupy some two-thirds of China) and that eastern China falls into three broad regions: the chernozem, degraded chernozem and grey forest soils of the northeast; the brown forest and grey brown loessial soils (and calcareous alluvium) of the north and centre; and the red and yellow soils of the south which include the lateritic soils of the tropical zone. Chinese expansion has been found from the original cradle area on the brown soils towards the red soils of the south and, later, the chernozems and leached soils of the northeast.

The region under study comes within both the arid zone and the humid tropics, as defined by geographers and climatologists. Meigs (1953) mapped the arid zone on the basis of the system of Thornthwaite (1948) with refinements from certain definitions of Emberger (1955) for the very arid regions, and thereby provided the basis for the UNESCO programme of research and interchange of information which was in operation from 1951 to 1962. Wallén (1966) has discussed the causes of arid zones, in relation both to macro-climatic influences (general circulation of the atmosphere, topography and proximity to the sea, and the climatic elements which need to be studied in relation to these — radiation, temperature of the air and of the soil, precipitation, hot winds and sandstorms) and micro-climatic influences which influence plant life more directly (water balance, evaporation and evapo-transpiration). See also 2.5 and Gates (1972) for discussion of the climatic types proposed by Hinds, Köppen & Thornthwaite; the rating scheme for evaluation of human climatic index from Maunder (1962), and the classification of microclimates on the basis of radiation, air temperature, wind and percentage humidity.

Evaporation may be determined if accurate knowledge is available of all other factors contributing to the heat balance of the evaporating surface (short-wave radiation, long-wave radiation, outgoing radiation from the surface, reflected short-wave radiation and sensible heat transfer upwards and downwards from the surface). As the determination both of the outgoing radiation and of the sensible heat transfer is most difficult, results are not always satisfactory. The most promising method for the determination of evaporation (Wallén, 1966), is the accurate measurement of the moisture content at two levels, and the vertical velocity of the air in between.

In arid zone research on agriculture and irrigation in particular, it is necessary to know the water balance conditions over large areas. Because of the prohibitive costs of the instruments which would be required at a large number of stations, climatological approaches based on empirical formulae for calculating evaporation and water need from simple measurements of common meteorological elements must therefore be adopted. Wallén refers particularly to the methods of Thornthwaite (1948) and Penman (1948) for the calculation of potential evapotranspiration (defined as the rate of evaporation from an extended surface of short green crop, actively growing, completely shading the ground, of uniform height and not short of water) from macrometeorological data. For arid and semi-arid regions with a summer rainfall regime (western Monsoon Asia), Thornthwaite's formula -- being derived from data in such climates — may be applicable; for other parts of the world (e.g. the Mediterranean semi-arid regions) Penman's more complicated formula, requiring a knowledge of radiation conditions, moisture content of the air, cloudiness and wind, is recommended for more careful analysis. Ojo (1969) has used a modification of Penman's formula in West Africa, obtained by substituting the theoretically calculated net radiation, using ordinary meteorological parameters such as temperature, cloudiness and sunshine for his net radiation formula. The vegetation zones can be delimited by using the P-PE index. The distribution of PE increases from the coast inland, in a zonal pattern. The water surplus/water deficit is a function of the movement of the Intertropical Convergence Zone. The forest/savanna border is delimited by P-PE = 0, that between Northern and Sudan Savanna by approximately −750 mm. isoline, that between Sahel Savanna and Sudan Savanna by approximately −1550 mm. isoline, and that between Sahel Savanna and Southern Savanna Zone by approximately −3000 mm.

92

The humid tropics, a term applied to low-lying areas up to an altitude of about 1000 m., have been defined (Garnier, 1958, 1960 on the basis primarily of West African data) as the area where 1. the mean monthly temperature for at least eight months of the year equals or exceeds 20°C.; 2. the vapour pressure and relative humidity for at least six months of the year average at least 20 millibars and 65 per cent respectively; and 3. the mean annual rainfall totals at least 1000 mm. and for at least six months precipitation is 75 mm. per month. The emphasis on the average state of the atmosphere as expressed by temperature and atmospheric humidity means that a map of the basic humid tropics is divided into two parts: those areas which experience the necessary conditions of temperature and atmospheric humidity for twelve months of the year (the 'core' areas); and those which experience them to the limits given above (the 'peripheral' areas).

Precipitation is regarded by Garnier (op. cit.) as a secondary factor, used to make three further subdivisions of the main region: areas that have 1000 mm. of rain per year and twelve months with 75 mm. per month; and areas that do not qualify in either of these respects. The addition of precipitation as a primary factor in delimiting the humid tropics would greatly reduce the total area regarded as humid and tropical throughout the year, as defined in atmospheric terms alone. The only extensive core area would be among the islands of southeast Asia. By limiting the definition of tropical to places with high average temperatures, areas are excluded that show characteristics found only within the tropics in a locational sense. Garnier considers that there is scope for a separate study on the 'cool' humid tropics as distinct from the 'hot' humid tropics with which he has been primarily concerned.

Ramage's book (1971) on monsoon meteorology in the International Geophysics Series is designed for senior undergraduate and graduate students of meteorology and climatology who have studied theoretical meteorology for at least a year, and for professional meteorologists working or expecting to work in the monsoon area. The author himself states that his book 'is constructed around synoptic systems common in the monsoon area. Differing sequences and frequencies of these systems are related to both regional and interannual variations in the character of the monsoons. Changes in the mix of synoptic systems through the transition seasons must be comprehended too, as essential parts of the dramatic annual cycle which is the monsoon hallmark. Using synoptic models as building blocks, a monsoon area structure of linked diversity becomes possible.'

From synoptic and aerologic data, and from many pictures from meteorological satellites, Flohn (1970) has outlined the seasonal varying large-scale flow patterns above the Indo-Pakistan subcontinent, the rain frequency distribution, the development of summer monsoon winds and rainfall patterns and the role of the Tibetan highlands in these processes. Hantel (1970) has studied monthly charts of surface wind vergence over the tropical Indian Ocean, notes the possibility of the existence of a double ITCZ in certain places at certain times, and describes the threefold monsoonal rhythm in terms of the first and second harmonics of the annual march of temperature.

Ramaswamy (1967) discussed problems relating to the Indian southwest monsoon in his Prince Mukarram Jah lectures to the Indian Geophysical Union, under the heads of: origin, onset, establishment and withdrawal of monsoon; basic causes of strong monsoon in Assam; meteorological situations leading to floods in Himalayan rivers west of 78° E; mean monthly amplitudes of microseisms in relation to normal wind regime over the Indian seas; climatic trends in the monsoon; palaeoclimatology of the monsoon; extra-territorial influences; weather modification; forecasting, short, medium and long range.

From the days of the Harappans to the present, the monsoon in no two years years in no two situations has been alike. For example, the basic causes of the large-scale deficiency in the monsoon rainfall in 1965 and 1966, which so seriously affected foodgrain production, have to be sought in the middle and upper topospheric westerlies and easterlies and in the abnormalities created by them, several of which actually or nearly broke all previous records (Ramaswamy, 1969). The fact that so many abnormalities occurred during the eight monsoon months (four in 1965 and four in 1966) seems to explain why such large deficiencies occurred in two successive years; the deficiencies did not occur due to insufficiency of water vapour in the monsoon air. For the future, it is essential to know why the high-level westerlies and easterlies exhibited these unique features. More observations and research may begin with the simulation of the upper tropical westerlies in massive and costly laboratory experiments (Ramaswamy, op. cit.). A more extensive study of the nature and characteristics of instability in the middle-latitude westerlies to the north of India during the southwest monsoon period is also urgently required.

3.4.2 The region and its constituent parts

With regard to climatic classifications in general, Smith (1968) noted in his summary of discussions at the Symposium on Agroclimatological Methods (UNESCO, 1968b) that there was a general reaction against them, and a full realization of their serious limitations: 'Accuracy had of necessity to be sacrificed in the simpler classifications, and if greater complexity was introduced to remove the obvious inaccuracies, the ensuing system became impracticable. The basic weakness is that meteorology, with its multiform and inter-related facets, does not lend itself to division into rigid compartments, in contrast to a botanical classification. Furthermore, the biological meteorological relationships are in themselves complex, and also for the most part imperfectly established, so that for agroclimatological purposes the use of existing climatic classifications is severely restricted.

'It was, however, suggested that there was a place for the better type of classification in the realm of teaching. Moreover a new approach to systemization based on individual crop requirements could be envisaged. This would seek to establish the delineation of, say, a wheat climate, subdivided into zones

94

appropriate to the many wheat varieties; the overlapping of such biologically defined zones would be inevitable, and would make simple map presentation difficult, but this indeed is the actual state of ecological affairs. There was considerable dislike expressed regarding classification for classification's sake, but a realization that a degree of codifying was desirable in planning practices.'

The geographical textbooks provide a background to the climatic characteristics within the region. For example, Robinson (1966) recognized eight mean types of climate in what he calls Monsoon Asia, which comprises more exactly monsoonal and equatorial Asia, with overlaps to the temperate and continental land to the north, thus: equatorial, tropical monsoon, temperate to monsoon (China type), hot desert, mid-latitude continental (steppe), humid continental (Manchurian type), temperate desert, high plateau (Tibetan type).

Dobby (1961, chapter 3) traces the origin of the term, monsoon, to the Arab expression 'mausin' or 'mausim', meaning a season of winds such as occur around the Arabian Sea, blowing for six months of the year as north-easterlies and for the other six months as southwesterlies. The effects in Monsoon Asia of the seasonally opposed winds differ widely from point to point, according to the origin and character of the moving air rather than to the peculiar seasonal reverse. The 'monsoon climate' is much more than a climate of seasonal winds, produced by a mechanism which is not yet fully understood, owing to the scarcity of observing stations: 1. a permanent high-pressure belt of anticyclones associated with the Tropic of Capricorn over northern India and the Pacific; 2. a cold-season anticyclone over Central Asia which induces outflows of cold continental air, descending into eastern China and converging with the moist, maritime air blowing out from the north Pacific high-pressure belt but blocked to the west because of the height and breadth of the Tibetan and Himalayan relief systems; and 3. the intertropical front, the phenomenon of the tropical zone, with variations of movement over Monsoon Asia which account for much of the peculiarities of local raininess (Dobby, op. cit.).

Spencer & Thomas (1971, chapter 7) discuss 'climatology and sensible climate' in relation to temperature patterns, the monsoon as a climatic agent, mechanics of weather change, moisture and its distribution, varieties of weather and climatic regions (maps of Thornthwaite's classification and of a modified Köppen classification). Since neither of these satisfy the authors, they have evolved a map of thermal comfort (their Fig. 7.1) — an attempt to differentiate seasonal weather patterns as human reactions perceive them.

Kazutake Kyuma (1971) of the Center for Southeast Asian Studies, Kyoto University, applied the Thornthwaite classification to the climates of south and southeast Asia. As, however, the number of climatic types classified on the basis of thermal efficiency, humidity and seasonal distribution of water surplus and deficiency amounted to 42, a delineation of climatic regions was most difficult to achieve. The author has now adopted his numerical taxonomic method evolved for Japan in a study of south and southeast Asia (Kyuma,

1972). The objective has been to define climatic regions as a basis for considering the classification of alluvial soils in the rice-growing tracts.

The following nine climatic regions can now be recognized (Fig. 3.5):
I Straits–Sunda; II Malay-northern Borneo; III Oceanic Sumatra-west Java; IV Southwest-facing coastal; V Southern Indochina-southern India; VI Middle Vietnam; VII Central India-northern Indochina; VIII Tonkin-Assam; IX Lower Indus.

The Philippines are so complex climatically that it has been impossible to define climatic regions on the basis of data from only a few stations. The mean monthly temperatures and rainfall for the sample stations in each region are given in the author's Fig. 3.

Kyuma (op. cit.) gives some details of the climates in each of his groups, as follows, and compares the final result with that obtained by using the Thornthwaite method:

I. Humid equatorial climate with very small temperature fluctuation; minimum rainfall of ca. 100 mm. in August and maximum rainfall of ca. 250 mm. in January.
II. Humid to perhumid equatorial climate with very small temperature fluctuation; minimum rainfall of ca. 150 mm. in July and maximum rainfall of ca. 400 mm. in November–December.
III. Perhumid equatorial climate with very small temperature fluctuation; minimum rainfall of ca. 200 mm. in September and maximum rainfall of over 400 mm. in November–January.
IV. Tropical monsoon climate with small temperature fluctuation; minimum rainfall of less than 10 mm. in January–February and maximum rainfall of over 550 mm. in July.
V. Tropical monsoon climate with small temperature fluctuation; minimum rainfall of ca. 20 mm. in February and maximum rainfall of ca. 200 mm. in September–October.
VI. Tropical monsoon climate with moderate temperature fluctuation; minimum rainfall of ca. 50 mm. in March–April and maximum rainfall of ca. 500 mm. in October–November.
VII. Tropical to subtropical monsoon climate; minimum rainfall of less than 10 mm. in December and maximum rainfall of ca. 300 mm. in July–August.
VIII. Subtropical monsoon climate; minimum rainfall of ca. 30 mm. in December and maximum rainfall of ca. 300 mm. in July–August.
IX. Subtropical arid climate; slightly monsoonal with a rainfall maximum in July–August (60–70 mm.), but for the rest of the year rainfall does not exceed 30 mm.

Other regional studies among many include the recognition by correlation of indices of five macro- and ten micro-agroclimatic regions in Rajasthan (Sen, 1972); the nature and influences of the southwest monsoon on the climate of Sri Lanka in general (Domrös, 1971) and in the southeast quadrant in particular (Wikkramatileke, 1956). Spate & Learmonth (1967) have published as their Fig. 2.9 a simple, empirical and descriptive map of the South Asian subcontinent, which indicates the distinction between the more continental north and the peninsula. Shanbag (1956) considers that the concept of Thornthwaite, on which the Spate/Learmonth map is based, has little relation to observations on growing plants. Shanbag proposes a formula related to actual plant growth

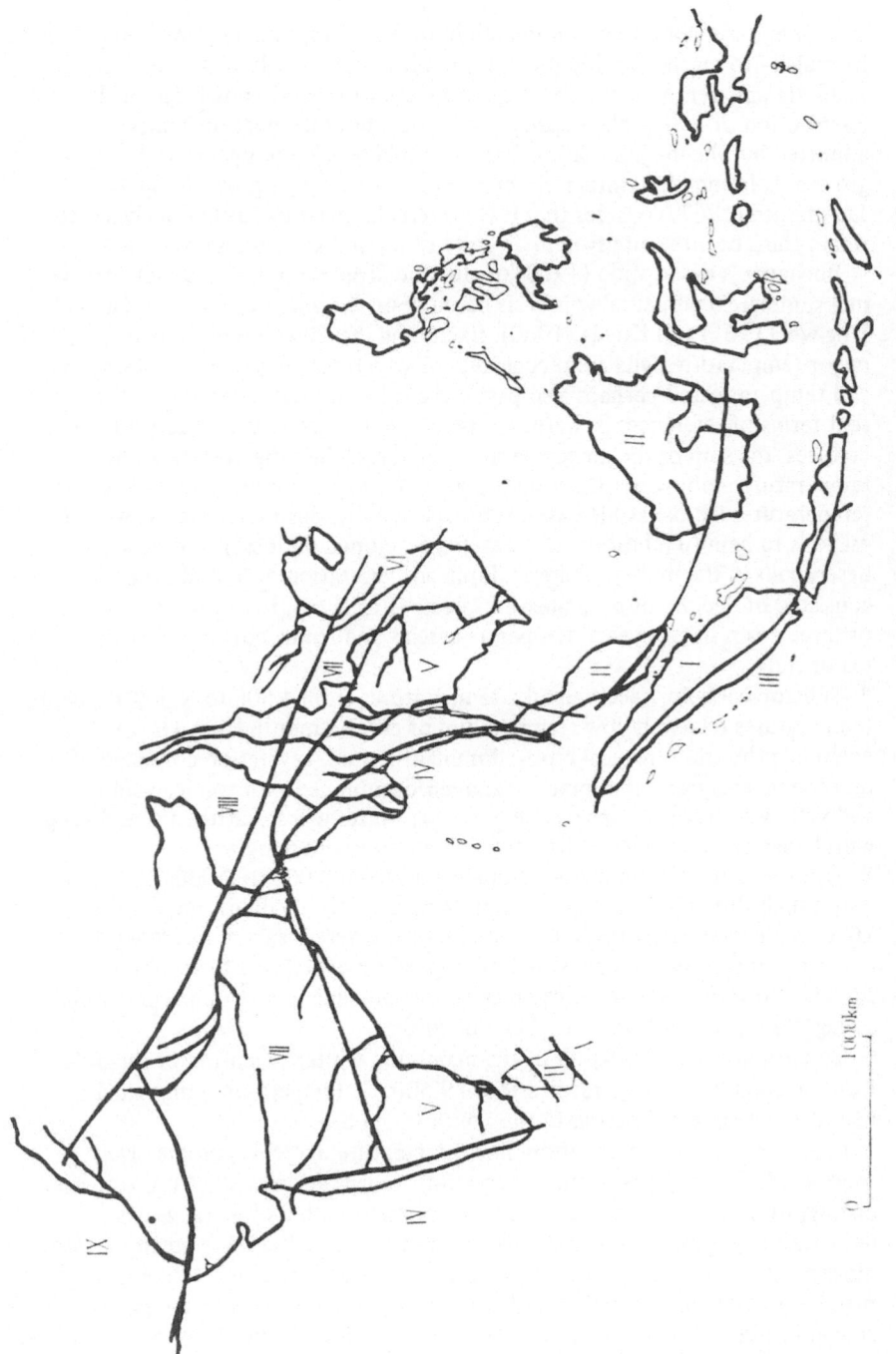

Fig. 3.5. Numerical classification of the climates of south and southeast Asia; see text page 96 for key (Kyuma, 1972).

at different temperatures, and related in turn – it appears by purely empirical formula – to mean monthly precipitation for each month of the year. The maps are apparently attractive to geographers with good knowledge of the distribution of Indian plant species and crops, but one great difficulty, admitted by Shanbag, is that his basic formula relating temperature to plant growth, is founded on rather old experiments dealing with maize. Spate & Learmonth (1967) consider that it is preferable, pending further analyses, to retain the Thornthwaite system, in spite of its well-known disadvantages.

Buchanan's description (1970) of climatic diversity in China, from 'perpetual summer to perpetual winter' is based upon the analyses made by Huang Ping-wei (1961) and Kovda (1960). Excluding the Tibet/Chinghai plateau, six major temperature belts are recognized, of which the first two, cold-temperate and temperate, and perhaps also partly the third, do not come within the present terms of reference: 3. warm temperate belt – accumulated active temperatures, the sum of the air temperature on days when the average 24-hour temperature is above 10°C., from 3,200 to 4,500°C., winters cold but summer temperatures almost as high as in subtropical belt; soils vary from brown forest soils to heilutu (chestnut chernozems developed on loess), and brown desert soils in the more arid areas; dominant vegetation of humid areas consisted of deciduous broad-leaved forests (Fig. 3.10), two crops per year or three crops in two years, temperate cereals and fruits, cotton, no citrus, tea or tung;
4. subtropical belt – accumulated temperatures from 4,500 to 8,000°C., winter temperatures relatively low (temperature of coldest month from 16° to 0°C.) excluding the truly tropical crops; dominant natural vegetation consisted of monsoonal evergreen and mixed evergreen/deciduous forests on leached red and yellow soils; two crops of rice per year, winter wheat, citrus, tea and tung, with bananas, pineapples, lichees, etc., along southern margin;
5. tropical belt – accumulated temperatures from 8,000 to 9,000°C., temperatures high throughout the year, mean temperatures of coldest month above 16°C. with an average absolute minimum over several years of not lower than 5°C.; lowland vegetation consisted of tropical monsoon rain forests on lateritic soils; no winter wheat, three crops of rice possible, plus tropical crops such as sugarcane, coconut palm, rubber and coffee;
6. equatorial belt – lies south of the maximum winter extension of the polar front, accumulated temperatures above 9,500°C., temperature range small; includes the scattered islands in the South China Sea.

In eastern China, precipitation, like temperature, increases progressively southward. On the basis of the evaporation/transpiration ratio, it is possible to distinguish four major zones. In this environment with its vast ranges in temperature and humidity, and where conditions may fluctuate greatly – and disastrously – from year to year, the Chinese agriculturist and animal husbandman have to fit their techniques of land use and husbandry into the patterns shown in Figs 3.3 and 3.4 (see also Newman & Jen-yu Wang, 1959 on the defini-

98

tion of agricultural seasons in middle latitudes, based on the responses of various field crops to the distribution of daily maximum and minimum temperatures, within weekly periods).

Buchanan (1970) concludes the section on climate thus: 'the peasants of Old China (were) helpless when confronted by the variability of China's climate. One of the major results of the social and political changes of the last two decades has been to create a new relationship between man and climate in China, a relationship in which man, as a member of a tightly-knit group possessed of increasing technical competence, has initiated a transformation of the vegetation cover and the hydraulic situation which is establishing him as an ecologic dominant.'

3.4.3 Methods and objectives in agroclimatology

The views of WMO were presented by P. M. Austin Bourke in his introduction to the Symposium on Agroclimatological Methods (UNESCO, 1968b) which was designed as a critical review of methods of measurement, analysis and presentation in agroclimatological studies, and their practical application. The following fifteen paragraphs are relevant sections of that address:

'In considering aims and objectives, we must necessarily think in terms of agrometeorology as a whole rather than the more restricted aspect, agroclimatology . . . It is true that when we come to consider the regional surveys we shall be concerned primarily with agricultural climatology, but even here the fundamental difference is in the field of application; the work is oriented towards assisting the long-term strategy of agriculture, whereas the rest of agrometeorology is mainly directed towards day-to-day tactics. But the raw material is the same in both cases — the meteorological and biological observations, and the interrelationship between the two. If an adequate network of observing stations has been established for general agrometeorological purposes, then the basic data exist for an agroclimatic analysis; if, in the course of season-to-season agrometeorological research, good practical models have been evolved of the relationship between weather and particular crops, farm animals, pests and diseases, then these models provide the essential link to convert the climatic analysis into practical agricultural terms. In other words, progress (or lack of it) in agroclimatology is linked with progress (or lack of it) in agrometeorology as an entity. When, as in this paper, we are concerned about objectives and their realization, we cannot usefully segregate the part from the whole.

'The opening paragraphs of the World Meteorological Organization's Guide to Agricultural Meteorological Practices provides a good summary of the scope of our speciality: Agricultural meteorology is concerned with the interaction between meteorological and hydrological factors, on the one hand, and agriculture in the widest sense, including horticulture, animal husbandry and forestry, on the other hand. Its object is to discover and define such effects, and thus to apply knowledge of the atmosphere to practical agricultural use. Its field of interest extends from the soil layer of deepest plant and tree roots through the air layer near the ground in which crops and woods grow and animals live to the highest levels of interest to aerobiology, the latter with particular reference to the effective transport of seeds, spores, pollen and insects.

'In addition to natural climate and its local variations, agricultural meteorology is also concerned with artificial modifications in environment (as brought about, for example, by wind-breaks and shelter-belts, irrigation, glass-houses, etc.); in climatic conditions of storage, whether indoors or in field clamps; in environmental conditions in animal shelters and farm buildings; and during the transport of agricultural produce by land, sea or air.

'The text goes on to consider the relationship between weather and agriculture under six main heads: soil, plants, farm animals, diseases and pests of crops and animals, farm buildings and equipment, and artificial modifications of the meteorological and hydrological regime.

'All this is relevant and excellent, but − with the exception of the phrase "to apply knowledge of the atmosphere to practical agricultural use" − it largely skirts the question of what constitutes the ultimate goal of agrometeorology. The same is true of other standard texts which I have consulted; presumably the objective of agrometeorology is considered to be so self-evident that it is superfluous to spell it out.

'I have therefore been forced to make my own definition: The task of the agrometeorologist is to apply every relevant meteorological skill to helping the farmer to make the most efficient use of his physical environment, with the prime aim of improving agricultural production, both in quantity and quality.

'The words "farmer" and "agricultural" are here used in the broadest sense, recognizing that horticulture, animal husbandry and forestry all come within the ambit of our subject. With this understanding, this statement of the ultimate objective would, I think, win general acceptance.

'The definition recognizes that the agrometeorologist does not himself intervene in agricultural production; his role is confined to passing on helpful information, either directly or through other advisers of the farmer. This implies that there are important problems of communication, presentation, interpretation and education. It is essential to promote an informed awareness of the importance of weather, both in agricultural research work and in practical farming, and an understanding of the potential and the limitations of agricultural meteorology − what we can and cannot do.

'Between the outbreak of the first world war and the end of the second, meteorologists virtually abandoned agriculture in favour of more immediate and rewarding pursuits. Left to their own devices, most agricultural research workers understandably fell back on the simpler meteorological elements − temperature and rainfall − and often considered these solely in terms of monthly averages and totals. Small wonder that relatively little progress was made in relating biological and meteorological events.

100

'But we should ask ourselves whether, in some cases at least, the pendulum has not in recent years swung too far from simplicity to complexity ... in these days it is normally easier to get financial sanction for complexity and impressive instrumentation than for the tedious and unspectacular drag of reducing and interpreting data. Nor is further automation anything like a complete answer; we are groping our way through complicated and largely unmapped territory where there is no clearcut route with which to programme a computer. In exploring this largely virgin country, there is no real substitute for the flexible probing human brain, with time to wonder and puzzle, and capable of those sudden flashes of inspiration which are essential to progress.

'I have come to think of late that we have been perhaps placing the wrong emphasis in our weather observational programme for agricultural meteorology. The WMO Guide deals virtually exclusively with specialized "agricultural meteorological stations" which have the task of recording both meteorological and biological data. . . . I would, however, suggest that the primary need at most agricultural experiment stations is less a comprehensive standardized observing station, comprising many elements which are irrelevant to the experimental programme, but rather an observing station primarily geared to precise needs.

'On the subject of plant diseases, I may note in passing that changes in the international synoptic weather code have in recent years lost us quick access to relative humidity values, so essential in forecasting epidemics. The inclusion of air and dew point temperatures to the nearest degree Celsius is no adequate substitute.

'The lack in most countries of a regular programme of biological observations adequate to provide a sound basis for practical agrometeorological help to farmers has been, I think, one of the most serious factors in retarding progress. It has made it even more difficult to establish correlations between weather factors and agriculture, and to verify tentative hypotheses. It is true that a comprehensive programme of phenological observations is operated in some parts of the world. The results are undoubtedly often of scientific interest, but, in so far as they centre on observations of wild plants, birds and animals, one must question their utility, either direct or indirect, within the narrow criterion of giving practical help to farmers.

'More immediately relevant to the operational needs of the agrometeorologist is a biological observing programme on cultivated crops and farm animals ... It is surprisingly difficult to assemble precise information on even such a matter as the variation from year to year of the dates of planting and emergence of crops, for which agrometeorological advice is required. Such data are of more than mere seasonal importance; they are vital to determine whether or not there are significant trends in, for instance, planting dates — the kind of gradual change which can be masked for a long time by year-to-year fluctuations.

'The difficulty of absolute verification means that even the better

101

models are also open to vigorous criticism. Thus much of our time and energy is spent in fruitless controversy, with one side arguing that a formula or procedure falls short of absolute accuracy and the other unwisely claiming for it something of the infallibility of a physical law. The fact is that in the difficult world of agrometeorology, we can hope at most for something little better than a rule of thumb, a practical approximation which may, if we are lucky, reach to within, say ± 20 per cent of the truth.'

Weather/yield relations are difficult to quantify, even on an empirical basis, because of the very complex nature of the problem, states L. P. Smith (1968) in his compte rendu des débats, after the Symposium on agroclimatology. 'Under certain restricted circumstances, where one independent variable is of paramount importance, working formulae can be proposed, but extrapolation to different circumstances is not always advisable. Meteorological parameters such as temperature sums have in the past been extensively used but with only partial success. Although such methods might continue to have some limited use, when they are intelligently applied, other workers have found considerable promise in the use of accumulated potential evaporation, which, being essentially an integration of energy supply, would appear to be the more logical.

'The zoning of farming methods, followed in greater local detail by the zoning of crops, leads finally to the zoning of crop varieties. An example of zoning in the case of sorghum showed that independent research by two scientists arrived at almost identical conclusions regarding the weather factor. There is an immense field of research opening here which should be exploited in future years, but progress will largely depend on the availability of reliable data and the interim solution can only be arrived at by carefully planned and interpreted pilot trials. Poorly designed trials can only waste both time and money. Intelligent use could be made of growth chambers.'

R. O. Slatyer (Australia) reviewed the proceedings of the Symposium on Agroclimatological Methods by concentrating on two main points: the role of the plant climatologist compared with the animal climatologist; and the effectiveness of climatological area surveys.

The main common ground for plant and animal climatologists is the need to understand the basis of physiological response of certain species to different levels of radiation, temperature, etc., and a need to manipulate the environment or the organism to minimize production losses due to climatic conditions.

The plant climatologist should attempt to understand the pattern of energy and water cycling in relation to growth, development and yield, as is being done by several research groups within CSIRO, to provide a stronger basis for the mathematical simulation of plant growth and production in different environments. In this way, it will become possible to indicate and perhaps avoid specific crop hazards, and to predict crop yields.

Three main types of agroclimatic surveys may be recognised: the broad

102

climatic classification of the Thornthwaite or Köppen type may be useful in roughly and rapidly comparing major zones in different continents; in the intermediate category are surveys which show specific hazards, comprising soil/water balance studies, useful only if applied with careful reference to the cropping system and environment for which they were devised; third, the survey based on sophisticated growth and evapotranspiration models.

The first is too coarse, the third may provide much more detail than is required, the intermediate type is best for regional purposes. Slatyer suggests that, for the intermediate-level surveys which are based on an evaluation of the soil/water balance, 'at least four sets of water balance models should be constructed to cover typical cropping systems in regions under study. The first of these could represent a normal cropping mode, probably one in which cropping takes place during the normal rainfall season and the land is uncropped for the remainder of the year. The second could be a natural or sown pasture mode in which growth would again be concentrated during the rainfall season. The third model could be an irrigation mode and could provide estimates of water requirements at points in the region where irrigation might be feasible. The fourth could be a model based on the experience and ingenuity of the investigator and could involve altering the normal cropping system to avoid climatic hazards or to capitalize on alternative crops. It could, for example, explore the possibility of fallowing land during the normal cropping period and growing an alternative crop on stored soil water. . . models should not be set up uncritically, but should be made as realistic as possible and the results should be tested against field observations before regional applications are undertaken. . . .

'In looking to the future, one can envisage a rapid increase in the quantity and applicability of the data coming from the integrated teams of physicists, physiologists and agronomists. It behoves the scientist concerned with area surveys to familiarize himself with the work of these groups, and to progressively incorporate the relationships they develop into his own studies. The more sophisticated and realistic his own models are, the easier it will be for him to utilize this information and the more valuable will be his own surveys' (R. O. Slatyer in UNESCO, 1968b).

3.4.4 Bioclimatology

The French Institute at Pondicherry has made extensive studies of bioclimates in India and adjacent countries, using as the basis of their techniques the ombrothermic diagram developed by the Toulouse school of Professor H. Gaussen (P. Legris, personal communication, 1974). On a graph are marked, on abscissa, the months of the year; and on the ordinates, to the right the scale of precipitation in millimetres, to the left the temperature in degrees Centigrade at a scale double that of the rainfall. A month is considered dry when its precipitation is less than twice the mean temperature:

$$P \leqslant 2T$$

dry if $\qquad$ $P < T$
sub-dry $\quad : 2T < P < 3T$
sub-humid $: 3T < P < 4T$
humid $\qquad : 4T < P < 5T$
per-humid $: \qquad P > 5T$

During a dry month, all the days need not be biologically dry. There may be some rainy, foggy or misty days and relative humidity may also be quite high. To consider the effect of these secondary factors which reduce the severity of drought, Bagnouls & Gaussen (1953) proposed a xerothermic index. This is an evaluation of the relative intensity of dryness in terms of biologically dry days during the dry season. Later they proposed a hygrothermic index, an evaluation of wetness during the rainy season (Bagnouls & Gaussen, 1964). The suitability of these two indices to explain the vegetation types of India has been demonstrated (Meher-Homji, 1968a).

Gaussen (1954) suggested that it might be possible to indicate the ecology of a place by means of a formula. The important factors to be considered are temperature, precipitation, length of dry season, length of the growth period and nature of the soil. The detailed bioclimatic maps of the Indian subcontinent and of southeast Asia prepared at the French Institute take into account four factors: 1. amount of precipitation; 2. rainy season; 3. length of dry and rainy seasons; 4. mean temperature of the coldest month (Legris & Viart, 1961; Labroue, Legris & Viart, 1965; Gaussen, Legris, Blasco et al., 1967).

A classification of biological climates based on the rhythms of the seasonal march of precipitation and temperature followed the ecological formulae (Bagnouls & Gaussen, 1957). Following the principles of this classification and the ecological formulae, Meher-Homji (1960) delineated the bioclimatic types of the Indian subcontinent and further established their analogous types in the world. The same method has been used by Gupta (1964) in establishing the homologous types of the western Himalaya in the Alps and Pyrenees, and by Shyam Sunder (1966) for comparison of the climates of Mysore State with those of tropical Africa and America.

Legris (1972) has presented a formula of 'hydric balance': P-X(ds): where P is precipitation, X is xerothermic index and ds is the deficit of saturation of the dry season. A good correlation is shown with the vegetation of peninsular India when the values of the hydric balance are plotted against the mean temperature of the coldest month. Legris further emphasized the use of this formula in palaeoecology, by assessing the critical differences in the hydric balance and the temperature separating two vegetation types and pointing out the change in the value of precipitation and temperature that could lead to the annihilation of one vegetation type and the invasion of the other in the gap created by the former.

Recently, the stress is on the variability in the bioclimatic factors. Legris & Blasco (1969) observed that it is the calculation of the average from a large number of years which makes the dry season of variable duration and intensity disappear from the average climate–diagram of the South Indian hill stations. The actual dryness lasts for 1 to 4 months and the spell of consecutive dry days extends from 30 to 80 days, coupled with very low relative humidity.

Meher-Homji (1968b, 1971a, 1974a) laid stress on the variability of regime, besides precipitation amount and length of the dry period. For certain stations of the northwest part of the Indian subcontinent, not only the concept of the average year breaks down, but that of the probable year comes close to it in view of strong inter-yearly variability.

The mediterranean regime of Pakistan is not of a pure type as in west Asia, but one of transitional type where 'mediterranean' years are interspersed among 'tropical', 'bixeric' years and years having no seasonal rhythm of precipitation. The degrees of 'mediterraneity' recognized on the basis of percentage of summer rainfall are supported by data of vegetation and floristic elements. The proportion of intermingling of tropical and mediterranean floras varies according to the degrees of mediterraneity (Meher-Homji, 1973). (See also Selod, 1961, on the bioclimates and vegetation, and Hamid, Siddiqui & Hassan, 1969, on agro-climatic classification of Pakistan.)

The climate of certain areas of southern India has been analyzed in considerable detail on the basis of daily rainfall data to make them of practical utility from an agroclimatological point of view (Legris, Couget & Meher-Homji, 1971; Couget, 1971; Meher-Homji, 1974b).

Other studies on these and related aspects include Mueller-Dombois (1968) on an ecogeographic analysis of a climatic map of Sri Lanka, and Perera (1968) on climate/vegetation correlations in Sri Lanka.

3.4.5 Assessment of climate in integrated surveys

The Australians regard climate as a most significant physical resource factor, since it sets the broad pattern of biological activity. However, their conclusion that, although the use of land may modify the microclimate, the macroclimate pattern may be regarded as a permanent immutable feature is apparently now open to some doubt (2.5). The climate at a particular location may differ significantly from the zonal mean. Individual components of climate fluctuate considerably, and these fluctuations are often as important in relation to productivity as are mean climatic values. Precise assessments of climatic components require physical recordings at many sites over a long period of time (30 years – see 2.5). Where such data are not available, extrapolations may be made from existing data, or by interpretation from other factors of the environment (Christian & Stewart, 1968).

The key members of a CSIRO Land Research survey team are the geomorphologist (land form), soil scientist and plant ecologist. An assessment of the climate is considered as an integral part of each survey, but it is not essential for a climatologist to spend a long period in the field. Specialist discussion of climate follows the main land-system descriptions, which relate to genetic factors, overall resources and man-made factors.

According to Zonneveld (1972), differences in the total character of land forms in the valley and in the adjacent hills or mountains are partly due to climate. Aerial photographs rarely provide direct information on climate; satellite photos are valuable because they show cloud forms and overcast patterns that are susceptible of interpretation. Temperature recording (infra-red scanning) supplied data on surface temperature, but only on a very short-term basis. Climatic data for land survey must therefore be obtained indirectly, 'calibrated with the help of climatic stations recording data over a period of a year.'

Plant ecologists regard vegetation — types or even individual species — as a lesser indicator of climate. The 'life zones' of Holdridge (1966, 1971) are a kind of land unit in which climate is the main guiding principle, land forms and vegetation serving only as properties and characteristics for classification and mapping. Climatic data measured at climatic stations are extrapolated over wide regions with the aid of mapping units based on vegetation and land form.

3.4.6 Assessment of environmental potential for agriculture

The term 'agroclimate' is defined by van Eimern (1968a) as the synopsis of all atmospheric conditions and weather processes near the soil surface which influence plant and animal production either immediately or indirectly and which are typical for a location, an area or a larger region, considered in a characteristic distribution of the most frequent, mean and extreme values during a long period. This definition is generally valid for the agroclimate, not only with respect to the macroclimate, but also to local, topo- and microclimate.

The topoclimate is the climate of a local area in which topographic characteristics such as soil, vegetation, settlements, lakes and rivers are decisive for the climate and which can be presented on maps of topographic scale. Considerations of agrotopoclimate are restricted to those factors of the topoclimate which affect plant and animal production in all forms of land use. Special instruments, tools and devices are necessary for the study of topoclimates (van Eimern, 1968b).

There are many different types of land use and agriculture and each requires quite different observations. To define the special needs for agriculture, it is necessary to describe the toal needs, both as to observation and density requirements, to see how these fit into present observational networks, and then to determine additional requirements (Shaw, 1971). The U.S. Federal Plan for a National Agricultural Weather Service defines the user requirements in

Table 3.3. United States of America: Examples of requirements of cotton producers in terms of forecast information on weather (Shaw, 1971).

Activities	Important Parameters	Operational Requirement	
		Description	Forecast Period
1. Planting	1. Soil moisture	Less than 80% field capacity to 4 in. depth	Up to 2 weeks
	2. Soil temperature	Greater than 65° F.	Do.
	3. Temperature (air)	Greater than 50° F.	Up to 36 hr.
	4. Precipitation	Less than 0.05 in.	Up to 12 hr.
	5. Wind	Less than 25 m.p.h.	Do.
2. Defoliation	1. Soil moisture	Between 50% and 80% field capacity	Up to 48 hr.
	2. Temperature (air)	Between 50° and 85° F.	Up to 36 hr.
	3. Precipitation	None	Up to 24 hr.
	4. Wind	Less than 10 m.p.h.	Up to 18 hr.
	5. Dew	Presence and period	Up to 12 hr.
	6. Cloudiness	Less than 0.7 cover	Do.
3. Harvesting	1. Soil moisture	Less than 90% field capacity to 6 in. depth	Up to 48 hr.
	2. Temperature (air)	Variable	Up to 36 hr.
	3. Precipitation	None	Up to 12 hr.
	4. Wind	Less than 20 m.p.h.	Up to 36 hr.
	5. Humidity	Less than 70% relative humidity	Up to 12 hr.
	6. Dew	Presence and period	Up to 24 hr.
4. Transportation			
5. Agri services (buying)	Each of these would have special requirements as above.		
6. Marketing			
7. Others			

Source for items 1, 2 and 3: Federal plan for a National Agricultural Weather Service (Environmental Science Services Administration, 1967).

respect of special meteorological data. An example of requirements for cotton producers is shown in Table 3.3. The parameter meteorological requirements which should ideally be measured at all categories of stations (but will not be for a long time, even at a network of principal stations in the United States) relate to: atmospheric composition, atmospheric profiles, clouds, dew, heat budget, humidity, hydrology, precipitation, pressure, radiation, temperature and wind.

Workers in South Africa have evolved a method for analysing climate which is based upon the assumption that the environmental variables may be transformed in such a way that the transformations vary linearly with the environmental potential index (EPI). It is recognized that, although edaphic, biotic and other factors are important in determining agricultural potential, climate may be regarded as the dominant factor, and a numerical index reflecting this potential alone may be of great value (see original papers for method of calculating EPI: de Jager, 1971; Scotney & de Jager, 1971). It is admitted that the exclusion of soil is a weakness, but this is due to the paucity of long-term data on crop yields on individual soil series. Oceanic influences along the coastal belt are also not accommodated.

In attempting to relate the indices to the potential yields of several adapted crops grown under dryland conditions, the EPI was derived using the linear transformation and the three environmental variables, soil moisture deficit, available heat, and rainfall variability; inclusion of data for solar radiation would have been desirable if they had been available.

The selected moisture factor was based on the calculation of potential evapotranspiration, expressed as the difference between annual water need and annual precipitation. The heat factor was calculated on a yearly basis, the summation of positive temperature above a base or zero point of vital activity being accepted as the working basis; for maize, a base of 5° C. was chosen. To obtain the yearly heat units, mean monthly temperatures above 5° C. were multiplied by the number of days in each month, and the results summated. The coefficient of variation of annual rainfall, calculated over a period of at least 20 years, was used.

The ranges of the variable factors, based on data from selected stations in Natal and Transkei, were: moisture: from −250 to + 250 mm. per annum; heat: from 3,000 to 6,000 degree days; rainfall: from 15 to 30 per cent coefficient of variation.

The EPI, the degree of intensification in terms of farming systems, and the potential yield of several crops and of natural veld corresponding to five main rating classes, are given in Table 3.4. Once the computed EPI for a particular farm is known, reference to tables of this nature will indicate the range of yields which might under normal conditions be expected. A rating of 3 or below is required for the economic production of most crops.

It is not intended that the technique should indicate the type of crop to be grown, but rather those areas in which adapted crops will yield satisfactory

Table 3.4. South Africa: Relations between environmental potential and estimated potential yields of rain-grown crops and natural veld (Scotney & de Jager, 1971).

Rating class	Environmental potential	Degree of intensification of farming system	Dryland Yield Potential				Wattle		Natural veld
			Maize	Potatoes	*Eragrostis curvula* dry matter		Bark	Timber	
			Mg/ha^{-1}	Mg/ha^{-1}	Mg/ha^{-1}		Mg/ha^{-1}		ha/mlu
1(0–1)	Very high	Very intensive	> 5,0	> 30	> 8		> 20	> 84	> 2
2(1–2)	High	Intensive	4,0–5,0	22–30	6–8		16–20	64–84	2–2,5
3(2–3)	Moderate	Semi-intensive	3,0–4,0	15–22	4–6		12–16	40–64	2,5–4
4(3–4)	Low	Extensive	1,0–3,0	7–15	2–4		< 12	< 40	3–5
5(4–5)	Very low	Very extensive	< 1,0	< 7	< 2		–	–	5–6

returns. Generally, the better the index, the wider the range of crops that can be grown. The method also indicates those areas where consistently high yields may be expected, and where expenditure on fertilizers might be more economic. Indices in the table between 3 and 5 generally indicate regions most suitable for irrigation. Indices may also be used to define the limits of vegetation types, and their relative carrying capacities. One of the main advantages of this speculative approach is that assessment of land potential in terms of climatic factors becomes more qualitative. With sufficient data, basic information could also be used in the detailed planning of a farming enterprise.

Proceeding from studies in Tohoku, Japan, near the northern limit of rice cultivation (Hanyu et al., 1966) on the determination of suitable regions and periods for the direct seeding of rice, Hanyu, Uchijima & Sugawara (1966) applied the index of climatic productivity for expressing the quantity of ripening of paddy rice. This index, first proposed by Murata (1964), used the average of daily mean air temperature and that of daily amounts of insolation during the productive period of 40 days, from 10 days before to 30 days after heading. However, on the basis of data obtained in Hokkaido & Tohoku, Hanyu & Uchijima (1970) consider that an index should be based on elements covering the full period of growth; for this purpose, these workers analysed the relation between yield, daily mean air temperatures and the total duration of sunshine in the northern districts during June and July, periods of transplanting and heading. In another paper (Uchijima & Hanyu, 1967), it is found that the important limiting factors in increasing the value of the index are low temperature in Hokkaido and Tohoku, high temperature in western Kyushu, and insufficient sunshine in Kanto plain. More recent work has been concerned with the relation between temperature for the safe cultivation period and the stability of yield in Hokkaido (Hanyu, 1971; Hanyu & Ishiguro, 1972).

The contrary approach to that of the agroclimatologist concerned with assessing the environmental potential of an area for agricultural production is that of the plant and animal physiologists. Climatic factors in relation to animal production are discussed in 4.2.3. For plants, one would have to consider the vast literature and experience on actual and potential production of agricultural crops, in relation primarily to photosynthetic efficiency (Alberda, 1962), or the total ability of crop plants to accumulate dry matter to the absolute upper limit (Loomis & Williams, 1963). The environmental factors influencing crop growth in a humid tropical climate have been fully reviewed, with extensive bibliographies, by Williams & Joseph (1970), with special reference to use of moisture and light, temperature limitations to growth and production, and the photoperiodic reactions of common tropical species (rice, sugarcane and other annuals, coffee and non-photosensitive crop plants).

Primault (1970) considers that the methods of presentation of climatological figures for use in agricultural planning must be changed completely; this task can be undertaken only by the use of electronic computers. The punch cards to be used should contain data for at least the past 30 years, and for

110

some elements such as precipitation, the intrinsic variability is such that even a 30-year period is insufficient for determining the variance curves. The following quotation is of relevant paragraphs in the English summary of Primault's article:

'Efforts towards finding relationships between meteorological factors and agriculture date from antiquity, but it was only in the 18th century that a systematic approach provided a long series of observations. These led to the general conception of climate through the comparison of mean values reflecting in various regions one or more meteorological elements (temperature, humidity, sunshine duration, wind, etc.). Frequently one parameter was taken separately to describe the suitability of regions for growing a certain crop (e.g. $> 9°C$. for vine), which resulted in errors. Annual averages were then abandoned and monthly data were used to establish climatograms in which several factors were combined, usually two: temperature and precipitation. Other workers sought to integrate these factors by defining their influence on physiological plant processes (e.g. the geo-botanical approach of Braun-Blanquet (1951), in which plant associations are determined by soil conditions and the local climate, and Ellenberg (1954), who proposed that climatic influences be determined by phenological observations and comparison of dates at which plants in various regions reach the same stage of development). However, the development of plant associations is repeatedly obscured through human interference, while phenological data must be taken over a long period of time to avoid too much influence being given by exceptional years. On the other hand, the climatograms are often also inadequate to compare conditions in various places; they tend to be over-simplified and do not sufficiently reflect the variability of the factors considered (see author's Figs. 1 and 2).

'In agriculture the average weather is rather unimportant since it occurs so rarely. Moreover, a linear relationship between a single meteorological factor and the development of a given plant species is also very rare. For example, growth is enhanced linearly by the temperature as soon as this exceeds the zero-threshold of development; but beyond a certain maximum, growth cannot be raised by higher temperatures, although these are not yet detrimental to the plant. Nevertheless, temperature can kill, at both ends of the scale, by burning or freezing. Mean values do not give any clue as to the risks involved in these extreme conditions.

'For the planning of agriculture more precise data are needed. . . . Usually, however, the data collected by the observation stations have been worked out on the rather long-term basis of whole months. For agricultural purposes, where an evaluation of profitability is the aim, shorter periods (ten, seven or five days) would be needed and for each of these periods the figures would have to be classified according to the following criteria:

a. The highest and lowest values recorded in the period considered should be given. They represent the total amplitude.

b. Growers are aware that adverse weather conditions may occur and in

111

assessing the profitability they must include this risk, for which a bad crop once in five years is an acceptable estimate. Since failure càn be caused by either high or low extremes, 10 per cent is eliminated on either side of the period or date so that the remaining 80 per cent between the useful maximum and the useful minimum represents the useful amplitude.

c. The mean value of the series does not indicate the normal, which occurs exceptionally. However, it can be assumed that once every two years, the figures remain within the expected limits. Therefore the normal amplitude in which a value can fluctuate without becoming exceptional is situated in the first and third quartiles of the series. By adding the figures for the medial and average values a very complete picture of the possible fluctuations of each meteorological element can be obtained.

'The complicated procedure described is necessary because in climatology the distribution curves are practically never of the normal (Gauss) type. For temperature the curve is rather normal, but for precipitation it is asymmetrical and can rise *ad infinitum* while for nebulosity it shows maxima at the two extremes and as to insolation it is again asymmetrical. It is very difficult to reproduce the frequency of the various values in graphs since these rapidly become indistinct. The only solution is therefore to give all these figures in table-form (cf. author's tables 1, 2 and 3).

'The figures so far available should be complemented. Efforts have been made to determine the duration of the growing season as influenced by changes in the weather, but more precise data are needed for the beginning and end of this phase. Meteorologists can calculate these by applying the same criteria of total, useful and normal amplitudes (cf. author's table 4). Moreover, some elements should be considered to determine the frequency especially of extreme temperatures (cf. author's table 5). Finally, agriculture should be provided with facts on the duration and number of periods in which certain temperature thresholds were exceeded (see author's Fig. 3).

'Plant Protection Services can use the general agricultural data so far discussed, but they have special requirements to know to what extent weather factors act upon the development of pests and diseases. To this end biometeorological values must be worked out for which certain basic figures must be known, such as temperature sums above a certain threshold, and insolation and precipitation sums. Such figures . . . will allow the interpretation of climatic factors so as to assess the risk of pests and diseases in a given region.'

It is not always realized in agroclimatological assessments that the actual and potential conditions governing plant and animal biology are operating at ground level. It is thus necessary to consider also micrometeorology in relation to the scientific disciplines concerned and in regard to the practical problems it is ultimately designed to help solve (WMO, 1972). Micrometeorology is the science concerned with the detailed examination on a microscale of the physical and meteorological processes taking place within the ecosphere – the zone lying principally within the boundary layers, between the top of a plant or animal and the bottom of the roots in the soil, the zone between atmosphere and earth wherein take place the biological processes. Micrometeorology must not be confused with mesoscale investigations of a local or topoclimatological nature which are sometimes referred to as microclimatology. Because of the part played by external influences such as radiation, rainfall or the transport of airborne particles, some degree of macroscale meteorology must be considered, and it is both impossible and undesirable to define precise dividing lines. Considerations of this type have to be made by a wide variety of scientists, especially during the planning of programmes of experimental research, and when formulating the details of specific projects of research.

The WMO working group suggests that directors of research, or university professors with research commitments and their scientific staff will find that WMO Technical Note no. 119 will enable them a. to take a comprehensive view of all the physical processes and the part they play in the scientific disciplines concerned with the soil, the atmosphere, plants and animals, b. to recognize the extent to which interdisciplinary collaboration is advisable, and c. to select lines of research in which relatively independent action can be taken. The Note will also show how the examination of the physical processes involved in the full understanding of a particular type of practical problem will indicate both the type of meteorological factor which has to be considered, and also what type of fundamental research is most necessary (Table 3.5 (a) to (g)).

3.4.7 Dry seasons/droughts/floods/variability/probability

The agronomist would wish to be able to tell his dryland farmers for how many years in ten, twenty or fifty years the precipitation might be, or might be expected to be, 50 per cent, 25 per cent, or 10 per cent of normal. Such information would be of great value to designers of farming and cropping systems, to breeders of adapted crop cultivars, and to those responsible for planning the storage of basic reserves of foodgrains harvested in the good years for consumption in the bad. The Rajasthani cultivator knows the fluctuations of his climate by experience and tradition, and so always keeps a two-year reserve of foodgrains sufficient for his family buried under or near the house.

The Symposium on Agroclimatological Methods (UNESCO, 1968b) agreed that no climate is constant. It is therefore necessary to consider all acceptable statistical techniques to ensure the essential inclusion of variations about the theoretical and much over-used average. The Symposium was not, however, very hopeful of being able to foretell the persistence or disappearance of climatic trends in the near future, although it was fully realized that a small change in, say, rainfall frequencies, might have large effects on the production potential of a regional zone. There is greater hope for seasonal forecasts, which could be of great practical application (papers to the Symposium on the use of avail-

able data, pp. 25–128). WMO Technical Note 81 (Thom, 1966) gives an introduction to the basic principles for the making of climatological predictions,
using the mathematics of climatological analysis (see also below in relation to
precipitation probability).

It is necessary to distinguish between the aridity of dry seasons and the aridity of droughts. The dry season is a normal component of the seasonal pattern
in a low-rainfall monsoonal climate. The economic ecosystem in such areas is
geared to this regular and fully expected annual cycle. Crises arise when there
is a deficiency of rainfall in what is normally the short season of rains. In the
equatorial zone which some have called 'everwet' and which may be the subject
of a UN interagency project (3.4.9.), droughts of economic significance may be
quite long and intense, but may occur at any time of the year. The rural
people are therefore not so well prepared for them, they may suddenly arise
in the middle of the season of cultivation of a main food or cash crop, and the
economic results may be much more severe.

There have been a number of studies of Asian drought of a more academic
meteorological, ecological or plant geographical nature. For example, Ruprecht
(1970) notes that, although the Thar desert in northwest India and Pakistan is
affected by the summer southwest monsoon, it is one of the driest regions of
the earth. The three-dimensional motion above this area has been studied. From
the vertically integrated equation of continuity, the vertical velocity at the
0.6, 1.5 and 2.1 km. levels was computed. Ruprecht shows that a divergence
in the lower layers, caused particularly by the heat low over Baluchistan (thermal wind), produces a subsidence of the air masses, with the result that con-
vection is substantially reduced. This subsidence corresponds well to the
pattern of the three-dimensional circulation of the tropical easterly jet.

The ecology of drought has been the subject of a series of papers from the
French Institute, Pondicherry; ecological and phytogeographic definitions and
significance of drought (Meher-Homji, 1964, 1965), delimitation of arid and
semi-arid climates in India (Meher-Homji, 1967), application of pluviothermic
quotient and xerothermic index — measure of the duration and intensity
of the dry season in number of biologically dry days — and hygrothermic
index — similar evaluation of the wet season — to the definition of vegetation
types in the Indian subcontinent (Meher-Homji, 1968a), and a review of the
problem whether the Sind-Rajasthan desert of the south Asian subcontinent
is of recent origin and which factors have been responsible for its creation
(Meher-Homji, 1973). In the last paper (73 literature references), it is concluded that, although man may have caused subsequent deterioration in local
ecological conditions due to abusive land-use, the origin of the desert itself
appears to be linked to meteorological phenomena 'especially the retreat of
the Ice Age over 10,000 years ago'.

Whyte (1975b) carries this further back in tectonic and orogenic history,
relating the origin and subsequent trends of aridity in the subcontinent to the
arrival and junction of the South Asian Plate, and to the progressive uplift of

114

Table 3.5. (a) to (g). Process analysis in micrometeorology – for detailed explanation see original document (WMO, 1972).

(a) Division of radiation processes group (R) into sub-groups

Soil science	RS1	The optical and physical properties of the soil in regard to radiation processes; problems of reflection, absorption and transmission.
Atmospheric science	RA1	Distribution of quantity and quality in time and space.
	RA2	Surface temperatures.
Plant science	RB1	The response to radiation; processes of photosynthesis and morphogenesis.
	RB2	Plant geometry; the space and time distribution of surfaces.
	RB3	The optical and physical properties of plant surfaces; leaf optics.
Animal science	RZ1	The response to radiation; photoperiodicity.
	RZ2	Behavioural response to radiation; effects of mobility and choice.
	RZ3	Animal geometry; radiation load.
	RZ4	Optical properties; reflection and absorption.
	RZ5	Surface temperatures.

Within these sub-groups, it is thought that the most important subject about which least is at present known is RA2: the surface temperatures of vegetation. Processes coming in the sub-groups RS1, RB2, RZ1, RZ2 and RZ4 are probably of a much lower category.

(b) Division of momentum processes group into sub-groups

Soil science	MS1	Wind erosion processes; soil 'blowing', dust and sandstorms.
	MS2	Gas exchange between soil and air.
Atmospheric science	MA1	Aerodynamic analysis; form drag and skin friction.

115

	MA2	Sampling and measurement in time and space; fluctuations with time.
Plant science.	MB1	Crop response to wind; growth, response of stomata.
	MB2	Crop response to wind; geometrical change with wind, leaf flutter deformation, mutual shelter, orientation in relation to dispersion and deposition.
	MB3	Crop response to wind; lodging, mechanical damage.
Animal science	MZ1	Animal geometry; exposure.
	MZ2	Animal response to wind; mobility and choice.

The sub-groups with the least claim to priority are thought to be MS1, MS2 and MB3.

(c) Division of heat-transfer processes group into sub-groups

Soil science	HS1	Conduction of heat in the soil; effects of soil type and moisture content, conduction to surface air.
	HS2	Distribution of temperature in time and space in the soil.
Atmospheric science	HA1	Free convection in the atmosphere.
	HA2	Forced convection.
	HA3	Distribution of convection in time and space.
Plant science	HB1	Temperature response; photosynthesis, respiration, morphogenesis, translocation of heat.
Animal science	HZ1	Temperature response; thermal regulation, behaviour and response to stress.
	HZ2	Production and reproduction.

Of these sub-groups, HB1 is thought to have the greatest importance; HS1, HA2 and HA3 are of least immediacy.

116

(d) Division of water-transfer processes group into sub-groups

Soil science	WS1	Infiltration of water into the soil.
	WS2	Internal redistribution of water within the soil.
	WS3	Distribution of soil moisture in time and space.
Atmospheric science	WA1	Transport of water; distribution of sources, sinks and fluxes in space and time.
	WA2	Conditions of surface wetness.
	WA3	Dewfall.
Plant science	WB1	Plant anatomical properties; stomatal control.
	WB2	Pathway characteristics of water transfer within the plant.
	WB3	Distribution of water status in time and space.
	WB4	Effect on crop physiology and morphogenesis of internal water deficits.
	WB5	Response of pests and diseases to water status and moisture conditions.
Animal science	WZ1	Response of parasites and diseases to water status and moisture conditions.
	WZ2	Water-balance factors, internal and external.

It should be noted that WA2, concerning surface wetness, is also a concern of the other three sciences and could, with equal validity, be included under their respective headings.

(e) Division of carbon-dioxide transfer group into sub-groups

Soil science	CS1	Upward flux of carbon dioxide from the soil.
Atmospheric science	CA1	Carbon-dioxide flux estimates and processes.
	CA2	Measurement of vertical profiles of carbon dioxide.
	CA3	Distribution of carbon dioxide in time and space.

117

Plant science	CB1	Plant response: photosynthesis.
	CB2	Plant response: respiration.
	CB3	Transpiration stream transport: role of stomates.
Animal science	CZ1	Soil biology respiration.

(f) Division into sub-groups

Soil science	OS1	Distribution in time and space of sources of matter such as chemicals, or other additives to the soil.
	OS2	Distribution in time and space of sinks of particulate matter.
Atmospheric science	OA1	Release, diffusion, transport and deposition processes.
	OA2	Distribution in time and space of matter pertaining to air quality, whether in solid, liquid or gaseous form.
	OA3	Distribution of atmospheric conditions influencing the viability of seeds or spores and the survival of insects or pests.
Plant science	OB1	Release and absorption processes in plants; effect of height and geometry.
	OB2	Response of plant.
	OB3	Response to environment of disease spores, pests, etc.; host microclimate infection conditions.
Animal science	OZ1	Release and absorption processes in animals; effect of height, geometry, and mobility.
	OZ2	Response of animal.
	OZ3	Response to environment of disease and parasites; host microclimate infection conditions.

Included within the OS sub-groups should be the effect of transfer by water of matter such as dissolved chemicals; this can be regarded as already having been considered in WS1 and WS2.

Sub-groups OB3 and OZ3 are already partly covered by HB2 and HZ3 respectively.

118

(g) The complete matrix may now be presented as follows:

Process		Science		
	Soil	Atmosphere	Plant	Animal
Radiation	RS1	RA1	RB1	RZ1
		RA2	RB2	RZ2
			RB3	RZ3
				RZ4
				RZ5
Momentum	MS1	MA1	MB1	MZ1
	MS2	MA2	MB2	MZ2
			MB3	
Heat	HS1	HA1	HB1	HZ1
	HS2	HA2		HZ2
		HA3		
Water	WS1	WA1	WB1	WZ1
	WS2	WA2	WB2	WZ2
	WS3	WA3	WB3	(WA2)
	(WA2)		WB4	
			WB5	
			(WA2)	
Carbon dioxide	CS1	CA1	CB1	CZ1
		CA2	CB2	
		CA3	CB3	
Other material	OS1	OA1	OB1	OZ1
Transfers	OS2	OA2	OB2	OZ2
	(WS1)	OA3	OB3	OZ3
	(WS2)			

Not every entry in this matrix should be regarded as of equal importance, but attempts have been made in the text to indicate which processes are of first-order significance.

the Himalaya, its associated ranges and the Tibetan plateau.

Agriculture in areas in India where the annual rainfall is higher than 1,200 mm. may be considered to have the same production potential as irrigated areas (IARI, 1970). Areas receiving an annual rainfall below 40 cm. can be classified as arid and need specialized techniques for improving production. The problem zones for dry farming can be taken as those areas which have an annual rainfall between 40 and 120 cm., or 160 million hectares out of a total cropped area of 330 million hectares. These dry areas are characterized by low and uncertain rainfall, high annual evaporation (200 to 350 cm.) and high summer temperatures (daytime maxima of up to 49°C. at some places).

The amount and distribution of rainfall received during the monsoon (kharif) season depend to a great extent on the number and intensity of the low pressure systems like depressions and cyclonic storms that move inland from the Arabian Sea and the Bay of Bengal, their track and extent of travel over land, and the number and durations of the 'break' situations in the monsoon season when little or no rain is received over most of the country except the extreme north (sub-Himalayan ranges). The distribution of rainfall in the rabi season in the north depends on the number and frequency of western disturbances, their location and intensity. These in turn are governed and influenced by the upper air atmospheric circulation patterns (IARI, op. cit.).

The scale of rainfall analysis, whether it should be limited to the annual, seasonal or weekly periods, depends upon the specific problem one is considering. There are examples where the monthly rainfall might have been in deficit, but floods might have occurred in the area in the same month due to a brief spell of heavy rain on two to three consecutive days. Monthly rainfall observed at one station may be equal to the normal amount, but there may have been a period with no rain during the first fortnight. Annual or seasonal rainfalls are useful parameters in areas where rainfall is received in almost equal amounts in each of the months during the rainy season. In view of the high variability of seasonal rainfall from year to year in India, climatic data need to be analysed for shorter periods of a fortnight or even a week. But the lower the rainfall and the shorter the period considered, the greater is the variability in rainfall pattern. It should be noted that the severity of drought depends not on the total amount of rainfall alone, but on the distribution of rainfall and the phases of the crop (IARI data on maize) affected by dry periods.

It will be seen from Table 3.6a,b that at the IARI area in New Delhi four continuous dry days (less than 6 mm. in 24 hours) are common between 7 July and 27 October, a continuous dry week not so common until the third week of September.

Forecasts of weather probability or precipitation probability are made available to the farmer in one form and to land planners in another (Chisholm, Frey & Haggett, 1971). The WMO Guide to Agricultural Meteorological Practices (WMO, 1963, with supplements) discussed the value to the farmer of the general forecasts of agricultural weather and their degree of probability. 'It is a very rare situation in which the forecaster can confidently state that the probability is one hundred per cent that the forecast will be correct in every detail. If a categorical forecast is issued when the probability of the forecast event occurring is only slightly better than 50 per cent, the forecast may convey the incorrect message that, in the forecaster's estimation, the event will probably occur. A farmer, having no method of evaluating the categorical forecast in terms of probability, will either accept the forecast event as a certainty and plan his operations accordingly, or he will make some arbitrary allowance for the general uncertainty of weather forecasts that may not be

120

Table 3.6a. Frequency of continuous dry days (rainfall less than 6 mm/24 hours) during the weekly periods of July–October at New Delhi (IARI, 1970). (Based on data for the period 1940–1969)

Number of years of occurrence in 30 years

Number of continuous dry days in the week	Week Nos (1st week ends on 7th July . . . 17th week ends on 27th Oct. of each yr.																
	1	2	3	4	5	6	7	8	9	10	11	12	13	14	15	16	17
Four	27	19	15	18	20	17	18	20	24	21	24	25	28	29	29	30	30
Seven	16	9	5	8	6	11	6	8	13	14	12	18	19	22	26	28	29

Table 3.6b. Probability of occurrence of specified number of continuous dry days (rainfall in 24 hours < 6 mm.) at New Delhi (IARI, 1970).

Month	No of continuous dry days (<6 mm/day)	Probable frequency
July	7	Once in 3 years
	20	Once in 6 years
August	7	Once in 5 years
	20	Once in 15 years

121

appropriate in this particular case. This places the farmer at a distinct disadvantage. Many farming operations must be done when a crop is at a certain stage of maturity or else a loss in the quality or size of the crop will result. The cutting of hay is an example of such an operation. If the farmer has a forecast that is stated in terms of the probability of the occurrence of hazardous weather, he can make a rational decision on whether to go ahead or delay the cut.'

The Environmental Data Service of the National Oceanic and Atmospheric Administration of the U.S. Department of Commerce has carried out studies of climatic characteristics and probabilities of precipitation in Asian countries and subregions. It is understood that ECAFE* and WMO also plan to map rainfall probabilities (daily basis) in the ECAFE region, using the computer and other facilities available at the Philippine Weather Bureau. Data cards will be punched as far as possible in the individual countries concerned.

A study by the U.S. Department of Commerce (Mooley & Crutcher, 1968) investigates a. whether the monthly monsoonal rainfall at Indian stations can be characterized by the incomplete gamma distribution function, b. the length of the period required to permit stabilization of the estimates of the distribution parameters — between 80 and 100 years inferred, and c. the correlation of month-to-month rainfall during the monsoon season — essentially no linear relation.

Crutcher, Nash & Kropp (1969) contributed a note on the climatology of Thailand and southeast Asia, as part of a training programme designed to present some new techniques of climatological analysis relating to temperatures, dew points, winds, precipitation, monsoons and typhoons (monthly averages or totals, with means and standard deviations). Techniques of correlation and regression analyses and harmonic and spectrum analyses are also included in the programme.

It would be possible to recalculate the data used in the compilation of the NOAA Atlas no. 1, on precipitation probability for Eastern Asia (Yao, Barger & Crutcher, 1971), in order to meet the needs of the agronomists. This NOAA study presents a partial picture of the water budget for eastern Asia, partial in that it shows the probabilities of precipitation by month, not for regimes within a month or for periods longer than one month. Precipitation amounts for ten levels of probability are presented for each of the twelve months, but only five of these are analysed and shown as maps (0.99, 0.80, 0.60, 0.40, 0.10). 'Parameters of the incomplete gamma distribution for individual months can be combined to produce approximate frequency distributions for n-month total precipitation. These maps and tables provide useful basic guidance information, though other information must be available for a good depiction of the water budget and associated budgets of solar energy, wind and other elements for better understanding of man's environment and its potential' (Yao, Barger & Crutcher, op. cit.).

*now ESCAP

122

In the technical report from ESSA on the characteristics and probabilities of precipitation in China, Yao (1969a) introduces the following monthly precipitation probabilities:

zero to a trace

1 inch or more	7 inches or more
3 inches or more	10 inches or more
5 inches or more	14 inches or more

'The probabilities of zero to a trace of precipitation reverse the situation found for other levels of precipitation. The probabilities are distributed so that there is a high probability during the winter and near zero probability from April to October. The patterns of probability distribution for different levels of precipitation are generally similar. That is to say the centers of the high and the low probabilities are located roughly in the same general area during the same period for all precipitation levels. The probability densities of these levels of precipitation are however different. In general, Taiwan Strait is in a permanent low probability area. The Yunnan Plateau is in a low probability center during the winter and spring, and Central China is in a high probability center except from July to September, when the probability is low. The North China Plain is in a low probability region except from July to September when the probability there is high. The patterns of precipitation probabilities are closely associated with the general circulation over the Chinese continent.'

These conclusions are presented in more agroclimatic terms elsewhere (Yao, 1969b). 'Spring drought and summer flood and drought pose continuous threats to crop production in the North China Plain. High moisture stress during the spring is shown by the low ratios of actual evapotranspiration (ET) to potential evapotranspiration (PE). Uneven precipitation distribution, together with the low preseasonal soil moisture storage, is responsible for spring drought. Precipitation probability analysis for the months of May and June indicates only a 10–30 per cent chance of an adequate moisture supply in this region during spring. Variability of precipitation amount and intensity share the responsibility for summer flood and drought. Precipitation probability alone for the months of July and August indicates that, on the average, the North China Plain has only a 30–50 per cent chance of adequate moisture for crop growth. However, the probability of an optimum amount of precipitation decreases considerably when rain intensity is taken into consideration. Near the two centers of maximum intensity, in July and August, rainfall reaches 22–25 mm. per rainy day. By comparison, growing-season precipitation in the winter wheat region of the Great Plains of the USA is less intense and more evenly distributed' (Yao, op. cit.).

3.4.8 Analogous/homologous climates

Some workers have introduced the principles and techniques of applied bioclimatology in order to establish the analogous climates of the world between which it might be advantageous to exchange seeds and other propagating material of economic plants. Meher-Homji (1963, 1971b) has applied the methods of Gaussen to establish the homoclimatic counterparts elsewhere for the stations in the dry tracts of the Indian subcontinent. Attempts to define homoclimates may be grouped in three categories (references to literature cited in Meher-Homji, 1971b): 1. physical and meteorological methods using climatic formulae, indices and coefficients (Transeau, 1905, Penck, 1910, Köppen, 1900, 1918, Lang, 1920, de Martonne, 1926, Thornthwaite, 1933, 1948, Nuttonson, 1947, 1951 — see below, Howe, 1960, Kaushik et al., 1969, Subrahmanyam & Sastry, 1969). 2. diagrammatic representation: hydrothermic figure (Raunkiaer, 1908); clima-diagram (Chaptal, 1933), agro-climatic diagram (Azzi, 1954), hythergraph (Taylor, 1920), ombrothermic diagram (Bagnouls & Gaussen, 1953), and its modified version — klima-diagramme (Walter, 1955). 3. biological criteria to reflect climatic analogies: floristic, ecological, vegetational (growth forms of Humboldt, 1805, and Grisebach, 1884; life forms of Raunkiaer, 1908); only vegetation types based on the characteristics of physiognomy and structure are shown to reflect climatic similarities (Meher-Homji, 1953).

One of the simplest methods of defining analogous climates would be to compare the curves of mean monthly precipitation and temperature of various stations on a graph. These values are the results of the averages of several years and do not show inter-yearly variability (Meher-Homji, 1971b). Analogy by means of diagrams alone is not considered to be possible, and a classification of biological climates becomes necessary (Fig. 3.6). Within the framework of this broad classification based mainly on the rhythms of precipitation and temperature, a further climatic analogy may be achieved by introducing resemblances in the temperature, rainfall and lengths of the dry season. It must be remembered that bioclimate is only part of the complete ecology of a species, and a knowledge of the total ecological requirements is essential.

Nuttonson has defined the agroclimatic analogues in North America in respect of China (1947, 1970), and Japan (1949), and in the northern hemisphere in respect of the Ryukyu Islands (1952). The purpose is 'to formulate an agronomic and horticultural approach to ecology and to promote research along the lines of crop ecology as it affects plant adaptation, plant introduction, and the exchange of varietal plant material among the various agricultural areas of the world.' Climatic analogues are defined as 'areas that are sufficiently alike with regard to the length-of-day conditions and some of the major weather characteristics affecting crop production, particularly during the growing period, to offer a fair chance in transplanting plant material from one area to a climatic latitudinal counterpart.'

Available meteorological data from weather stations in the Asian countries and from comparable stations in North America (China and Japan) or within the same latitudinal belt in the Northern Hemisphere (Ryukyus) have been used as basic source material. Elements of comparison are mean monthly and

124

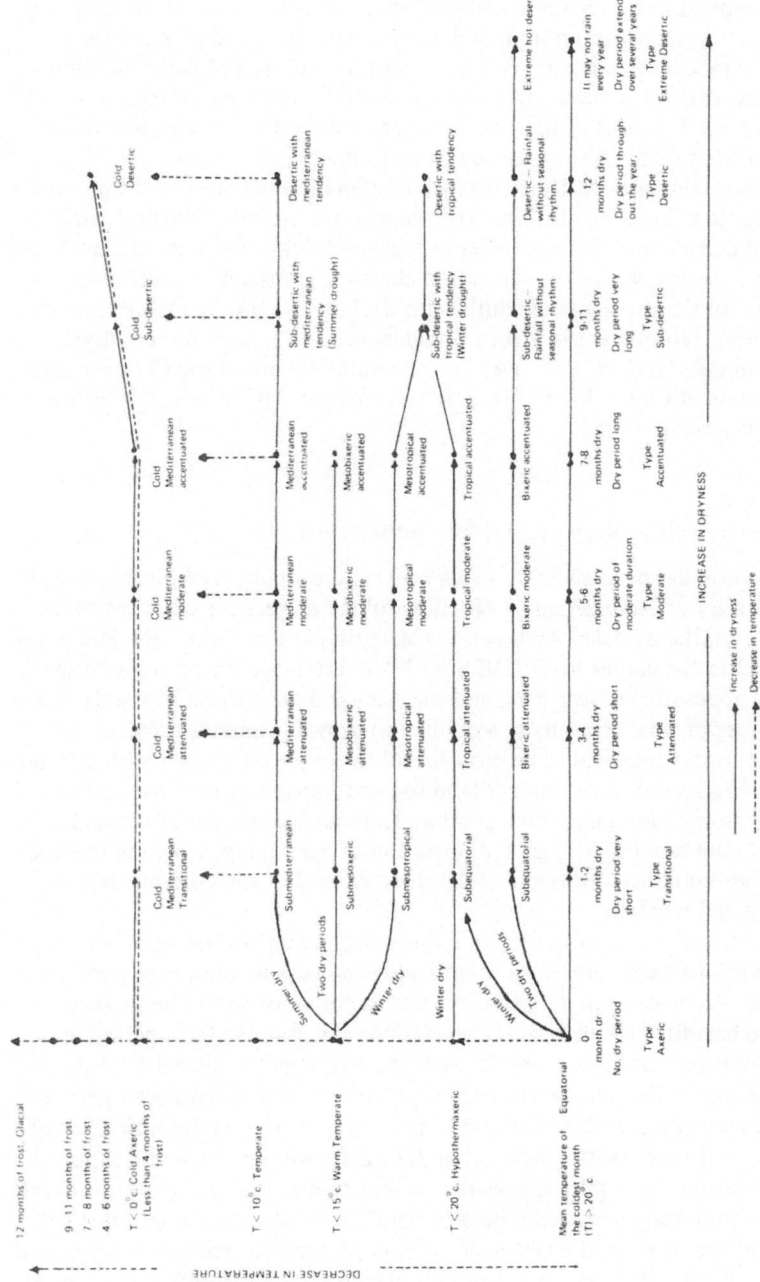

Fig. 3.6. Diagrammatic representation of the classification of biological climates of Bagnouls & Gaussen (1957), from Meher-Homji (1971).

125

yearly temperatures, maximum and minimum temperatures, average monthly, seasonal and yearly precipitation and relative humidity, and precipitation effectivity indexes and ratios (Thornthwaite). In addition to determination of year-round analogous areas, separate studies relate to the April/September and October/April fractions of the year to determine supplementary climatic or thermal analogues for spring-planted and autumn-planted crops.

When one compares Nuttonson's map of climatic analogues of the provinces of the People's Republic of China with their counterparts in North America, one notes considerable discrepancies in bioclimatology. The Great Plains States of North America, where the vegetation climax is grassland, are introduced as analogues for the provinces of south-central China, where the former climax before devegetation was evergreeen broad-leaved forest, mixed mesophytic forest, and deciduous broad-leaved forest, among others (Wang Chi-wu's map of the forests of China, 1961 (Fig. 3.10) also Whyte, 1972a, Fig. 4; Fig. 7 in Chen Cheng-Siang, 1970).

3.4.9 An agroclimatology project for southeast Asia

Consideration has been given by FAO/WMO (Consultant, G. W. Robertson) to the possibility of establishing an agroclimatology project for what has been called the southeast Asian Archipelago (Malaysia, Indonesia and the Philippines) as a sequel to the earlier FAO/UNESCO/WMO Inter-Agency projects on the semi-arid zones of the Near East, and the semi-arid zone south of the Sahara. The purpose of this third survey would be to provide climatological inform- ation in a form which would be most useful to people concerned with different aspects of agriculture: planning of land use and transportation; discussions regarding storage and marketing; plant and animal breeding/selection and testing of new varieties; design and operation of machinery; crop and livestock disease; agronomic practices and techniques; extension and advisory services; education and research.

The great variability of rainfall in monsoonal and equatorial Asia from day to day, week to week, month to month and even year to year, is of great con- cern to all. An analysis in depth of this one factor alone would be of great economic benefit. Provided daily observations are available for a period of twenty years or more, it is possible, by using a systematic numerical analysis technique, to provide answers to such problems as: how often does a period of a number of days with no rain occur in ten years; what is the variability of the onset of the wet season; how often during the wet season would supple- mental irrigation be required; does the current trend in rainfall (towards excess or deficit) indicate a persistent climatic trend? It would also be desirable to provide information on the factor of equal importance to rainfall in the humid tropics — global radiation. There are indications that global radiation is not as abundant as may have been thought, and may be limiting for many crops in a

126

degree far greater than most agronomists realize. Its influence on the production of crops such as rice, sugarcane or pineapple may be 'very subtle indeed', and careful experimentation is called for.

A few stations may have sufficiently long-term records of daily rainfall and possibly daily extremes of temperature for fairly reliable estimates to be made of certain averages and their variability. Global radiation has not yet been measured for a sufficient number of years to provide a reliable estimate of its long-term average, trends and variability. A few-long term records of daily duration of bright sunshine may be converted to global radiation by appropriate equations.

The consultant proposes that the study and analysis of such data as may be available or can be collected and the subsequent publication of results be done on an area to area basis. It may be possible to define some fifty such areas in the three countries mentioned above.

3.5 Water sources, resources and irrigation potential

For ten years from 1965, UNESCO is supporting the International Hydrological Decade (IHD), to be conducted in association with United Nations and its Specialized Agencies, with the major objectives:
'Hydrology is the science which deals with the waters of the earth, their occurrence, circulation and distribution on the planet, their physical and chemical properties and their interactions with the physical and biological environment, including their responses to human activity. Hydrology is a field which covers the entire history of the cycle of water on the earth.
'The rapid advances made in industry and agriculture, the accelerating growth of human populations, and the desire to secure higher standards of living, have resulted in increased use of water by man, both in agriculture and for industrial, municipal and domestic purposes, to the extent that the availability of water as well as the control of excessive water have become a critical factor in the development of many regions of the world.
'The absolute necessity of increasing the degree of rational management of water as a vital element of the human environment is therefore recognized. For this it is essential that an adequate body of reliable data should be available in each country, and that water use and management should be based on scientific principles established for all the branches of hydrology. In order that a progressive approach may be maintained as new problems arise, further scientific research in hydrology and extensions and improvements of basic data are also essential.
'The overall objective of an international programme in the field of hydrology is to accelerate the study of water resources and the regimen of waters with a view to their rational management in the interest of mankind, to make known the need for hydrological research and education in all countries, and

to improve their ability to evaluate their resources and use them to the best advantage. That is, the programme will focus on science but will give strong consideration to utilitarian factors ' (UNESCO, 1964).

The fields of research in hydrological phenomena are defined: The air/land interface – precipitation, evaporation and transportation, soil moisture; Surface run-off; Ground water – occurrence and movement, hydrology of calcareous terrains and karst topography; Dynamics of lakes and reservoirs; Geomorphological and geochronological problems – sediment transport and production, archaeohydrology, estuarine, deltaic and coastal problems; Snow, ice, periglacial, glaciers and ice caps; Quality and chemistry of water; Influence of man's activities on the hydrological processes.

It is the last item which is most relevant to the present discussion of the hydrological cycle, as one of the major components of the environment which it is desired to assess in arriving at a most productive and at the same time conservative form of land use. A Working Group on the Influence of Man on the Hydrological Cycle within the scope of the IHD was sponsored by FAO. This produced a first report (UNESCO, 1972a) giving guidelines to policies for the safe development of land and water resources, particularly in relation to forest lands (3.6.3), grasslands (3.6.4), arable lands, irrigation and salinity, swamp drainage, urbanization and water pollution, landslides and road construction, and location of danger areas in very large catchments.

Subsequently, the UNESCO/FAO Working Group decided that this first report should be complemented by another having a different approach. The second volume (FAO, 1973) stresses the fact (frequently disregarded in practice and in planning exercises) that man's efforts to control the water cycle invariably involve factors other than hydrology and engineering: ecological, sociological, economic, cultural and political considerations and forces.

Part One of the FAO document (by J. C. I. Dooge, 1973) shows how the heat of the sun supplies the enormous energy needed to keep the cycle turning. Some 500 million million tons of water are raised annually from the surface of the earth to the atmosphere, to be redistributed around the globe as vapour, to return as rain, hail and snow, and to begin the cycle again (Figs. 3.7 and 3.8).

A statement of the biological significance of water in the ecosystem has been given in 2.4. In Part Two of the same document (FAO, 1973), A. B. Costin & J. C. I. Dooge (1973) consider the components of the hydrological cycle related to land use (population interception and storage of water, surface run-off, soil water storage and evapotranspiration, and deep percolation), control of hydrological processes, increasing of available water through engineering, and the effects of agricultural and forestry practices (see 3.6.3 for precipitation, evaporation, infiltration and erosion under forest covers, and 3.6.4 for the hydrological situation under grass covers).

UNESCO has published a series of studies within the scope of the International Hydrological Decade or on subjects related thereto:

128

THE HYDROLOGICAL CYCLE

A continuous exchange
of water through
precipitation and
evaporation

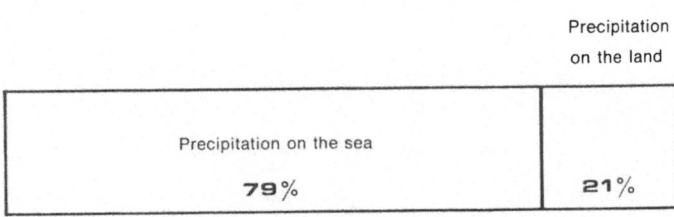

Precipitation

on the land

Precipitation on the sea

79%

21%

Water Potentially Controllable and Controlled by Man

Of the total annual world precipitation 79 percent is lost to man immediately since it falls on the sea. From the remaining 21 percent falling on land, 13.4 percent is lost as a result of evaporation, leaving 7.6 percent to run off in the rivers of the world and through the ground. Of this amount it is estimated that half is beyound the control of man. Of the other 3.8 percent which is potentially controllable by man 3.5 percent is not controlled at present, leaving the amount of water actually under control at only 0.3 percent of the total annual precipitation. For the most part the unused but potentially usable supply is located far from where it is most needed. To meet future demand for water without harmful effects will require that the necessary large scale investments and engineering are based upon knowledge and understanding of the place of water in the whole environmental equilibrium.

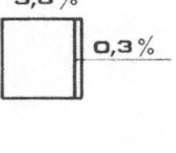

3,8%

0,3%

Evaporation

from the land

Evaporation from the sea

86,7%

13,4%

Fig. 3.7. The hydrological cycle (FAO, 1973).

129

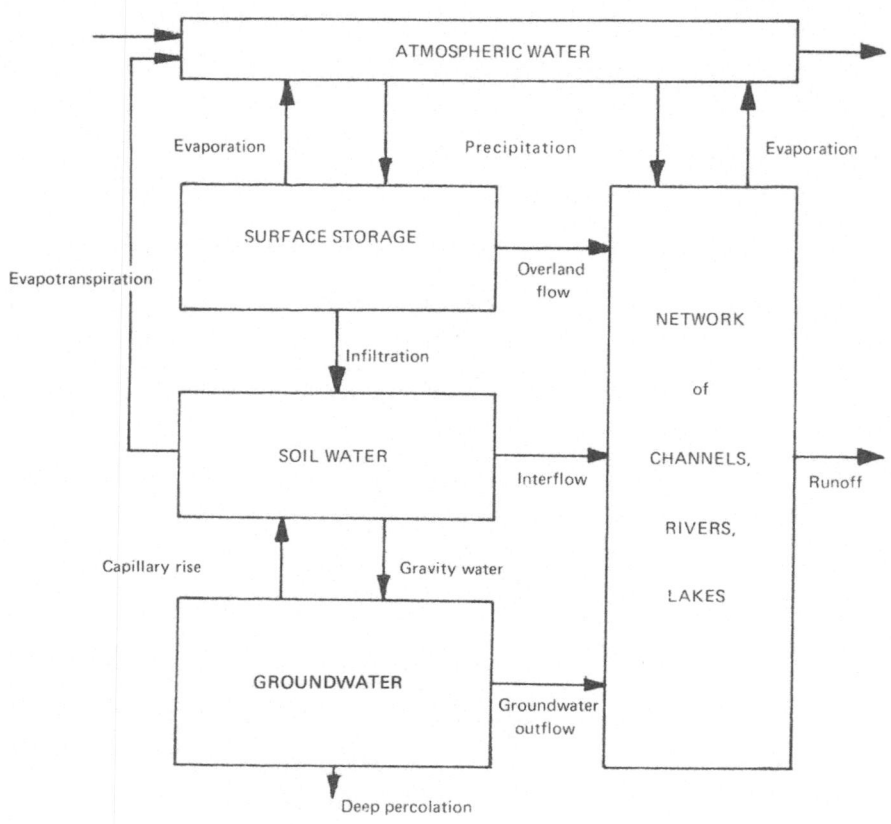

Incoming water vapour

Outgoing water vapour

ATMOSPHERIC WATER

Evaporation

Precipitation

Evaporation

SURFACE STORAGE

Overland flow

Evapotranspiration

Infiltration

SOIL WATER

Interflow

NETWORK

of

CHANNELS,

RIVERS,

LAKES

Runoff

Capillary rise

Gravity water

GROUNDWATER

Groundwater outflow

Deep percolation

Fig. 3.8. The hydrological cycle over a catchment area (Dooge, 1973).

Irrigation, Drainage and Salinity: An International Source Book (FAO/ UNESCO, 1973).
A synthesis of current scientific knowledge, a guide to practical experience obtained in irrigation and drainage methods in relation to salinity and alkalinity of arid lands; problems related to the control of soil humidity in saline conditions.
Ground-Water Studies: An International Guide for Research and Practice (UNESCO, 1972b)
The organization of, and methods for undertaking ground-water investigations, the procedures for establishing ground-water balances, the forecasting of ground-water conditions, selection and location of ground-water stations,

130

choice of instruments and equipment, and methods for collecting, processing and analysing data.

Studies and Trends of Research in Hydrology, 1965–1974 (UNESCO, 1972c) Proceedings of the International Conference on the Practical and Scientific Results of the International Hydrological Decade, and on international co-operation in hydrology, UNESCO Headquarters, Paris, December, 1969.

World Water Balance: Proceedings of the Reading Symposium, July 1970 (International Association for the Hydrological Sciences/UNESCO/WMO 1972). Distribution. Geographical features. Research requirements: accuracy and adequacy of observations, theoretical development, formulation of physical and mathematical models of large-scale water balance problems.

Mathematical Models in Hydrology: Proceedings of the Warsaw Symposium, July 1971 (UNESCO/WMO/IASH 1974b). Stochastic hydrology, parametric hydrology, and water resource systems.

Design of Water Resource Projects with Inadequate Data (UNESCO/WMO/IASH, 1974a).

Teaching Aids in Hydrology (UNESCO, 1972d).

Lal Bahadur Sastri stated in his inaugural message to the Sixth Congress of Irrigation and Drainage: 'The effective utilization of our water resources may mean the difference between poverty and plenty in India.' Sukhatme, Erus & Memoria (1970) state that, among the principal factors which help to improve productivity in agriculture, water is of paramount importance, and then proceed to illustrate how the yield response of crops to fertilizers can be increased by the application of adequate water at the right time. The availability of sufficient water and the avoidance of excess are essential if the high-yielding varieties of foodgrains are to maintain their characteristics.

In the contribution of the FAO Land and Water Development Division to the volume in commemoration of the 25th Anniversary of the United Nations and the 20th Anniversary of the International Commission on Irrigation and Drainage (ICID, 1970), it is noted that more than 10,000 litres of water are needed to produce the daily food for a single person. Food crops are the largest consumers; to produce one ton of wheat up to 5,000 tons of water are required, or double that amount if irrigation losses are considered. In spite of the fact that the high exploitation costs for water development restrict the economics, irrigation is continually spreading; in the Near East, crop production from irrigated land represents 60 per cent of the total agricultural output, but the irrigated harvested area comprises only 20 per cent of the total harvested area. The following activities of the Land and Water Development Division are relevant in the present context (ICID, op. cit.).

Inventory of resources and development plans — estimation of long-term requirements for agriculture, survey and appraisal of resources, policy and planning, river basin development.

Resource development, use and conservation — soil and water conservation, surface and ground-water development and engineering (see Burdon, 1971, on

131

exploitation of groundwater for agricultural production in arid zones), land reclamation, drainage, flood control, reclamation of saline and alkaline soils, methods of water use and irrigation including fertilizer application, and the integration of engineering undertakings and agricultural activities (paper on farm water management in the Far East presented by C. E. Houston to the FAO Seminar on Water Management and Control for Agriculture, Tokyo, July, 1972).

Irrigation as the basis for increased production was discussed at the 9th FAO Regional Conference, Bangkok, 1968 (see also Kyoto University, 1966; CENTO, 1966). 'Although the Far East Asian region is generally blessed by relatively abundant rainfall, its distribution is subject to great seasonal variations, and water control, in the form of irrigation, drainage and flood control, plays a key role in raising and stabilizing agricultural production. The importance of controlled water supplies is further enhanced by the dominance of rice in the cropping pattern of the region. The lack of proper irrigation results in moisture shortage, even during the rainy season when periodic droughts occur, and it also limits or prohibits dry season cultivation. At the·same time in many locations, super-abundance of water causes floods and water-logging, detrimental not only for crop production but also for the introduction of modern agricultural methods using high cash inputs. Among these inputs water control is one of the more expensive and the fullest benefits from irrigation and other facilities for water control can be achieved only when they are fully integrated with the use of other inputs. Conversely, except for certain areas where crop cultivation without water control is possible, correct water control is the prerequisite infrastructure which enables the use of other inputs. This is especially so for crops with high water requirements and for high-yielding varieties which require highly sophisticated farming techniques. Although major emphasis is placed on major engineering works, there is also an essential role for correct water management, associated with other farming inputs, based upon a comprehensive distribution system down to field level, not only to expedite the use of costly water, but also to ensure its fullest use and to reduce avoidable losses.'

In 1968, an IBRD/FAO Technical Working Party discussed differences in approach in land classification with special reference to feasibility studies for irrigation projects. It was agreed that semi-detailed and reconnaissance soil surveys are useful in but not sufficient for feasibility studies. Detailed systematic soil surveys as carried out by FAO can supply the soil information necessary for land classification for feasibility studies for irrigation development. with supplementary advice from special consultants. The Working Party recognized that land classifications made in feasibility studies that may be considered by IBRD are based upon U.S. Bureau of Reclamation standards and specifications. It will be necessary to consider how the land classification system for irrigation projects adopted by the Bureau of Reclamation of the U.S. Department of the Interior applies to the socio-economic situation in Asia (from FAO Soils Bulletin 22, 1974).

The selection of lands for irrigation in the United States involves social, economic and physical factors and construction of irrigation projects is generally very costly. Consequently, the selection of irrigable lands depends basically on economic criteria; the feasibility of a project is determined by overall costs and benefits, which vary from project to project. According to the Bureau, lands to be irrigated should have a favourable 'payment capacity' which is defined as the residual amount of funds available to defray the cost of irrigation water after all other costs have been met by the farm operator. Institutional factors, managerial levels, farm practices, cost-price relationships, markets, social conditions, climate and other factors are considered in determining the payment capacity for each project and part of a project. In general:

$$Y = -a + bX_1 - cX_2 - dX_3$$

where

Y = Payment capacity (U.S. dollars)
X_1 = Productivity rating (per cent)
X_2 = Land development cost (dollars)
X_3 = Farm drainage cost (dollars)
a, b, c and d = Constants derived from farm budget analyses

'A large investment may be made to reclaim a saline-sodic soil which after improvement would yield a net farm income of U.S. $200 per acre; in another climate and economic setting where net income after improvement would only be U.S. $30 per acre, a soil having similar saline-sodic conditions would be regarded as non-irrigable.

'Physical, chemical and biological evaluations of project areas are very important in the Bureau of Reclamation procedures, particularly for characterization of climate, soil, topography and drainage. In general, as climate favours higher farm incomes, greater expenditures can be made for land forming, farm distribution systems, leaching salt and exchangeable sodium, profile modification practices, and farm surface and subsurface drainage. When considered in terms of land-class-determining factors such as uneven microrelief; soil texture, structure, and depth; exchangeable sodium and soluble salt levels; permeability of substrata; and depth to groundwater barriers, then more severe deficiencies involving such factors can be tolerated in climates favouring high incomes than in those resulting in lower incomes.'

Six land classes are recognized and defined by the Bureau of Reclamation, four on arable land, two on non-arable land. In a given project area, specific limits of soil properties and other parameters are set up to segregate the different classes. Flexibility in the limits relating to each class is required from one project area to another.

Although the Bureau of Reclamation's irrigation suitability classification sets up specific limits for classes and subclasses, the specifications are not absolutely rigid, and can be modified from one project area to another. In Thailand, for example, a special class set up for arable lands for wetland rice included: 'Lands that are highly suitable for paddy rice production under irrigation, being capable of producing sustained and relatively high yields of rice at reasonable cost. The surface horizon soil should be of medium to fine texture associated with subsoil characteristics and drainage conditions that provide for optimum soil submergence, but the build-up of reduction products or salinity should not be toxic to plant growth. These lands should occupy a position with project facilities to assure adequate runoff of surface water. These lands shall have relatively high net farm income potential.'

The feasibility report for the Pa Mong area in the Mekong River drainage basin is an example of adaptability of the Bureau's irrigation suitability classification. The Stage 1 study determined '1. lands in Laos and Thailand that would be of the most suitable quality for a first-stage irrigation development along the Mekong River a short distance downstream from the Pa Mong dam site. 2. the area that could be served by an economically located and sized gravity conveyance system originating at Pa Mong Dam, and 3. the scale of development that could be utilized under the economic and social conditions that are expected to prevail in the region with project development.'

Analysis of soils in the context of the assessment of the potential for irrigation was considered in 1969 by a small working party composed of representatives from Departments and agencies concerned in the States of Queensland, New South Wales, Victoria and West Australia, and from the CSIRO Divisions of Land Research, Irrigation Research, and Soils. The CSIRO Division of Soils was asked to prepare a document indicating preferred methods of analysis. The resulting publication (Commonwealth Bureau of Soils, 1974) is based on Australian experience, and particular attention has therefore been given to the analytical problems associated with expansive clay soils.

Zonneveld (1972, p. 71–2) discusses the extent to which land units in integrated surveys should include information about hydrology which can be used as a basis for evaluation. A hydrological study of land begins with a delimitation of the catchment area, following definition of the divides (watersheds) with the help of geomorphological features. Only some hydrological features can be linked to land units, e.g. conditions occurring on and in the soil, soil moisture, groundwater, springs and floods. Land units dominated by hydrological factors are coastal and riverain flood plains and marshes. Deep lying water resources may be very important, but are less clearly connected with land units.

The CSIRO Division of Land Research discusses water resources in relation to integrated surveys in the following terms (Christian & Stewart, 1968): 'The water resources of an area are determined by rainfall *in situ*, evaporation and surface and sub-surface drainage. The water that enters surface or sub-surface drainage systems may migrate and hence the water resource of an area is influenced by adjacent areas as well as by its own rainfall, evaporation and drainage characteristics. Added to this is the control and influence on water resources that man can exercise through engineering and through the use and misuse of land. Water as a resource has many uses, for human or stock consumption, domestic and urban purposes, industry and for power production. As a resource it must be assessed in terms of these uses according to circumstances. The magnitude of the resource may be measured by suitable recording instruments and some indications can be obtained from the drainage patterns of the landscape, vegetation and other factors. This complex and dynamic nature of water resources, with its variations in distribution, quality, place, season and the interactions with other features of the land surface, makes assessment of water resources a complex matter not adequately covered by locality descriptions and measurements alone. The position becomes clearer when' the purpose of water use is defined, but the usefulness of water will be dependent upon the nature and magnitude of other resources.'

The FAO emphasis noted above (Ninth Far East Regional Conference) was continued at the FAO/UNDP Seminar on measures to accelerate benefits from water development projects by improved irrigation, drainage and water use at the level of farm and field (FAO, 1971b). The discussions during the Seminar can be divided into five subjects: 1. water development and use — planning, design and execution; 2. agronomic consideration given to water development schemes; 3. legal and institutional aspects of water development and use; 4. socio-economic consideration, and 5. research, education and training aspects. These five elements are so closely interrelated that the Seminar recognized the need for an integrated approach towards the optimal use of land and water resources for the socio-economic benefit of the regions, countries and rural population.

The discussion is summarized (FAO, op. cit.) under the following heads:
Planning, design and execution of water development projects
Logical sequence of planning; maxi- versus mini- and extensive versus intensive projects; use of groundwater; system of water distribution; drainage needs; pilot scheme; irrigation methods and practices; modification and renovation of existing systems; land preparation (see also Toyoda, 1967 on supplemental irrigation on slope land in Taiwan; also Boaz et al., 1971 on village irrigation schemes).
Agronomic considerations: Upland cultivation (farmers of region not familiar with cultivation of upland crops under irrigation); land capability and soil characteristics (standardized method with local modifications essential for soil survey, soil interpretation and land classification); multiple cropping (requiring heavy farm labour, additional inputs and intensive extension work); fertilizer and other inputs (see also Kung, 1971, on irrigation agronomy in monsoon Asia, 5.3.6). Institutional and legal aspects; Socio-economic considerations; Research, education and training.

135

UNESCO sent two specialists from the International Institute for Land Reclamation and Improvement, Wageningen, (Schulze & de Ridder, 1974) to Egypt to study problems of the rising water table and accompanying salinity phenomena in the West Nubarya Project area, some 50 km. southwest of Alexandria. It has been recommended that a systems approach be adopted in the study of control of the water table, and that the investigations should not be regarded as individual undertakings, but as component parts of an integrated programme. The following interrelated subsystems are significant to the purpose (Fig. 3.9): a. the agricultural production subsystem; b. the surface-water distribution subsystem; c. the groundwater subsystem; d. the artificial drainage subsystem.

An interdisciplinary team in the fields of geology, irrigation technology, soil science and crop science, three from the Center for Southeast Asian Studies and one from the Faculty of Agriculture in Kyoto University studied the environmental determinants affecting the potential dissemination of high-yielding varieties of rice in the Chao Phraya river basin of northern and central Thailand (Fukui, 1971a). The main factors governing the spread of these varieties are topography and potential water resources, which determine the relative difficulty of gravity irrigation and drainage. These conditions are most favourable in the traditional irrigation of the inter-mountain basin (ca. 320 x 1,000 ha.) but here adverse factors are the smaller demand for marketable surpluses due partly to remoteness from the urban markets, and the fact that glutinous rice is the main product of northern Thailand.

The topography of the water-deficient foothills is suitable for irrigation and drainage, but water resources are limited (rice land ca. 1,310 x 1,000 ha.) and fear of drought will prevent adequate investment in fertilizers for the full yield potential of the varieties to be realized. Gravity irrigation and drainage are possible but more difficult in the higher elevation areas of the barrage irrigation area; high-yielding varieties could potentially be disseminated in the area (ca. 80 x 1,000 ha.) where rice growing by transplanting is the current practice.

Where gravity irrigation is not possible due to flat topography, states Fukui (op. cit.), the spread of high-yielding varieties is practically impossible, even where flood water is controlled by conservation irrigation, a method which has been well developed and will be further extended in the basin, on the less-flooded delta and the canalled lowland. In these areas, complete poldering is essential for the wide dissemination of high-yielding varieties, but the construction of polders, although technically feasible, is not economically practicable for a low value crop such as rice. As a drainage system at higher sites is possible only by sacrificing lower areas, further spread of high-yielding varieties seems totally impossible. Only by drastically modifying the water flow of the whole Chao Phraya River system could the general water supply and conditions for rice production of the lowlying plain be improved.

Thus one finds a contrasting situation in the non-deltaic and deltaic areas of Thailand; the methodology is being applied by these Japanese workers for

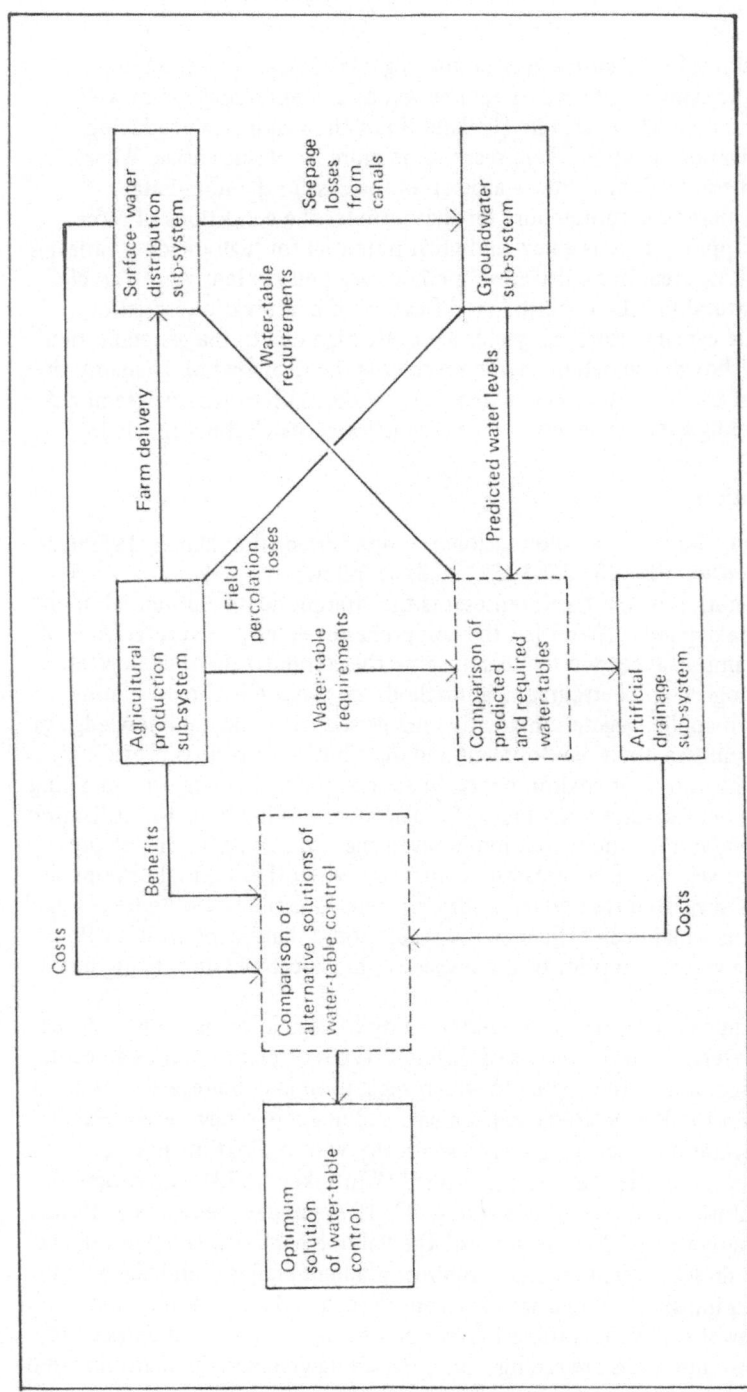

Fig. 3.9. Major interrelated subsystems affecting the optimal solution of water-table control (Schulze & de Ridder, 1974).

137

the subdivision of river basins in Asia into regions of rice cultivation, in the belief that the same two broad categories will be evident elsewhere as well. Although the non-deltaic area in Thailand is not the major rice-producing region, cultivation practices there seem to be more labour-intensive. Where the inherent soil fertility is low — as is often the case in the non-deltaic region — farmers tend to use more fertilizer. Under the conditions of more intensive cropping, there is a new and great potential for high-yielding varieties.

In the deltaic area, rice cultivation methods are quite extensive and stable, thanks to natural flooding and the modification of conservation irrigation. Except in the canalled lowland, yields are quite high due to the generally rich soils. Yet, although the deltaic region represents the rice bowl of Thailand, the possibility of the introduction of intensified methods up to the level required by high-yielding varieties seems to be extremely remote (Fukui, op. cit.).

3.6 Vegetation

A resolution adopted at an international symposium held in March, 1959 (reported in *Nature* 184: 23—25, 1959) reads as follows:

'The vegetation of the Earth represents the vital productive potential upon which all life depends. Therefore, the comprehensive study of vegetation is of the utmost importance, and for this purpose the combination of ecological, phytosociological and cartographical methods are required. The present-day methods of mapping vegetation greatly enlarge our fundamental knowledge of plant communities, their development and distribution as well as providing a deep insight into their environments. In applied phytosociology the mapping of vegetation constitutes a solid basis for assessment of habitats, for utilization of vegetation, and for the evaluation, or even the forecasting, of any change or damage to vegetation by erosion, wind, water and other natural or human factors. It is suggested that no large-scale technical measures should be planned or carried out which may influence the vegetation or landscape without first mapping the vegetation prior to the respective technical measures being put into effect.'

There is an extensive literature on the study of vegetation as a pure science, with little reference to the merits of different types of plant cover as an economic resource, nor to the degree to which vegetation may be regarded as an indicator of a total integrated environment. Volume 5 of a new series, Handbook of Vegetation Science, dealing specifically with ordination and classification of communities has been published (Whittaker, 1973), and reviewed by A. W. Johnson in *Science* (184: no. 4137, 1974) under the heading 'Plant ecology: gradients and discontinuities.' The volume itself is divided into three main parts: direct gradient analysis, indirect gradient analysis, and classification (American, Finnish, Scandinavian, Russian, French and numerical systems).

The reviewer states that ordination is a process by which communities or samples of communities are arranged in order along gradients of environmental

138

change, or alternatively the samples are ordinated along compositional gradients, with those communities most similar to each other being placed adjacent on a scale describing the range of similarity values among all such samples. This approach grew largely out of dissatisfaction with rigid views on the discontinuity of communities and from increasingly improved information on the nature of species.

In classification the emphasis is on combining samples that are similar to each other. In ordination, the differences between them are used to arrange the samples in a linear or multidimensional framework that will reveal something of the relations between the samples and their environments. On the degree of mathematical sophistication which is frequently involved in these studies, the editor and one of the contributors state that there is often an inverse relation between mathematical complexity or elegance, and research effectiveness of ordination procedures. To which the reviewer adds: Amen. 'This conclusion will be greeted with some enthusiasm by those ecologists who tend to be suspicious that the kind of analysis that is most gratifying to the mathematician may be irrelevant or misleading to the ecologist.'

Although the controversy between the various schools of classification has largely ended, the traditions have not; about half the book is given to fairly detailed descriptions of the six or seven classification systems still in relatively wide use. 'Some of the systems have rather wide use, while others have become entrenched among plant scientists in local areas. A few of these scientists, such as the European advocates of the Braun-Blanquet system, seek to promote wider use of their methods . . . many will point out that ordination and classification are considerably behind the cutting edge of modern community studies. None the less, this volume is valuable both as a historical summary of the development of two techniques . . . and as a handbook of the techniques themselves' (from review by Johnson, op. cit.).

Volume 8 of the Handbook of Vegetation Science, entitled Vegetation Dynamics, edited by R. Knapp, has also been published (1974). It consists of twenty-seven contributions, grouped in seven sections: Kinds of changes in vegetation; Methods of syndynamical analysis and of conclusions in vegetation dynamics; Cytogenetical, competitional, allelopathic and similar causes of vegetation dynamics; Classification of successions and of their terminal stages; Productivity and chemical changes (in ecosystems) in successional changes (including H. S. Beard: Vegetational changes on aging landforms in the tropics and subtropics); Examples of fluctuations (including R. T. Coupland on the North American grasslands); Synchronological vegetation dynamics.

The editor states that the theoretical and practical bases of environmental protection, of wildlife management and of various aspects of nature conservation are fundamentally related to the results of studies on vegetation dynamics. The concluding chapters deal with fluctuational changes occurring within short periods, and the history of plant communities over millenia or longer periods of time.

139

3.6.1 Vegetation as indicator of environment and as resource

The plant resources of part of the region are discussed fully in Natural Resources of Humid Tropical Asia (UNESCO, 1974), by:

R. A. de Rosayro: vegetation of humid tropical Asia — classification of vegetational types and mapping; quantitative studies; forest inventory

Tran Van Nao: forest resources — types: tropical evergreen, moist deciduous, dry deciduous, mangrove and rear mangrove, coniferous — quantitative evaluation of area, growing stock and annual increment; distribution of major species — Dipterocarps, teak (*Tectona grandis*), bamboo, tropical conifers, mangrove; management systems; production, biomass, economic yield, balance of needs and resources, factors hindering or favourable to production

P. Legris: vegetation and floristic composition of humid tropical Asia — bioclimates; floristic affinities and phytogeographic division; taxonomy and flora; new genera and species; palynology; herbaria and research; organizations; bibliography of literature and vegetation maps

R. O. Whyte: grasses and grasslands

M. Jacobs: botanical panorama of the Malesian archipelago — biogeographical concept; Indonesia, Malaysia, Brunei, the Philippines, Singapore, eastern Timor, eastern New Guinea, and perhaps Solomon Islands; life forms of Malesian plants; climaxes and seres; altitudinal zones; threat to dwindling forests; floristic subdivision of Malesia; the Asian/south Malesian drought disjunction; origin of mountain flora; palaeo-plant geography; present state of taxonomy

The UNESCO/FAO map of the vegetation of the Mediterranean zone (UNESCO, 1970) overlaps into the monsoon region in its sections on the Indus and Gangetic basins, and on the eastern parts of Iran and Afghanistan. There is an extensive literature on the arid and semi-arid tracts of western Monsoon Asia. Bibliographies to literature on the floras and vegetation of monsoonal and equatorial Asia and the western Pacific include those covering the Indo-Pacific area published in the *Flora Malesiana Bulletin* (Rijksherbarium, Leiden); the bibliography of Indochinese botany from 1955–1969 (Vidal, 1972); bibliography to floras of southeast Asia (Reed, 1969); bibliographies on plant ecology (autecology, synecology, vegetation, succession, forestry, range, soil and land survey, arid zone climatology, flora) in India (Meher-Homji & Gupta, 1971), in Pakistan (and Bangladesh when East Pakistan) (Gupta & Meher-Homji, 1971) and in Sri Lanka (Meher-Homji & Perera, 1972).

An introduction to the European schools which have dealt with tropical vegetation as a resource may be found in the writings of Walter (1973) who describes the vegetation of the earth in relation to climate and eco-physiological conditions, with reference to the zones of the evergreen tropical rain forest, of tropical summer rainfall, and of subtropical semi-deserts and deserts; of the Montpellier school of phytosociology in the writings of J. Braun-Blanquet and L. Emberger, as presented by van der Maarel & Tüxen

(1972), Guinochet (1973), and by Long (1974, 1975) on the general principles and methods of phyto-ecological analysis, mapping and land management; and of the H. Gaussen/Toulouse school which has for many years been surveying and mapping the vegetation of India, from its centre at the Institut Français, Pondicherry.

The phytogeographical studies carried out by the Institut Français are at three levels of investigation (P. Legris, personal communication, 1974; Legris & Meher-Homji, 1968): small-scale for the inventory of natural resources over large areas, medium-scale for regional studies, and large-scale for the evaluation of primary productivity of some special ecosystems.

In these three cases, the components are analysed and correlations sought with environmental factors. This kind of integrated approach permits an evaluation of the potentiality of the land, the technique of study varying according to the scale. The aim is common to natural ecosystems as well as to those transformed by man.

The natural vegetation is studied from floristic, physiognomic and successional points of view. Man can create a number of landscapes like woodlands, thickets, savannas, scattered shrubs and pseudo-steppes by interfering with the original vegetation. In a given ecological region, these degraded stages tend to evolve towards a floristically and physiognomically definite maximum type of vegetation called the 'plésioclimax', if protected from the destructive activities of man and his domestic animals. The plésioclimax is theoretically defined as the state the natural vegetation would attain over a sufficiently long time without human interference. In practice it is recognized as the most developed formation found in an ecological region. It is as a rule a forest in a humid country, but scarcely even a thicket in an arid zone. As it is difficult to know with certainty the true climax, it is convenient to use the concept of plésioclimax. The successive physiognomic stages ranging from barren soil to the plésioclimax go to form a 'series of vegetation'. For the nomenclature and definitions of the physiognomic types, the conventions adopted at the Yangambi conference are followed by the staff of the Institut Français (see also Aubreville, 1956; Trochain, 1957). A series is named after two to four species of the plésioclimax stage; some are chosen for their abundance/ dominance, some for their characteristic value, and some for their economic importance. Every series is determined by a definite set of ecological factors: temperature, rainfall regime (amount and distribution), soil type.

The natural vegetation is sometimes regarded as an integrated expression of the various factors of the environment. However, Christian & Stewart (1968) state that native vegetation does not always exploit the natural environment as effectively as an introduced crop plant might. It is therefore not an accurate indicator of the potential productivity of arable agriculture, of the variation in productivity which may occur within an apparently uniform area. Only the broader differences in the environment of new areas can be predicted confidently from the vegetation, and then usually at the formation level. Major struc-

141

tural or floristic units may occur on several different land types and are therefore unsatisfactory for distinguishing potential land-use units.

According to Zonneveld (1970; see also his ITC lecture notes on vegetation
science and vegetation survey Chapter 8), vegetation science contributes in three
ways to the delineation of the attributes of natural resources: a. vegetation in
itself provides raw material for food, shelter, study and recreation; it also
has negative aspects in the form of pests and weeds; vegetation survey overlaps
widely with land-use survey; wild vegetation also acts as a reserve for gene
resources; b. vegetation is a reflection of the environment; where direct
measurement of environmental factors is restricted, vegetation can be used to
extrapolate the results of investigations over wide areas; the study of factors
such as climate, water, soil, fertility requires sequential observations over a
certain period of time − to a certain extent vegetation can summarize such
data; c. information from vegetation can be used in three ways; independent
vegetation mapping and subsequent evaluation of the data (pragmatic land
classification based on vegetation units); subordinate to surveys of other disciplines − soils, climate, hydrology; as an equal characteristic to other land
attributes in more holistic surveys, and in the subsequent evaluation of land
units for pragmatic use.

While engaged on the survey of various types of vegetation in north and
west Africa (Sahel), Rossetti (1963, 1967) evolved a technique of aerial photography which involved the use of a captive balloon filled with hydrogen, for
the taking of low-altitude vertical photographs. The balloon could be manipulated at altitudes ranging from 10 m. above the soil surface to about 800 m.
The greatest stability of the equipment could be obtained in light winds of
constant direction, when one worked at altitudes above 50 m. Above 100 m.
wind speed could exceed 20 km./hour and reach 30 to 35 km./hour, provided
conditions ensured the stability of the aerial equipment. At altitudes below
50 m. almost complete absence of wind is necessary. In this way it is possible
to take photos of good clarity at scales varying from 1/100 to 1/8000, by
means of negatives of 55 x 72 mm., on well delimited areas of soil surface
from a few dozen square metres to about 30 hectares. Work continued on
these lines and on the more common types of aerial photography of vegetation at the Centre National de Recherche Scientifique at Montpellier, and
with the commercial concern, Géotechnip at La-Celle-Saint-Cloud. Rossetti
and his associates presented a group of papers to the Second International
Symposium of Photo-Interpretation, Paris, 1966, including a study (Rossetti,
Kowaliski & Havé, 1967) on the characteristics of spectral reflection of some
plant species and their images on colour photographs, taken on the ground or
from the air. Kodak Ektachrome Infrared Aero film offers interesting possibilities for aerial detection of vegetation.

But now (Charles Rossetti, personal communication, 1974) the techniques
of aerial photography have become very sophisticated (radar, thermal infrared
detection, laser, etc.); the means required are outside the possibilities of a

private firm, and the entire field of remote sensing is now the responsibility of State-supported research organizations (see also 6.4).

Over much of the land surface of monsoonal and equatorial Asia, data from vegetation surveys may be used as an approximate indicator of the actual environment, but not directly for an assessment of potential. The natural and induced vegetation formations have been reduced over vast areas of land to a low stage in ecological succession. The soils associated with these degradation stages are also low in their own series, due to erosion, loss of structure and fertility — how many of the dryland cultivators of Asia are farming on sub-soils? The microclimates at ground level in these degraded types of vegetation, the ecoclimates or bioclimates in which most of the plants and animals, wild and domesticated, exist, are also low in that their own succession series, far below (more arid and with higher ambient temperatures at ground level) the undisturbed microclimates which one might expect to find associated with a particular macroclimate.

On such land, one may perhaps be able to recreate a mental picture of the original vegetation climax, and therefore also of the highest form of natural vegetation and of the original soil and climatic environment of which this vegetation was an approximate indicator. This recreation may be done on the basis of experimental evidence of the direction and limits of succession, and on some form of inspired guesswork based on circumstantial evidence from inaccessible mountains, remote valleys, Muslim cemeteries, temple courtyards, the former hunting grounds of maharajahs and the protected vegetation left around Chinese villages for geomantic (fung shui) reasons (Fig. 3.10).

3.6.2 Floras

The correct recognition of the genera and species in plant communities during vegetation surveys depends upon the availability of published floras and of herbarium specimens assembled by taxonomists and plant geographers. It is therefore unfortunate that these published floras contain the names, morphological descriptions and notes about habitats and local and general distribution of *all* plants which have been found in field collections in a specific area. It is fully recognized that, over much if not most of the land surface of monsoonal and equatorial Asia, the present vegetation cover represents a low seral stage in succession. Only if the modern communities are exposed to ecological/historical analyses (Whyte, 1972b, 1975b) is it possible to separate those genera and species which were present in the original climax vegetation from those which are modern intruders. If something approaching this distinction could be made in floras before publication, it would be much easier for the surveyors of vegetation to decide upon the correct place in succession and potential direction and rate of regeneration of the communities being studied. Information which might be used to distinguish between truly indigenous species, endemics and intruders in recent or modern times would also be of great value to those responsible for the conservation and utilization of genetic resources among wild ancestors of our cultivated plants.

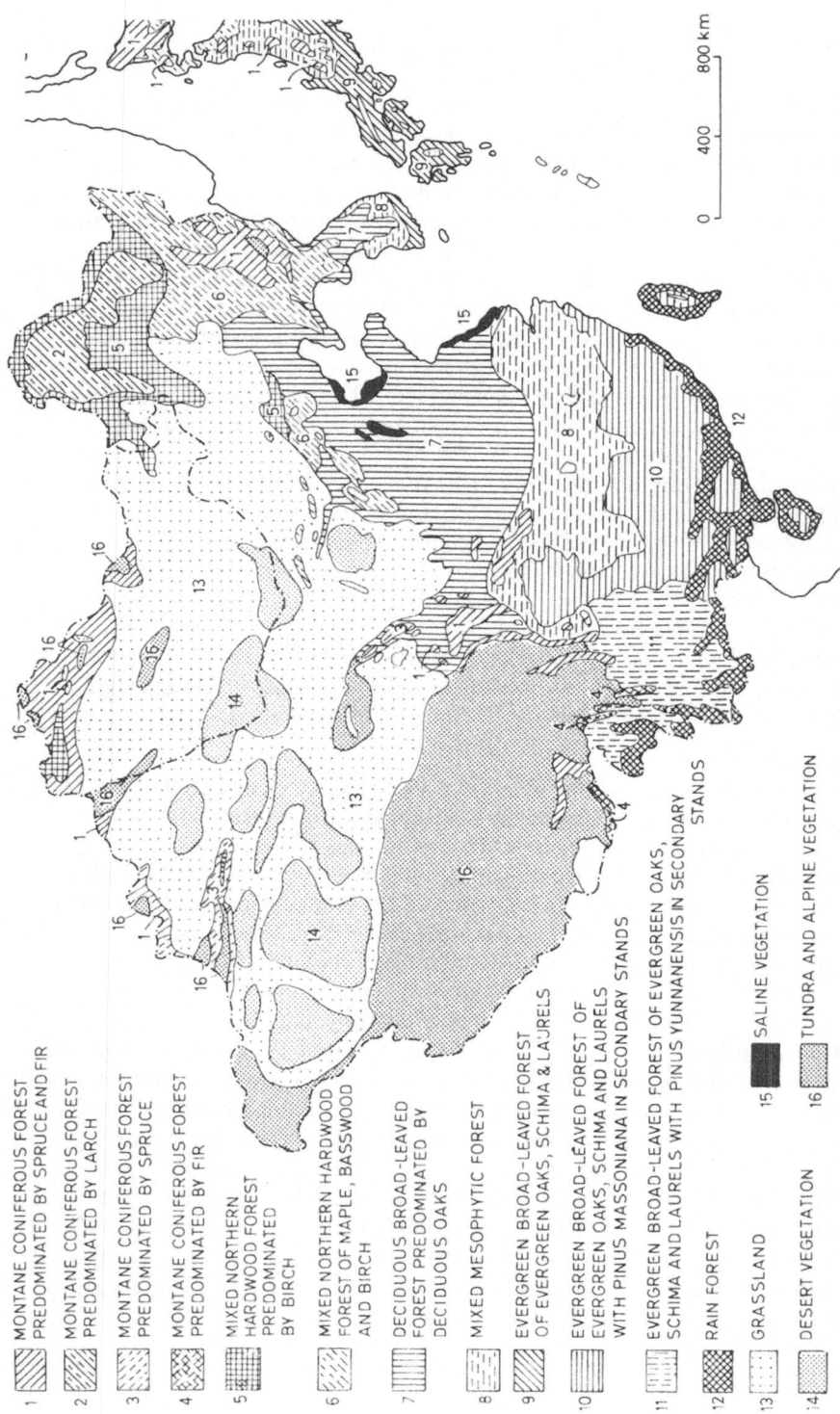

1 MONTANE CONIFEROUS FOREST
   PREDOMINATED BY SPRUCE AND FIR

2 MONTANE CONIFEROUS FOREST
   PREDOMINATED BY LARCH

3 MONTANE CONIFEROUS FOREST
   PREDOMINATED BY SPRUCE

4 MONTANE CONIFEROUS FOREST
   PREDOMINATED BY FIR

5 MIXED NORTHERN
   HARDWOOD FOREST
   PREDOMINATED
   BY BIRCH

6 MIXED NORTHERN HARDWOOD
   FOREST OF MAPLE, BASSWOOD
   AND BIRCH

7 DECIDUOUS BROAD-LEAVED
   FOREST PREDOMINATED BY
   DECIDUOUS OAKS

8 MIXED MESOPHYTIC FOREST

9 EVERGREEN BROAD-LEAVED FOREST
   OF EVERGREEN OAKS, SCHIMA & LAURELS

10 EVERGREEN BROAD-LEAVED FOREST OF
   EVERGREEN OAKS, SCHIMA AND LAURELS
   WITH PINUS MASSONIANA IN SECONDARY STANDS

11 EVERGREEN BROAD-LEAVED FOREST OF EVERGREEN OAKS,
   SCHIMA AND LAURELS WITH PINUS YUNNANENSIS IN SECONDARY
   STANDS

12 RAIN FOREST

13 GRASSLAND

14 DESERT VEGETATION

15 SALINE VEGETATION

16 TUNDRA AND ALPINE VEGETATION

Fig. 3.10. A theoretical reconstruction of the climax vegetation of the People's Republic of China, Korea and Japan, from which most of the forest types have either been eliminated entirely, or greatly reduced in area and successional status (from Wang Chi-wu, 1961).

3.6.3 Forests
Les forêts précèdent les peuples, les déserts les suivront — Jacquot

Types

Champion & Seth (1968) introduce their volume on the forest types of India (second edition) with the statement that, in an applied science such as forestry, the need for classification is emphasized by the general experience that most of the techniques devised for directing or controlling the development of a forest are applicable only within a limited range of conditions and forest types. A classification must therefore be available as a standard of reference for research workers and practitioners.

A forest type is defined as a unit of vegetation which possesses broad characteristics in physiognomy and structure sufficiently pronounced to permit of its differentiation from other such units, irrespective of physiographic, edaphic or biotic factors. Since forest types have a bearing on the practice of scientific forestry, greater importance has recently been placed on the main tree layers or on the most emergent vegetation, with the shrub and ground flora in a subordinate position. The main type-groups are subdivided into types on a geographic basis, since a recognizable type group varies somewhat with locality owing to differences in floristics and minor variations in climate and site occurring within the range associated with each group-type as a whole (Table 3.7 adapted from Champion & Seth, op. cit.).

In most types of vegetation, especially those of climax status, there are no sharp boundaries between adjoining types, as exist for example between tropical rain forest and savanna under certain specific conditions (Hills & Randall, 1968). There is normally a transition zone or ecotone, which may be of any width and is sometimes quite extensive, and which is intimately linked with the sharpness or otherwise of changes in the complex of other land resource factors (Champion & Seth, 1968). It is common practice to reduce this transitional zone to a minimum. 'As the same transition of site factors and vegetation may elsewhere be widely drawn out, perhaps at the expense of the conditions and vegetation selected as a type, the method is illogical and may render comparison with other areas impossible.' However, it has such advantage in local land use practice that it is generally adopted in forestry and agriculture.

The enquiries of FAO into world forest resources has led to the conclusion that existing forests and forests which can be established on potential forest lands are capable of providing a supply of forest products for a world population much larger than exists today. However, to do this, the following conditions must be fulfilled (Husch, 1971): a. the current devastation and uncontrolled exploitation must be reduced as much and as soon as possible; b. large areas of forests still considered inoperative or inaccessible must become a resource for useful products and services; c. all forests must be managed as renewable crops, protected from damage and rationally used, and d. large areas of land best suited to forest production should be either forested or reforested and brought into production.

145

Table 3.7. The forest types of India (from Champion & Seth, 1968, which for map see facing their page 48).

A. TROPICAL FORESTS
   Wet evergreen
   Semi-evergreen
   Moist-deciduous
   Littoral and swamp
   Dry deciduous
   Thorn
   Dry evergreen
All these main type-groups of tropical forest except the fourth and the last can be subdivided into southern and northern forms, the difference becoming less obvious the drier the type; the difference is to some extent in form but largely floristic.
B. FOREST TYPES DETERMINED BY SPECIAL SITE FACTORS
   Stages in natural succession or edaphic preclimaxes grouped under category 4 above.
   Littoral
   Tidal swamp
   Freshwater swamp
   Seasonal swamp
   Riverain fringes
C. MONTANE SUBTROPICAL FORESTS
   Subtropical broad-leaved hill forests
   Subtropical pine forests
   Subtropical dry evergreen forests
D. MONTANE TEMPERATE FORESTS
   Montane wet temperate
   Himalayan moist temperate
   Himalayan dry temperate
E. SUBALPINE FORESTS
   Stunted deciduous or evergreen forest, only in Himalayan and associated ranges, usually in close formation with or without conifers.
F. ALPINE SCRUB
   Moist (low but often dense scrub)
   Dry (xerophytic shrubs in open formation)

Inventory

Good forest management, correct silvicultural practices and the preparation of working plans are stated to be impossible without good data from an inventory correctly planned and operated. FAO has published a manual on the planning of a forest inventory (Husch, op. cit.) from which the following three paragraphs are quoted (see also R. A. de Rosayro & Tran Van Nao in UNESCO, 1974):

'Forest inventories are usually considered to be synonymous with timber estimates. In this sense a forest inventory is an attempt to describe the quantity and quality of forest trees and many of the characteristics of the land area upon which the trees are growing. It should be emphasized that forest inventories must include an evaluation of both tree characteristics and the land upon which they grow. Estimates of timber quantities divorced from the area on which the forests are growing have little meaning. A forest is not merely a quantity of woods, but associations of living plants which can and should be treated as a renewable crop.

'A complete forest inventory from a timber-estimate point of view should include a description of the forested area and ownership; estimates of volumes (or other parameters such as weight) of the standing trees, and estimates of growth and drain. In any specific inventory, there may be emphasis on, or elimination of, one or more of these items, depending upon the objective; but, for a complete evaluation of a forest area, and especially with a view toward managing it on a sustained-yield basis, all these elements should be secured.

'With the increasing use of forest areas for purposes other than the supplying of wood — recreation, watershed management, wildlife refuges or possible conversion to other land uses — the scope of the forest inventory must be widened. When these other values are important, their relationship to the forest and the land upon which it is situated must be observed, measured when possible, and the resulting data analysed. In some cases, a forest inventory will seek only wood or timber quantities and qualities; in other cases, both timber and nontimber information will be required and, with growing frequency, only nontimber information may be sought and may require different or additional observations and measurements. In many cases, a great deal of the information usually obtained in an inventory for timber can also be used for evaluating other forest associated values. For example, forest composition and topographic information, always necessary for a timber-oriented forest inventory, are also essential when evaluating forest recreational possibilities or watershed values. Similarly, forest site quality information can also provide essential information for consideration of forest land conversion to other land use. In this brief guide, the discussion will concentrate primarily on timber-oriented inventories. In planning forest inventories which require nontimber information, it is essential that specialists in these allied fields work cooperatively in planning and executing the inventory.'

The determination of present or potential growth is a complex and difficult operation. Under some conditions, preliminary estimates of growth may be prepared in a short time, but elsewhere and particularly in tropical zones, several years of repeated measurements may be needed to obtain data on growth rates. Commercial forest exploitation is frequently based on the estimation of timber of size and qualities suitable for immediate use, with little regard to potential productive capacity on a sustained-yield basis. The assessment of changes taking place in a forest over a period of time requires periodic or repeated inventories. Neglect of periodic assessments of the quantity of timber in the forest may conceal changes in growing stock and may seriously affect continuous management.

A Pre-Investment Survey of Forest Resources of India has been carried out in recent years by UNDP/FAO working in association with Government of India. Although a statement of the techniques adopted is not available, two officers of the Survey have proposed a standard classification scheme for land-use and forest cover types which is appropriate where aerial photo-interpretation is involved (Tomar & Maslekar, 1973; see also Husch, 1971, on the use of aerial photography in forest inventory).

The system of classification adopted in terrestrial surveys may require certain modification when aerial photographs are used. Certain types of information are difficult to recognize on aerial photographs, whereas broad features of landscape and vegetation, and the limits of forests and other types of vegetative cover are better judged on aerial photographs. The scale has to be broadened, from the very small area, say 0.1 hectare suitable for ground surveys, to 2 to 50 hectares for aerial surveys depending on the scale of photography. On aerial photographs, the forest may be classified by the following categories (Tomar & Maslekar, 1973): condition classes, forest type/species, stand and tree characteristics, conditions of site, extent of forest, conditions of terrain, and accessibility class.

A Forestry Inventory Manual for Cambodia was prepared in early 1962 to serve as a source of training material after the departure of the USAID forestry technicians; the text in English was to be translated into French (USAID, 1962). The inventory is based on the use of aerial photographs for mapping and field sampling. For the initial inventory, photographs at 1:40,000 scale covering the whole country are used for mapping; photos at 1:10,000 scale are used for locating field plots. Timber type, density and size class are delineated on the 1:40,000 photos and the delineations are then transferred to base maps. Maps provide area statistics for exploitable and non-exploitable forest land, timber types, present land use, reserved versus non-reserved lands, provinces and sampling units. Area statistics for such forest conditions as size class and stocking class are determined by proportions derived from field samples.

Volume statistics are based on field samples taken on a systematic grid. Sample plots are located in a rectangular grid pattern with a spacing of 16 km. in both N-S and E-W directions. Each sample plot consists of a cluster of four subplots. Periodic remeasurement of plots will be facilitated by photo and written records of plot locations and by permanent markings in the field.

Intensity of sampling and inventory techniques were set at levels considered feasible in relation to available funds, time and personnel. Experience in other countries indicate that these levels could be expected to provide a sampling error (on the basis of 2 chances out of 3) less than ±20 per cent per 30 million cubic metres of sawtimber growing stock. The estimates of area determined from the maps is considered free of sampling error.

Through remeasurement of plots, estimates essential for forest management at the local and national levels can be made of timber growth rates and changes that have occurred as a result of growth, mortality and cutting. The systematic grid is of particular value in this connection. The overall extensive inventory providing information for management at the national level should be planned on a cycle of about 10 years, with remeasurement of all original plots that can be relocated, or of new ones. New photos are necessary for each reinventory, but one photo cover at an intermediate scale should suffice.

Detailed inventories must be made of management units, after deciding on plot interval; this may be 8, 4, 2 or 1 km. depending on the allowable sampling error for volume estimates of the average management unit or working circle. When sufficient information is available on coefficient of variation, the number of plots required for the average working circle is computed by the formula:

$$n = \frac{t^2(cv)^2}{Ae^2}$$

where n = number of samples
   cv = coefficient of variation in per cent
   Ae = allowable sampling error (per cent) of volume for working circle
   t = a constant that reflects the reliability to be placed on the estimate.

The value of 1.000 may be used for *t* if an estimate is desired which has a 2 out of 3 chance of falling within the confidence limits indicated by the sampling error. If a precision of 9 out of 10 is desired, 1.645 should be used, if 19 out of 20, 1.960. Having determined the number of plots and knowing the area of the average working circle, the grid spacing may be determined (USAID, 1962).

The techniques adopted by the (British) Land Resources Division are given in the three-volume report of a forest inventory carried out at the request of the Government of Fiji (Land Resources Division, 1973), to obtain the following information: volume and area of accessible forest for timber production; identification of protection forest; mapping on a scale of 1:50,000 to show stratification of forest into forest types; preparation of volume tables.

The unit of survey was the river catchment. Those with sufficient forest cover to merit enumeration were recognized by preliminary airphoto interpretation, very steep hill ranges and areas of poor forest cover being excluded. The promising forests were divided into broad forest types on the basis of airphoto pattern and landform.

A base line was drawn on the 1:50,000 map down the main river of a catchment or parallel to a road or the coastline in the case of coastal catchments, and divided into 2,000 m. sections. Within each section the starting point of the line was chosen by reference to a table of random numbers. The lines were drawn parallel to each other, on a bearing that would cross the topography and, if possible, the forest types at right angles. These lines, which ran from the main river or from the forest edge to the watershed, were transferred from the map to the aerial photographs. In the field the starting points of the lines were marked on the ground by reference to the aerial photographs. Each line consisted of a continuous strip 20 m. wide of plots 50 m. long. Within the plot every tree over 35 cm. d.b.h. was enumerated by name giving the breast height diameter (or above buttress diameter) and length of merchantable bole. In every twentieth plot, a regeneration count was made, tallying the trees under five size classes. Only the regeneration of the recognized useful timber species was tallied in the first three size classes. The data from the lines were also used to finalize the division of the forest into distinctive forest types.

The volume of timber of a forest type was calculated by multiplying the enumerated volume by a factor representing the proportion of the total area to the sampled area. This can be written as follows:

Total volume of forest type A =

$$\frac{\text{Total area of forest type A}}{\text{Sampled area of forest type A}} \times \text{Sampled volume in forest type A}$$

The sampling error was computed by forest type for the trees grouped into five timber density groups. The formula for variance and sampling error, and the aim of the sampling error are given in appendix 2 to Volume 1 of the L.R.D. report. The aim of the sampling was to achieve a 20 per cent error limit in the production forest types.

Hydrology
A UNESCO/FAO Working Group considered man's influence on the hydrological cycle within the scope of the International Hydrological Decade (FAO, 1973). In the contribution by A. B. Costin & J. C. I. Dooge, the relation between forest management and the hydrological cycle was considered. 'It is now generally agreed that the presence of a forest cover in comparison with alternative vegetation does not affect the total amount of precipitation. However, there are local exceptions such as "occult" or "impingement" precipitation, and the interception of wind-blown snow. Occult precipitation is due to the collection on tree crowns and stems of fog-sized droplets moving horizontally into the vegetation. Other phenomenon are condensation of water vapour as dew on leaf surfaces, or freezing to form deposits of rime. . . .

'Under most climatic conditions, however, the effect of forest cover is to promote a higher rate of transpiration and hence to reduce the volume of total run-off. Accordingly, the removal of forest cover would normally have the effect of increasing the total run-off due to the reduction in evapotranspiration and also of increasing the flood peaks due to the decrease in infiltration capacity. The increases in water yield following forest clearances which have been measured in the field have been variable in amount. However, it would appear that the increases are approximately proportional to the percentage of area cleared and are in all cases less than an equivalent depth of 350 mm. of water per annum for total clearance, or a corresponding amount for partial clearance. In cases where adequate drainage is not available, the decrease in transpiration may result in a rise in the groundwater table which can cause serious problems if the groundwaters are saline. Cumulative effects on groundwater may occur as a result of changing vegetation patterns and associated rates of evapotranspiration (examples quoted from Australia). . . .

'High infiltration rates, with a corresponding reduction in surface run-off and soil erosion, are features of most forests in good condition. These attributes are due to the relatively stable and porous structure of the forest soils, and to the protection of the soils by the leaf mould and ground cover. High infiltration also increases the opportunity for recharge of groundwater, with the result that the flow of springs and streams tends to be more sustained; however, as noted above, total flow may be less due to the usually higher rates of evapotranspiration from forests than from most other types of vegetation. In areas of snowfall, forests further sustain streamflow by delaying the melting of the snow. These favourable conditions for infiltration which exist under many forests (the relatively stable soils, the protective ground cover and leaf mulch, and the usually higher rates of evapotranspiration) tend to operate in the same direction to minimize soil erosion as well as surface run-off. There are many instances of slopes steeper than the equilibrium slope for the soil becoming unstable after the forest cover is removed. Extensive mass movements may result, sometimes soon after clearing, but sometimes many years later, after the original root systems and surface cover have rotted away. However, it should not be thought that tree cover itself is an insurance against surface run-off and soil erosion. Removal of the ground cover and leaf mulch, as by grazing or fire, may have serious consequences.'

The need to conserve declining water reserves which has become apparent over the last few decades has led several investigators to conclude that plants with a high rate of transpiration endanger water resources, and the growth of such plants must not be encouraged. Some think that trees withdraw more water from the soil than other plant species and evaporate it excessively through the stomata of leaves. If from 10 to 40 per cent of rain is intercepted by the forest canopy and evaporates before it reaches the forest floor, and if trees transpire more than other cover types, then less moisture should remain in the forest soils after the rainless growing season than in the soils of neighbouring treeless areas where there is only seasonal vegetation growth. Gindel (1973) finds that the data from the arid zone contradict this supposition.

Conservation
Modern forest management in the South Asian (formerly Indian) subcontinent has until relatively recently aimed at reversing the ecological regressions of the past and at developing progressive seres modified and limited to suit the purposes of mankind. The demand now is for faster progress, for 'instant' timber (Champion, 1969). Economists now advise Governments to press their Forest Departments to replace naturally occurring species wherever possible by others of faster growth and those yielding the types of wood most in demand for wood-based industries. 'The foresters of Pakistan are being pressed to abandon their three major species, *Acacia arabica*, *Dalbergia sissoo* and *Pinus roxburghii*,

in favour of exotics, none of which have yet been grown successfully as crops under the local conditions. Again, fellings on a vastly increased scale are being made in the fir forests of the Himalaya under working plans depending for their acceptability on success in regeneration, artifical or natural, although so far no success has been obtained in regenerating even on a small scale, and it is difficult to be optimistic on results to date from experiments.

'The economists also press for examination of value production of forest land if put under other crops such as tea, coffee, cotton and rubber which all give quicker returns, even if they involve more capital outlay. It is left to the forester to point out the indirect but indisputably important returns from the forest, notably its protective value for soil and water, its recreational and amenity values, and its scientific value including wild life conservation – most of which values are difficult to quantify.

'Great risks are involved and are being taken in attempting big scale rapid changes without the necessary basic knowledge and experience. Ill-advised clearing of luxuriant rain forest may end up with rapid loss of topsoil and subsoil leaving exposed bedrock or an unproductive laterite pavement.

'It must be feared that the hand of man having progressed from being purely destructive to adopting conservative policies of protection and controlled use to meet human requirements is now reverting to the mainly destructive role by making drastic changes in the name of desirable development towards fuller use, without the necessary evidence that they will succeed – in fact against the indications of such evidence as is available. The right motto in this matter is surely *festina lente*, but this is not a motto that appeals to the modern politician!' (Champion, op. cit.).

The situation in India is causing increasing concern to professional foresters. In 1952, official policy stated that one third of the country should be maintained under forest, 60 per cent in the hills and 20 per cent in the plains. Over-cutting of non-reserved forests for timber and fuel, indiscriminate and excessive grazing and browsing of domestic livestock in forests (Whyte, 1964, Chapters 16, 17, 18), and the establishment of river valley projects by irrigation authorities without adequate consultation with Forest Departments (Pratap Singh, 1971), have all contributed to the reduction of the forest area and of its capacity to conserve soil and water or to prevent advancing desiccation. In the 1960's, foresters were saying that it was generally believed that the forest area had fallen from 33 to 25 per cent, but that the latter figure was actually 16 per cent. In the 1970's (Palit, 1970), it has been stated that the forest area in Rajasthan, for example (Prakash, 1972) represents only 11 per cent of the total land area, and that a further 7 per cent is degraded due to the uncontrollable action of anthropogenic influences.

The tropical rain forest is the most complex ecosystem on the earth, which can nevertheless be described in the same way as other self-supporting ecosystems: as an association of producing, consuming and decomposing organisms, all ultimately deriving their energy from sunlight (Richards, 1952, 1973;

see also Meggers, Ayensu & Duckworth, 1973). Today, however, the rain forest, like most other natural ecosystems, is rapidly changing. Satellite photos show that it is no longer a continuous belt of green extending 10 degrees or more north and south of the Equator, over the lowlands of the humid tropics of Central and South America, Africa, southeast Asia and the islands of Indonesia. Now this belt is fragmented and much reduced in area; in the past two decades, huge expanses have been felled for timber or replaced by plantations of oil palm, banana, rubber, cocoa and other crops, or for the system of shifting cultivation that demands a perpetual supply of uncultivated land. Sizable areas still stand in Amazonia, Africa, Borneo and New Guinea, but it is probable that by the end of this century, very little will remain. 'The destruction in modern times of a forest that is millions of years old is a major event in the world's history' (Richards, 1973).

The nature and magnitude of the consequences cannot be foreseen with precision. One effect which is probably already irreversible is a permanent alteration in the course of plant and animal evolution (Richards, op. cit., Stern & Roche, 1974). 'Biologists are generally agreed that much of the existing flora and fauna of the world, perhaps including man himself, originated in the humid tropics. The rain forest has for millions of years served as a factory and storehouse of evolutionary diversity from which plants and animals able to adapt to more rigorous environments have migrated to populate the subtropical, temperate and colder regions. This role the tropical rain forest can play no longer; the destruction of forests and other ecosystems has already cut the lines of communication and made these migrations impossible. . . .

'It is still uncertain (but see 2.5) whether or not the removal of all the world's rain forests would have any significant consequences for the global climate, even if they were replaced by bare rock and soil. As we have seen, the primary forests will probably be replaced by systems of impermanent cultivation, by artificial forests of much simpler ecological structure, and by secondary forests, scrub and savanna. This vegetation will also contribute oxygen to the atmosphere and will modify the microclimates at and near the surface, although the effects produced may be somewhat smaller than those produced by the original forest. . . .

'The tropical rain forest is a unique and rich community of plant and animal species that includes many of the most beautiful and bizarre forms of life. Only here are found such insects as the brilliant blue morpho butterfly of tropical America, countless striking and lovely birds, mammals such as the orang utan, the sloths and the scaly anteater, as well as magnificent orchids and trees. Tropical biota are so diverse and abundant, and they have received such limited scientific attention, that even among those species that have been named and described we know virtually nothing about the biological characteristics of a large majority. Much of the plant and animal life of the Tropics may thus become extinct before we have even begun to explore it' (Richards, 1973).

153

A Symposium on South East Asian Plant Genetic Resources held in Bogor, Indonesia, 20–22 March, 1975, recommended the conservation *in situ* of species of actual or potential use in forestry, horticulture or pharmacology (Whyte, 1975a, would include the perennial species of the Gramineae, especially of the genus *Oryza*, and the herbaceous Leguminosae, at their northern limits of distribution in southeast Asia and southern China). This conservation may be best achieved by setting aside selected large areas of virgin forest of a size indicated in MAB Project 8, and chosen to comprise the whole range of ecosystems occurring therein. The present destruction of these valuable resources of irreplaceable genetic resources is to be deplored.

FAO has also become aware of the increased tempo of forest exploitation in the tropics in recent years, and plans to convene a technical conference towards the end of 1975. See also the Goroka Symposium on the impact of man on humid tropics vegetation (Papua and New Guinea/UNESCO, 1960).

FAO refers to tropical moist forest, by which they presumably mean the tropical rain forest of most workers in the tropics, or the tropical wet evergreen of Champion & Seth (1968), with possibly other types of associated humid forests as well. Parallel to the massive exploitation there has not been growth in our knowledge of the composition and structure of these forests, of their behaviour following human intervention, of the appropriate silvics and silviculture of the many different forest communities, of suitable technologies for the harvesting and processing of tropical woods, and the costs involved. The criteria for decisions relating to land-use in the humid tropics are now being seriously questioned.

FAO concludes its preliminary note thus: 'There are certain functions – real or supposed – of the tropical moist forest which available methodologies find difficult to quantify. These have in general been neglected. They include the role of the tropical moist forest in mitigating the vulnerability of certain soils to the action of climate, in moderating the influence of intense rainfall on the discharge capacity of catchment areas, in maintaining breeding conditions apt for continuing speciation and hence species evolution (the 'gene pool' function) and in maintaining the global thermal and atmospheric balance.'

3.6.4 The Grass Covers

The communities and constituent species and the techniques which may be adopted in assessing them have been fully described elsewhere (Whyte, 1968, 1974a and c). It is therefore necessary to refer only briefly to certain aspects not already covered.

Ecosystems and communities – true and false
Following the study of the antiquity and evolution of the Gramineae of the region (Whyte, 1972b, 1975b), it becomes apparent that grassland workers should distinguish between true, false and artificial communities, and there-

fore also between true, false and artificial ecosystems. A community may be said to be true if it is composed entirely of species which are indigenous to that locality or habitat. The grass covers of Rajasthan are true, except where they are contaminated by relatively recent arrivals, such as perhaps *Panicum antidotale*, or species lower in the succession which have come from elsewhere to take up their position in the environments which have become unsuitable for the true indigenous species higher in the succession.

A community may be said to be false if it is composed entirely or largely of species which have arrived from elsewhere. The three or four species in such a community may each have arrived at different times in the vegetation history of the particular locality, for different reasons (natural migration or man-assisted movement), and from different places of origin. They come together on a site adapted to their individual autecologies. They therefore have to combine and compete with species with which they may not have been associated in their place of origin. Such false communities and ecosystems are characteristic of sites where grass covers have occupied the ground storey on land which was formerly under tropical rain forest or other grassless types of forest.

Artificial communities and ecosystems are the sown pastures and hay leys, and the cultivated fodder crops sown or planted pure or in mixtures, the composition of which is based on experimental or empirical knowledge regarding compatibility of species and related characteristics.

Two groups are working on grass ecosystems in Asia, one on temperate grasslands (Chairman, M. Numata, Chiba, Japan), and one on tropical grasslands (Chairman, R. Misra, Varanasi, India). The latter group held a meeting in Varanasi on 17 to 22 January, 1974, to assemble regional information which could be used as the tropical contribution (Misra, 1974) to the International Grassland Synthesis volumes to be published by the U.S. IBP Ecosystem Analysis Studies on the Grassland Biome. The agriculturist is, according to R. T. Coupland in his opening address, devoted to producing the maximum amount of food on a land surface. The biologists, on the other hand, are somewhat concerned about this because of the increasing inputs of N, P and K that are required in order to produce this food. The biologist's approach has been to find out more about the fundamentals of this production process in plants, and the use of this material in the natural and artificial association, so that one may better understand how this food production can be maintained at a sustainable level and what that sustainable level might be.

Thus, six working groups were formed at Varanasi to consider the tropical contribution to the Synthesis volumes under the heads: site description, primary producer system, consumer system, decomposer and nutrient cycling, systems analysis, and management and conservation. The few grassland practitioners that exist per hundred thousand square kilometers in Asia may be forgiven for considering how all this applies to the advice they are to give to the animal husbandmen; and whether the data that are laboriously and meticulously collected on biomass, litter, nutrient cycling, the higher rates of annual dry matter production in tropical grasslands as compared with temperate ones, and all the other criteria of the ecosystem specialist can really be accepted as reliable and permanent, on grass-dominant communities that vary so widely month by month and year by year, in accordance with climate, the biotic pressures, and the competition from the other constituent species. And the economic value of biomass as a nutrient resource is far greater in 100 kg. of desert grass and shrub from Rajasthan than in 100 kg. of tall jungle grass from the Terai or southeast Asia.

Supplementary to the techniques given in the earlier studies by the writer, reference should be made to the contributions from the Centre d'Etudes Phytosociologique et Ecologique, C.N.R.S., Montpellier, France, in the *American Journal of Range Management*, on sampling methods in dense herbaceous pasture which would be applicable in much of monsoonal and equatorial Asia (Long et al., 1972; Poissonet et al., 1972; Poissonet et al., 1973), also to the paper by Descoings (1971) on the concept and method of use of standardized data sheets for the analysis of the structure of intertropical herbaceous formations, also to papers on the methods of study and inventorization of the herbaceous stratum in palm savannas in Côte d'Ivoire (Descoings, 1972, Poissonet & César, 1972).

A grassland ecosystem model
Fig. 3.11 is a simulation model of the dynamics of a grassland developed as part of the systems analysis effort of the U.S./IBP Grassland Biome study at Colorado State University, Fort Collins. The model was developed to be representative of the grassland network sites and to address the effect of net or grass primary production as influenced by level and type of herbivory, moisture and temperature, and added nitrogen and phosphorus. ELM is considered a total system model as the abiotic, producer, consumer, decomposer and nutrient (phosphorus and nitrogen) components are all represented.

The abiotic model simulates the abiotic parameters via a water flow submodel and a temperature profile submodel which are stratified through the air, plant canopy and soil profile.

The producer model considers carbon and phenological dynamics of both above and below ground parts of a variable number of primary producers.

The decomposer model follows decomposition rates and microbe biomass in litter and dead material both above ground and below ground.

The mammalian consumer model and the grasshopper model simulate organismal, intrapopulation and interpopulation dynamics of consumers.

The nitrogen and the phosphorus models simulate nutrient flow through the system.

Each submodel is independently operational and also interacts with the other six submodels to give the total model (from *U.S./IBP Grassland Biome Newsletter* no. 14, June 1974).

156

Hydrology

'The hydrological effects of methods of grassland practices (A. B. Costin &
J. C. I. Dooge in FAO, 1973) depend greatly on the climate and on the level
of grassland management. Generally, a permanent grass cover has an effect on
the elements of the hydrological cycle which is intermediate between that of
a forest or dense scrub and that of bare soil or land cultivated for crops. Thus,
for example, the water yield and the tendency to erosion will be greater than
for forest cover but less than for ordinary farm land on comparable topography. The higher the level of grassland management, the closer will the
hydrological characteristics of the grazing area correspond to those of good
forest cover.

'In semi-arid areas, mismanaged grassland is particularly liable to produce
high flood peaks and relatively large amounts of soil erosion. In these areas the
native grass cover tends to be seasonal and at times offers minimal protection
to the whole of the soil surface. Thus the amount of livestock which can be
supported by such grazing land on a long-term basis may be quite limited.
Over-grazing of the area will intensify the problems of surface run-off and
erosion. The over-grazing of grassland can also result in a hydrological change
due to the compaction of a surface with inadequate cover due to the trampling of the grazing animals. This phenomenon is just as likely to occur in humid
as in arid areas and it further decreases the amount of infiltration and increases
the amount of surface run-off and the liability to surface erosion. Good
management of the pasture is not only of benefit for hydrological control
but is also in most cases capable of giving better economic returns. . . .

'It is possible to increase stock-carrying capacity several times, and yet also
reduce or eliminate flood run-off and surface erosion. The improvement of
grassland or the replanting of land to forest may, by reducing the run-off,
affect the water supplies downstream. The planting of land to trees or grass
to prevent siltation in a reservoir may accomplish that purpose, but may at
the same time reduce the chances of filling the reservoir with water. With more
integrated economic development in a grassland area, the concentration of
animals in confined spaces for winter feeding or before marketing may lead
to quite serious problems of waste disposal,' unless the cowshed wash is distributed into nearby crops of the gross-feeding African grasses, Para, elephant,
hybrid Napier, and Guinea, as is done in the Milk Colony at Aarey, Bombay
and other centres.

3.7 Faunas

Before man began to operate in an ever-increasingly destructive manner on the
biological ecosystems of the region, there was a rich faunal component of
large, medium, small and minute animals. Primitive man living in or on the

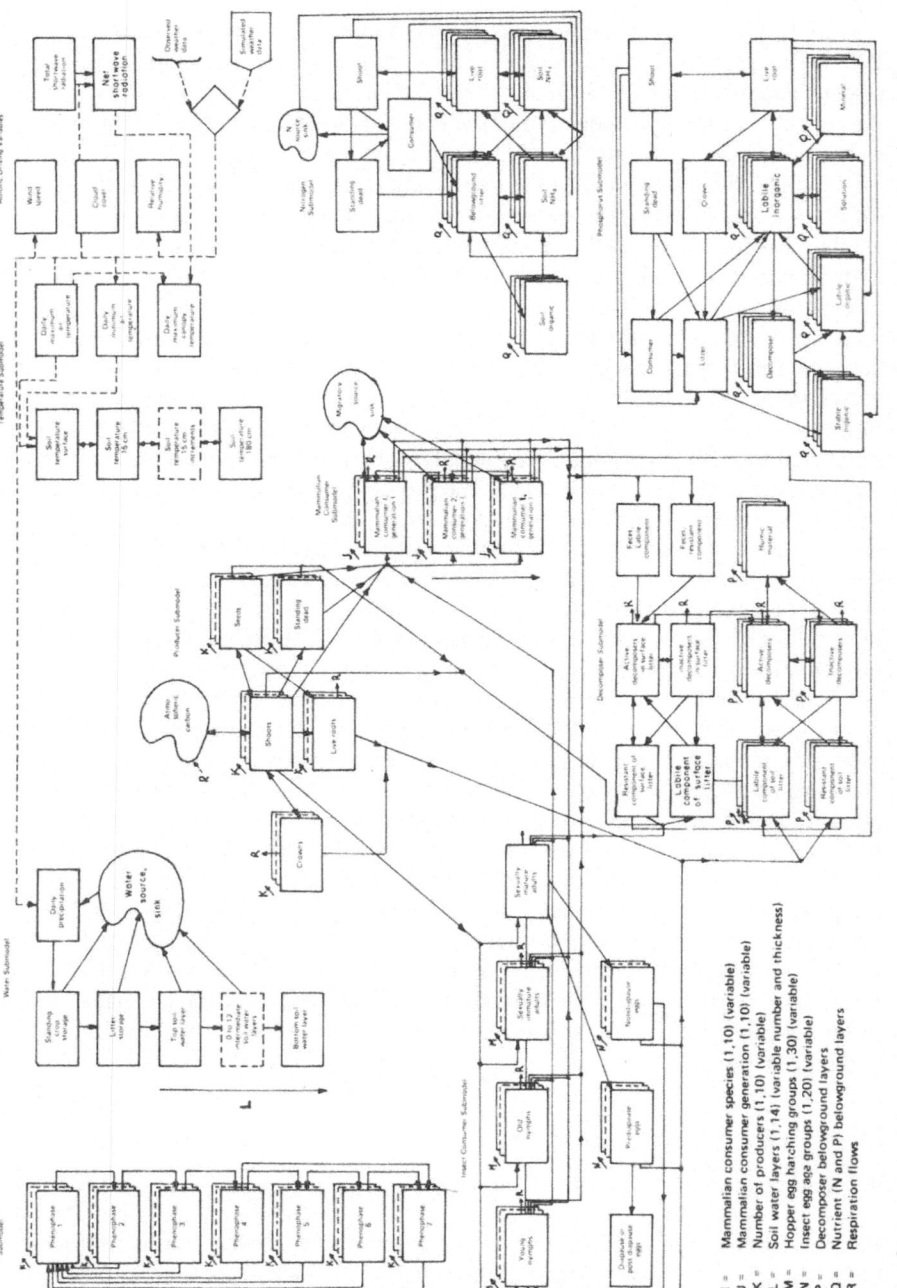

I = Mammalian consumer species (1, 10) (variable)
J = Mammalian consumer generation (1, 10) (variable)
K = Number of producers (1, 10) (variable)
L = Soil water layers (1, 14) (variable number and thickness)
M = Hopper egg hatching groups (1, 30) (variable)
N = Insect egg age groups (1, 20) (variable)
P = Decomposer belowground layers
Q = Nutrient (N and P) belowground layers
R = Respiration flows

Fig. 3.11. A grassland ecosystem model constructed by US/IBP Grassland Biome Study at Colorado State University, Fort Collins.

fringes of forest depended, and in a few cases still depends upon this fauna to provide him with an animal-protein balance to his otherwise vegetable diet.

In the arid and semi-arid zones, increasing pressures of humans and domestic livestock and the clearance of vegetation for cultivation has caused a drastic reduction in the former wildlife population. In the tropical rain forest environment, the former state of equilibrium between flora, fauna and primitive man in small numbers has been destroyed, again by increasing human populations and by the destruction of climax and other high seral stages of vegetation by shifting cultivation of progressively increasing intensity.

Simultaneous with this deterioration of the natural or indigenous fauna has been the increase of a predatory fauna which may have been indigenous, in biological balance with their associated types of vegetation, or which are intruders from foreign environments. Examples include many disease-disseminating organisms such as the anopheline mosquitoes (Whyte, 1974b, pp. 38–9), which transmit malaria, *Mansonia* and *Anopheles* which transmit Malayan filariasis, snails of the genus *Oncomelania* which are hosts to *Schistosoma japonicum*, causing Asiatic schistosomiasis, the predators on crops in the arid west of the region, the desert locust, and the many insect pests affecting arable and plantation crops, forest trees and domestic animals (N. C. Pant in UNESCO, 1974). Entomologists look with deep concern on the actions of the land scientists when they are directed to raising the potential of a desert region through the introduction of irrigation. For example, the new crops to be grown on the land commanded by the Rajasthan Canal will provide a lush environment and a rich new source of food for insect predators whose numbers were formerly held in check by the rigours of the desert ecosystems.

One is somewhat uneasy when reading Janzen's statements (1973) about the five major kinds of pest communities that may be encountered against the background of tropical ecosystems, biological and agro-, in particular, the suggestion that the 'conspicuous success of lowland rice monoculture in southeast Asia may be due, in part, to a generally depauperate insect community, as compared to that of other parts of the lowland tropics;' and the statement that 'animal communities in Borneo are drastically reduced when supported by tropical rain forest growing on nutrient-poor white sand soils.'

There is a need for an authoritative survey of the present state of knowledge on animals of all types in the land ecosystems of Asia. For general works on animal ecology, see under Charles S. Elton in the bibliography; see also Macfadyen, 1963, especially on the techniques for study of animal communities (Chapters 12 to 16) and on the ecosystem (Chapter 17); also Andrewartha, 1970, on the components of the environment (resources, mates, predators and pathogens, aggressors, weather, malentities) and on methods for estimating density, patterns of distribution and dispersal of populations of animals; also Lord Medway on the mammals of Borneo (1965) and Malaya (1969). P. Pfeffer (in UNESCO, 1974) has discussed the characteristics of the fauna of humid tropical Asia, the main biogeographic divisions and natural zones (coastal,

coastal and inland swamp, humid forests, savannas and open forests), their associated animal populations and improvement of biotypes or ecosystems. The Wallace Line (passing east of the Philippines, then between Borneo and Sulawesi and ending between Bali and Lombok in the Lesser Sunda Islands) is considered by Pfeffer to be a fairly clear-cut faunistic boundary for most land vertebrates, between the 'eastern region' (Indo-Malaysia or the southern edge of the continent and the Greater Sunda Islands), and the islands lying farther east, known collectively as Wallacea, as a halfway (and biologically impoverished) house for the fauna between the eastern region and Australia.

The review of the consumer organisms in the ecosystems of tropical savannas (Bourlière & Hadley, 1970) and of the adaptations of the consumers to the savanna environment is primarily concerned with Africa, as was also the international symposium of the British Ecological Society, Norwich, July 1970, on the scientific management of animal and plant communities for conservation. The meeting of the Tropical Ecology Group of the British Ecological Society, London, April 1974, on the unique range of habitats offered by tropical forests for mammals was concerned more with the faunal biology and ecology of these animals than with the total fauna as part of a land resource. F. Bourlière introduced a comparison of species diversity, morphological and behavioural convergences, vertical distribution, niche specialization and population dispersion, recruitment and turnover between mammals of the tropics of the Old and New Worlds, and discussed the role of mammals in maintaining the floristic complexity of tropical forests. Environmental pressures of a similar nature have shaped the mammalian communities in a similar way, despite different evolutionary histories. See also Willson (1973) on tropical plant production and diversity of animal species in tropical and temperate forests.

The assessor of land potential and the planner of new forms of land use need to know if it is desirable and practicable to consider the reintroduction or regeneration of the faunal components of ecosystems in which it is planned to revive an original vegetative cover, or one similar to it, through natural regeneration or artificial planting.

BIBLIOGRAPHY

ABROL, I. P. & D. R. BHUMBLA, – 1971 – Saline and alkali soils in India – their occurrence and management. In 'Soil Survey and Soil Fertility Research in Asia and the Far East,' pp. 42–51. FAO, Rome.
ABROL, I. P., K. S. DARGAN & D. R. BHUMBLA, – 1973 – Reclaiming Alkali Soils. Bull. no. 2. Central Soil Salinity Research Institute, Karnal, 58 pp.
AGARWAL, R. R., 1968 – Saline-Alkali Soils in India. Tech. Bull. (Agric. Ser.) 15. Indian Council of Agricultural Research, New Delhi. 227 pp.
AGRONOMIE TROPICALE, – 1968 – Le Colloque sur la fértilité des sols tropicaux, Tananarive, Madagascar, 19–25 Novembre, 1967. Agron. Trop. 23: 151–240 and 517–667.
AHMED, E., – 1969 – Origin and geomorphology of the Thar Desert. Ann. Arid Zone 8: 171–180.

ALBERDA, T., – 1962 – Actual and potential production of agricultural crops. *Neth. J. Agric. Sci.* 10: 325–333.

ANDREWARTHA, H. G., – 1970 – Introduction to the Study of Animal Populations (2nd ed.). Chapman and Hall, London.

ANUCHIN, N. P., – 1970 – Forest Mensuration. Keter Publishing House, Jerusalem. (Translation of original published 1960 by State Publishing House for the Wood and Paper Industry, Moscow, Leningrad). 458 pp.

ASHTON, P. S., – 1967 – Climate vs. soil in the classification of southeast Asian tropical lowland vegetation. *J. Ecol.* 55: 67–68.

AUBERT, G., – 1970 – Soils and the maintenance of their fertility as factors affecting the choice of use of land. In 'Use and Conservation of the Biosphere,' pp. 47–69. UNESCO, Paris.

AUBREVILLE, A., – 1956 – Essai de classification et de nomenclature des formations forestières africaines avec extension du système proposè à toutes les formations forestières du monde tropical. Centre Tech. Forestier Trop., Nogent-sur-Marne.

BAGNOULS, F. & H. GAUSSEN, – 1953 – Saison sèche et régime xérothermique. Documents pour les Cartes des Productions végétales 3. 47 pp.

BAGNOULS, F. & H. GAUSSEN, – 1957 – Les climats biologiques et leur classification. *Ann. Géogr.* no. 355: 193–220.

BAGNOULS, F. & H. GAUSSEN, – 1964 – Les climats tropicaux et végétation. *Adansonia* 4: 262–268.

BALI, Y. P., – 1971 – Use of soil surveys for planning agriculture. In 'Soil Survey and Soil Fertility Research in Asia and the Far East,' pp. 130–132. FAO, Rome.

BAULKWILL W. J., – 1972 – The Land Resources Division of the Overseas Development Administration. *Tropical Science* 14: 305–322.

BAUR, G. N., – 1964 – The Ecological Basis of Rainforest Management. Sydney. 510 pp.

BECKINSALE, R. P., – 1965 – Climatic change: a critique of modern theories. In 'Essays in Geography for Austin Miller,' pp. 1–38. University of Reading.

BHARUCHA, F. R., – 1955 – Afghanistan, India and Pakistan. In 'UNESCO Rev. Res. Plant Ecol.' pp. 19–39. UNESCO, Paris

BLACK, C. A., – 1968 – Soil Plant Relationships (2nd ed.). 792 pp.

BLASCO, F., – 1971 – Montagnes du Sud de l'Inde. *Inst. fr. Pondichéry, Trav. Sec. sci. tech.* 10: 1–436

BOAZ, M., A. OLMERT, M. YARON, M. GRINBERG, M. SHANI & A. GADIEL, – 1971– Village Irrigation Programmes : A New Approach in Water Economy. Irrigation and Drainage. Paper 4. FAO, Rome. 26pp.

BOURKE, P. M. Austin, – 1968 – Introduction: the aims of agrometeorology. In 'Agroclimatological Methods: Proceedings of the Reading Symposium,' pp. 11–15. Natural Resources Research 7. UNESCO, Paris.

BOURLIERE, F. & M. HADLEY, – 1970 – The ecology of tropical savannas. *Ann. Rev. Ecol. Syst.* 1: 125–152.

BRAUN-BLANQUET, J., – 1951 – Pflanzensoziologie (2nd ed.). Grundzuge der Vegetationskunde. Springer-Verlag, Wien.

BRAY, J. R., – 1971 – Vegetational distribution, tree growth and crop success in relation to recent climatic change. *Advances in Ecological Res.* 7: 177–233.

BRITISH ECOLOGICAL SOCIETY, – 1970 – Animal Populations in Relation to their Feed Resources (Symposium). Blackwell, Oxford.

BUCHANAN, K., – 1970 – The Transformation of the Chinese Earth. Bell, London. 336 pp.

BURDON, D. J., – 1971 – Exploitation of groundwater for agricultural production in arid zones. In 'Food, Fiber and the Arid Lands,' ed. W. G. McGinnies, B. J. Goldman & P. Paylore, pp. 289–300. University of Arizona Press, Tucson.

CENTO, – 1966 – Symposium on Hydrology and Water Resources Development, Ankara, Turkey. Office of U.S. Economic Coordinator for CENTO Affairs, Ankara. 484 pp.

CHAMPION, H. G., – 1969 – The effect of human population on the forests of the Indian sub-continent. In 'Problems of Land Use in South Asia. Yearbook of the South Asia Institute, Heidelberg University, 1968–69,' ed. U. Schweinfurth & M. Domrös, pp. 19–28. Otto Harrassowitz, Wiesbaden.

CHAMPION, H. G. & S. K. SETH, – 1968 – A Revised Survey of the Forest Types of India. Government of India, Manager of Publications, Delhi. 404 pp.

CHANG, Jen-Hu, – 1968 – Climate and Agriculture. An Ecological Survey. Aldine, Chicago. 304 pp.

CHAPMAN, J., – 1973 – A Manual on Establishment Techniques in Man-Made Forests. FAO, Rome, 114 pp.

CHEN, C. S., – 1970 – The Agricultural Regions of China, American Institute of Crop Ecology, Silver Spring, Md. 47 pp.

CHIA, Lin-Sien, – 1973 – Reliability of West Malaysian rainfall records: an objective evaluation. *J. Trop. Geogr*. 36: 1–7.

CHISHOLM, M., A. E. FREY & P. HAGGETT (eds.), – 1971 – Regional Forecasting: Proceedings Twenty-Second Symposium of Colston Research Society, University of Bristol, 1970. Butterworth, London. 480 pp.

CHORLEY, R. J. (ed.), – 1969 – Water, Earth and Man. A Synthesis of Hydrology, Geomorphology and Socio-Economic Geography. Methuen, London. 588 pp.

CHRISTIAN, C. S. & G. A. STEWART, – 1968 – Methodology of integrated surveys. In 'Aerial Surveys and Integrated Studies. Proceedings of the Toulouse Conference,' pp. 233–280. UNESCO, Paris.

CLARKSON, J. D., – 1968 – The Cultural Ecology of a Chinese Village; Cameron Highlands, Malaysia. University of Chicago, Department of Geography Research Paper 114. 174 pp.

COCHEME, J., – 1968 – FAO/UNESCO/WMO agroclimatology survey of a semi-arid area in West Africa south of the Sahara. In 'Agroclimatological Methods: Proceedings of the Reading Symposium,' pp. 235–247. UNESCO, Paris.

COMMONWEALTH BUREAU OF SOILS, – 1974 – Methods for Analysis of Irrigated Soils. Technical Communication no. 54, Commonwealth Agricultural Bureaux, Farnham Royal/CSIRO, Melbourne. 208 pp.

COMMONWEALTH SCIENTIFIC AND INDUSTRIAL RESEARCH ORGANIZATION, – 1964 – Some Concepts and Methods in Subtropical Pasture Research. Bulletin 47, Commonwealth Bureau of Pastures and Field Crops, Hurley, 254 pp. (390 references).

CONDON, R. W., – 1968 – Estimation of grazing capacity on arid grazing lands. In 'Land Evaluation – CSIRO Symposium,' ed. G. A. Stewart, pp. 112–124. Macmillan, London.

CONSEIL SCIENTIFIQUE POUR L'AFRIQUE AU SUD DU SAHARA, – 1956 – Rapport final de la Réunion de Spécialistes en Phytogéographie/Final Report of the Specialist Meeting on Phytogeography, Yangambi, Belgian Congo. Joint Secretariat, Bukavu. 379 + 32 pp.

COSTIN, A. B. & J. C. I. DOOGE, – 1973 – Balancing the effects of man's actions on the hydrological cycle. In 'Man's Influence on the Hydrological Cycle,' pp. 19–51. Irrigation and Drainage Paper no. 17. FAO, Rome.

COUGET, Y., – 1971 – Draft three-year Indo-French Co-operative programme for the agricultural development in the semi-arid zone of Andhra Pradesh, 1970–1972. Société d'Aide technique et de Coopération, Paris.

COUPLAND, R. T. & G. M. van DYNE (eds.), – 1970 – Grassland Ecosystems: Reviews of Research. Proceedings of the Second Meeting of the PT Grassland Working Group, IBP, Saskatoon and Matador, Canada, September 5 to 10, 1969. Range Science Department Science Ser. no. 7, Colorado State University, Fort Collins. 208 pp.

COX, I. H. (ed.), – 1970 – New Possibilities and Techniques for Land Use and Related Surveys. The World Land Use Survey, Occasional Paper no. 9. Geographical Publications, Berkhamsted. 138 pp.

CRUTCHER, H. L., S. B. NASH & D. K. KROPP, – 1969 – A Note on Climatology of Thailand and Southeast Asia. ESSA Technical Memorandum EDSTM 10. U.S. Department of Commerce, Silver Spring, Md. 166 pp.

DAY, T. H., – 1971 – Soil survey needs for irrigation project development. In 'Soil Survey and Soil Fertility Research in Asia and the Far East,' pp. 182–184. FAO, Rome.

DERBYSHIRE, E. (ed.), – 1973 – Climatic Geomorphology. Geographical Readings Ser. Macmillan, London, 296 pp.

DESCOINGS, B., – 1971 – Méthode de description des formations herbeuses intertropicales par la structure de la végétation. *Candollea* 26: 223–257.

DESCOINGS, B., – 1972 – Note sur la structure de quelques formations herbeuses de Lamto (Côte-d'Ivoire). *Ann. Univ. Abidjan sér. E (Écologie)* 5: 7–30.

DIXEY, F., – 1966 – Water supply, use and management. In 'Arid Lands,' ed. E. S. Hills, pp. 77–102. Methuen/UNESCO, London and Paris.

DOBBY, E. H. G., – 1961 – Monsoon Asia. University of London Press, 381 pp.

DOMRÖS, M., – 1971 – Der Monsun im Klima der Insel Ceylon. *Die Erde* 102: 118–140.

DOOGE, J. C. I., – 1973 – The nature and components of the hydrological cycle. In 'Man's Influence on the Hydrological Cycle,' pp. 2–18. Irrigation and Drainage Paper no. 17, FAO, Rome.

DUDAL, R. & F. R. MOORMANN, – 1964 – Major soils of south-east Asia. *J. Trop. Geogr.* 18: 54–80.

DUDAL, R., F. MOORMANN & J. RIQUIER, – 1974 – Soils of humid tropical Asia. In 'Natural Resources of Humid Tropical Asia,' pp. 159–178. Natural Resources Research 12. UNESCO, Paris.

DUFFEY, E. & A. S. WATT (eds.), 1971 – The Scientific Management of Animal and Plant Communities for Conservation. Eleventh Symposium of the British Ecological Society. Blackwell, Oxford. 652 pp.

EDNEY, E. B., – 1966 – Animals of the desert. In 'Arid Lands,' ed. E. S. Hills, pp. 181–218. Methuen/UNESCO, London and Paris.

EIMERN, J., van – 1968a – The topoclimate and its mapping for agricultural purposes. In 'Proceedings Regional Training Seminar on Agrometeorology, 13–15 May, Wageningen,' pp. 213–220. WMO, Geneva.

EIMERN, J., van – 1968b – Some hints on the use of instruments and equipment for topoclimatological work in the field. In 'Proceedings Regional Training Seminar on Agrometeorology, Wageningen,' pp. 289–293. WMO, Geneva.

EIMERN, J., van – 1968c – Method and techniques for the mapping of the topographical distribution of the air temperature. In 'Proceedings Regional Training Seminar on Agrometeorology, Wageningen,' pp. 319–342. WMO, Geneva.

ELLENBERG, H., – 1954 – Naturgemässe, Anbauplanung, Melioration und Landespflege. Landwirtschaftliche Pflanzensoziologie III. Stuttgart. 109 pp.

ELLENBERG, H. & J. LEBRUN, – 1970 – Natural vegetation and its management for rational land use. In 'Use and Conservation of the Biosphere,' pp. 105–122. UNESCO, Paris.

ELTON, C. S., – 1966 – The Pattern of Animal Communities. Methuen, London. 432 pp.

ELTON, C. S., – 1969a – Animal Ecology. Chapman and Hall, London. 242 pp.

ELTON, C. S., – 1969b – The Ecology of Invasions by Animals and Plants. Chapman and Hall, London. 182 pp.

ELTON, C. S., – 1969c – The Ecology of Animals. Chapman and Hall, London. 98 pp.

ELTON, C. S., – 1973 – The structure of invertebrate populations inside Neotropical Rain Forest. *J. Anim. Ecol.* 42:

EMBERGER, L., – 1955 – Une classification biogéographique des climats. Rec. des trav. Fac. Sci. de l'Univ. de Montpellier. 3–43.

FLENLEY, J. R. (ed.), – 1971 – The Water Relations of Malesian Forests. Trans. First Aberdeen–Hull Symposium on Malesian Ecology, Hull, 1970. University of Hull, Dept. Geogr. misc. ser. no. 11. 97 pp.

FLOHN, H., – 1970 – Elements of a climatology of the Indo-Pakistan sub-continent. *Bonner Meteorologische Abhandlungen* 14: 5–28.

FLOHN, H., M. HANTEL & E. RUPRECHT, – 1970 – Investigations on the Indian Monsoon Climate. *Bonner Meteorologische Abhandlungen* 14: 1–99.

FOOD AND AGRICULTURE ORGANIZATION OF THE UNITED NATIONS, – 1970 – Water development for agriculture. In 'International Cooperation in the Development of Water Resources for Agriculture,' pp. 168–182. International Commission on Irrigation and Drainage, New Delhi.

FAO, – 1971a – Soil Survey and Soil Fertility Research in Asia and the Far East. World Soil Resources Reports no. 41. FAO, Rome. 229 pp.

FAO, – 1971b – Village Irrigation Programmes: A New Approach in Water Economy. Irrigation and Drainage Paper no. 4. FAO, Rome. 26 pp.

FAO, -- 1973 – Man's Influence on the Hydrological Cycle. Irrigation and Drainage Paper no. 17. FAO, Rome, 71 pp.

FAO, – 1974 – Approaches to Land Classification. Soils Bulletin no. 22. FAO, Rome. 120 pp.

FAO/IAEA Division of ATOMIC ENERGY IN FOOD AND AGRICULTURE, – 1968 – Isotopes and Radiation in Soil Organic Matter Studies: Proceedings of a Symposium, Vienna, July 1968. FAO/IAEA, Rome and Vienna. 376 pp.

FAO/IAEA DIVISION OF ATOMIC ENERGY IN FOOD AND AGRICULTURE, – 1971 – Nitrogen-15 in Soil-Plant Studies: Proceedings of a Research Co-ordination Meeting, Sofia, December, 1969. FAO/IAEA, Rome and Vienna. 258 pp.

FAO/UNESCO, – 1973 – Irrigation, Drainage and Salinity: An International Source Book. Hutchinson/FAO/UNESCO, London, Rome, Paris. 510 pp.

FOSBERG, F. R., B. J. GARNIER & A. W. KÜCHLER, – 1961 – Delimitation of the humid tropics. *Geogr. Rev.* 51: 333–347.

FRANQUIN, P., – 1968 – Agroclimatic analysis in tropical regions. Soil-moisture conditions. *Cah. ORSTOM, Sér. Biol.* 5: 15–24.

FRANQUIN, P., – 1969 – Agroclimatic analysis in tropical regions. Rainy season and wet season. *Cah. ORSTOM, Sér. Biol.* 9: 65–95.

FRANQUIN, P., – 1970a – Modèles mathématiques de structures chez les végétaux. I. Principes de structure et de production de nombre. *Cah. ORSTOM, Sér. Biol.* 14: 77–126.

FRANQUIN, P., – 1970b – L'estimation par interpolation des valeurs d'une variable climatique. *Cah. ORSTOM, Sér. Biol.* 14: 127–147.

FRANQUIN, P., – 1972 – Modèles mathématiques de structures chez les végétaux. II. Relations de structure. *Cah. ORSTOM, Sér. Biol.* 17: 3–21.

FRENCH, N. R. (ed.), – 1971 – Preliminary Analysis of Structure and Function in Grasslands. Proceedings Symposium on The Grassland Biome, Colorado State University, September, 1971. Colorado State University Science Series, Range Science Dept., Fort Collins. 387 pp.

FUKUI, H., – 1971a – Rice culture in the central plain of Thailand. V. Possibility of higher yield viewed from the yield component surveys in farmers' fields. *Tonan Ajia kenkyu* 8: 518–533.

FUKUI, H., – 1971b – Environmental determinants affecting the potential dissemination of high yielding varieties of rice: a case study of the Chao Phraya river basin. *Tonan Ajia kenkyu* 9: 348–374.

GARNIER, B. J., – 1958 – Some comments on defining the humid tropics. *University College, Ibadan, Dept. Geogr. Res. Notes no.* 11: 9–25.

GARNIER, B. J., – 1960 – Delimiting the humid tropics. In 'ICSU Review,' vol. 2, pp. 210–218. International Council of Scientific Unions.

GATES, D. M., – 1972 – Man and His Environment: Climate. Harper and Row, New York, London. 175 pp.

GAUSSEN, H., – 1954 – Expression du milieu par des formules écologiques. Leur représentation cartographique. In 'Les Divisions Ecologiques du Monde. Colloques Intern. du CNRS,' Paris, and *Ann. Biol.* 31: 257–267.

GAUSSEN, H., – 1959 – The vegetation maps. *Inst. Fr. Pondichéry, Trav. Sect. Sci. Tech.* 1: 155–179.

GAUSSEN, H., P. LEGRIS, F. BLASCO et al., – 1967 – Bioclimats du Sud-Est Asiatique. *Inst. Fr. Pondichéry, Trav. Sect. Sci. Tech.* 3(4): 1–112.

GINDEL, I., – 1973 – A New Ecophysiological Approach to Forest-Water Relationships in Arid Climates. Junk, The Hague. 142 pp.

GOMEZ-POMPA, A., C. VAZQUEZ-YANES & S. GUEVARA, – 1972 – The tropical rain forest: a nonrenewable resource. *Science* 177: 762–765.

GOVINDA RAJAN, S. V., – 1971 – Soil survey organization and service for land use planning. In 'Soil Survey and Soil Fertility Research in Asia and the Far East,' pp. 176–181. FAO, Rome.

GRASSLAND RESEARCH INSTITUTE, HURLEY, – 1961 – Research Techniques in Use at the Grassland Research Institute, Hurley. Bull. 45, Commonwealth Bureau of Pastures and Field Crops, Hurley. 167 pp. (127 references).

GUINOCHET, M., – 1973 – Phytosociologie. Masson, Paris. 228 pp.

GUPTA, R. K., – 1964 – The bioclimatic types of western Himalayas and their analogous types towards the mountain chains of the Alps and the Pyrenees. *Ind. For.* 90: 551–577.

GUPTA, R. K. & V. M. MEHER–HOMJI, – 1971 – Bibliography on plant-ecology in Pakistan. *Excerpta Botanica* B 11: 183–208.

HAFFNER, W., – 1973 – Brachsysteme und zelgengebundener Anbau in Zentral-und Ostnepal. In 'Vergleichende Kulturgeographie der Hochgebirge des Südlichen Asien,' ed. C. Rathjens, C. Troll & H. Uhlig, pp. 40–42. Franz Steiner Verlag, Wiesbaden.

HALLAIRE, M., C. P. de BRICHAMBAUT & C. GOILLOT, – 1970 – Techniques d'Etude des Facteurs physiques de la Biosphère. Service de Publications INRA, Paris. 543 pp.

HALLAM, A., – 1973 – A Revolution in the Earth Sciences: from Continental Drift to Plate Tectonics. Clarendon Press, Oxford. 127 pp.

HAMID, Q., T. H. SIDDIQUI & S. J. -U HASSAN, – 1969 – Agro-climatic classification of West Pakistan. *Agriculture Pakistan* 20: 423–435.

HANTEL, M., – 1970 – Monthly charts of surface wind vergence over the tropical Indian Ocean. *Bonner Meteorologische Abhandlungen* 14: 32–79.

HANYU, J., – 1971 – The stabilities of temperature and rice yield during the cultivation period in Hokkaido. 1. The critical conditions of temperature for the safe cultivation of paddy rice. *Res. Bull. Hokkaido Nat. Agric. Exper. Station* no. 98: 1–11.

HANYU, J. & T. ISHIGURO, – 1972 – Temperature and rice yield stability during the cultivation period in Hokkaido. 2. Relation of the fluctuation of rice yield to the temperature during the safe cultivation period. *Res. Bull. Hokkaido Nat. Agric. Exper. Station* no. 102: 93–103.

HANYU, J. & T. UCHIJIMA, – 1970 – A consideration on the climatic productivity of paddy rice in the northern districts of Japan. *J. Agric. Meteorology, Tokyo* 25: 241–246.

HANYU, J., T. UCHIJIMA, T. SAITO & S. SUGAWARA, – 1966 – Agro-meteorological studies on the determination of suitable regions and periods for the direct seeding cultivation of rice plants in northern part of Japan. *Bull. Tohoku Nat. Agric. Exper. Station* no. 34: 1–25.

HANYU, J., T. UCHIJIMA & S. SUGAWARA, – 1966 – Studies on the agro-climatological method for expressing the paddy rice products. 1. An agro-climatic index for expressing the quantity of ripening of the paddy rice. *Bull. Tohoku Nat. Agric. Exper. Station* no. 34: 27–36.

HILLS, E. S., C. D. OLLIER & C. R. TWIDALE, – 1966 – Geomorphology. In 'Arid Lands,' ed. E. S. Hills, pp. 53–76. Methuen/UNESCO, London and Paris.

HILLS, T. L. & R. E. RANDALL (eds.), – 1968 – The Ecology of the Forest/Savanna Boundary. Proceedings IGU Humid Tropics Comm. Symp., Venezuela, 1964. McGill University, Savanna Res. Ser., Montreal. 128 pp.

HOLDRIDGE, L. R., – 1966 – The life zone system. *Adansonia* 6: 199–203.

HOLDRIDGE, L. R., – 1971 – Forest Environments in Tropical Life Zones. Pergamon Press.

HUANG Ping-wei, – 1961 – The Complex Natural Zonation of China. USSR Academy of Sciences, Geographical Series.

HUSCH, B., – 1971 – Planning a Forest Inventory. FAO Forestry & Forest Products Studies no. 17. FAO, Rome. 120 pp.

IGNATYEV, G. M., – 1968 – Classification of cultural and natural vegetation sites as a basis for land evaluation. In 'Land Evaluation,' ed. G. A. Stewart, pp. 104–111. Macmillan, London.

INDIAN AGRICULTURAL RESEARCH INSTITUTE, – 1970 – A New Technology for Dry Land Farming. IARI, Delhi. 189 pp.

INDIAN COUNCIL OF AGRICULTURAL RESEARCH, – 1961 – Handbook of Agriculture. ICAR, New Delhi. 761 pp.

ICAR, – 1971 – Soil and Water Research in India in retrospect and Prospect (a Review). Tech. Bull. (agric.) no. 22. New Delhi. 391 pp.

INTERNATIONAL ASSOCIATION OF HYDROLOGICAL SCIENCES/UNESCO/WMO, – 1972 – World Water Balance: Proceedings of the Reading Symposium. UNESCO, Paris. 169 pp.

INTERNATIONAL COMMISSION ON IRRIGATION AND DRAINAGE – 1970 – International Co-operation in the Development of Water Resources for Agriculture. ICID, New Delhi.

INTERNATIONAL UNION FOR THE CONSERVATION OF NATURE – 1970 – Eleventh Technical Meeting, New Delhi, November, 1969. Vol. 1. First and Second Sessions: Commission on Ecology. 282 pp. Vol. 2. Third Session. Survival Service Commission. 132 pp. Vol. 3. National Parks in southern Asia. 134 pp. Vol. 4. Environmental conservation education. 156 pp. Vol. 5. Creative conservation in an agrarian economy. 98 pp. IUCN publications new series, nos. 17, 18, 19, 20 and 21. Morges, Switzerland.

JAGER, J. M. de, – 1971 – Theory for the assessment of environmental potential. *Agrochemophysica* 3(4): 67–69.

JAIN, S. K., – 1970 – Floral composition of Rajasthan: a review. *Bull. Bot. Survey India*: 12: 176–187.

JANZEN, D. H., – 1973 – Tropical agroecosystems. *Science* 182: 1212–1219.

JEWITT, T. N., – 1966 – Soils of arid lands. In 'Arid Lands,' ed. E. S. Hills, pp. 103–125. Methuen/UNESCO, London and Paris.

KANWAR, J. S. & I. C. MAHAPATRA, – 1971 – Soil fertility research in India. In 'Soil Survey and Soil Fertility Research in Asia and the Far East,' pp. 77–84. FAO, Rome.

KASSAS, M., – 1966 – Plant life in deserts. In 'Arid Lands,' ed. E. S. Hills, pp. 145–180. Methuen/UNESCO, London and Paris.

KNAPP, R. (ed.), – 1974 – Vegetation Dynamics: Handbook of Vegetation Science, Part 8. Junk, The Hague. 368 pp.

KOVDA, V. A., – 1960 – Soils and Natural Environments of China. U.S. Joint Publications Research Service, Washington.

KOZLOWSKI, T. T. & C. E. AHLGREN (eds.), – 1974 – Fire and Ecosystems. Academic Press, New York. 542 pp.

KUNG, P., – 1971 – Irrigation Agronomy in Monsoon Asia. FAO, Rome. 106 pp.

KYOTO UNIVERSITY, Center for Southeast Asian Studies, – 1966 – Water Resource Utilization in Southeast Asia. Symposium Series III. Kyoto University. 236 pp.

KYUMA, K., – 1971 – Climate of south and southeast Asia according to Thornthwaite's classification scheme. *Tonan Ajia kenkyu* 9: 136–158.

KYUMA, K., – 1972 – Numerical classification of the climate of south and southeast Asia. *Tonan Ajia kenkyu* 9: 502–521.

LABROUE, L., P. LEGRIS & M. VIART, – 1965 – Bioclimats du sous-continent Indien. *Inst. fr. Pondichéry Trav. Sect. Sci. Tech.* 3(3): 1–32.

LAMB, H. H., – 1972 – Climate: Present, Past and Future. Vol. 1. Fundamentals and Climate Now. Methuen, London/Harper and Row, New York. 613 pp.

LAND RESOURCES DIVISION, Overseas Development Administration, – 1973 Fiji Forest Inventory. Vol. 1. The Environment and Forest Types. 94 pp. + map. Vol. 2. Catchment Groups of Viti Levu and Kandavu. 62 pp. Vol. 3. Catchment Groups of Vanua Levu. 88 pp. + map. Land Resource Study no. 12, Tolworth Tower, Surrey.

LEGRIS, P., – 1965 – La végétation de l'Inde: Ecologie et flore. *Inst. Fr. Pondichéry. Trav. Sect. Sci. Tech.* 6: 1–589.

LEGRIS, P., – 1972 – A new formula of hydric balance. *Trop. Ecol.* 13: 12–26.

LEGRIS, P. & F. BLASCO, – 1969 – Variabilité des facteurs du climat: Cas des montagnes du Sud de l'Inde et de Ceylan. *Inst. Fr. Pondichéry. Trav. Sect. Sci. Tech.* 8: 1–95.

LEGRIS, P., Y. COUGET & V. M. MEHER-HOMJI, – 1971 – Drought climatology of Anantapur district, A. P. In 'Proceedings 58th Indian Science Congress,' vol. III. p. 379.

LEGRIS, P. & V. M. MEHER-HOMJI, – 1968 – Vegetation maps of India. In 'Proceedings Symposium Recent Advances in Tropical Ecology,' pp. 32–41. International Society for Tropical Ecology, Varanasi.

LEGRIS, P. & M. VIART, – 1961 – Bioclimates of South India and Ceylon. *Inst. Fr. Pondichéry. Trav. Sect. Sci. Tech.* 3: 165–178.

LETTAU, H. Evapotranspiration climatonomy. 1. A new approach to numerical prediction of monthly evapotranspiration, runoff and soil moisture storage. *Mon. Weath. Rev.* 97: 691–699.

LIKENS, G. E., F. H. BORMANN & N. M. JOHNSON, – 1969 – Nitrification: importance to nutrient losses from a cutover forested ecosystem. *Science* 163: 1205–1206.

LONG, G. A., – 1974 – Diagnostic Phyto-écologique et Aménagement du Territoire. I. Principes généraux et Méthodes. Recueil, Analyse, Traitement et Expression cartographique de l'Information. Masson, Paris. 256 pp.

LONG, G. A., – 1975 – Diagnostic Phyto-écologique et Aménagement du Territoire. 2. Application du Diagnostic phyto-écologique. Masson, Paris.

LONG, G. A., P. S. POISSONET, J. A. POISSONET, P. M. DAGET & M. P. GODRON, – 1972 – Improved needle point frames for exact line transects. *J. Range Management* 25: 228–229.

LOOMIS, R. S. & W. A. WILLIAMS, – 1963 – Maximum crop productivity: an estimate. *Crop Science* 3: 67–72.

MAAREL, E. van der & R. TÜXEN – 1972 – Grundfragen und Methoden in der Pflanzensoziologie. Junk, The Hague. 521 pp.

MACFADYEN, A., – 1963 – Animal Ecology (2nd ed.). Pitman, London. 344 pp.

MAMEDOV, O., – 1971 – Burma: Agrochemical Study of Burmese Soils. FAO UNDP TA Report no. 2956. FAO, Rome. 19 pp.

MARKS, P. L. & F. H. BORMANN, – 1972 – Revegetation following forest cutting: mechanisms for returning to steady-state nutrient cycling. *Science* 176: 914–915.

MAUNDER, W. J., – 1962 – Rating scheme for evaluation of human climate index. *Weather* 17:9.

MEDWAY, Lord, – 1965 – Mammals of Borneo: field keys and an annotated checklist. *J. Malayan Br. Roy. Asiatic. Soc.* 36: 1–193.

MEDWAY, Lord, – 1969 – Wild Mammals of Malaya and Offshore Islands. Oxford University Press, Kuala Lumpur.

MEDWAY, Lord, – 1971 – The Quaternary mammals of Malesia: a review. In 'The Quaternary Era in Malesia. Transactions Second Aberdeen-Hull Symposium on Malesian Ecology,' ed. Peter & Mary Ashton, pp. 63–77. University of Hull, Dept. Geography Misc. Ser. no. 13.

MEGGERS, B. J., E. S. AYENSU & W. D. DUCKWORTH, (eds.) – 1973 – Tropical Forest Ecosystems in Africa and South America: A Comparative Review. Smithsonian Institution Press, Washington. 350 pp.

MEHER-HOMJI, V. M., – 1960 – Les bioclimats du sub-continent indien et leurs types analogues dans le monde. Thèse Science, Toulouse.

MEHER-HOMJI, V. M., – 1963 – Les bioclimats du sub-continent indien et leurs types analogues dans le monde. *Inst. Fr. Pondichéry, Trav. Sect. Sci. Tech.* 7: 1–386.

MEHER-HOMJI, V. M., – 1964 – Drought: its ecological definition and phyto-geographic significance. 1. Ecological definition. *Tropical Ecology* 5: 17–31.

MEHER-HOMJI, V. M., – 1965 – Drought: its ecological definition and phyto-geographic significance. 2. Phytogeographic significance of drought. *Tropical Ecology* 6: 19–33.

MEHER-HOMJI, V. M., – 1967 – On delimiting arid and semi-arid climates in India. *Indian Geographical J.*: 42: 1–6.

MEHER-HOMJI, V. M., – 1968a – Application of pluviothermic quotient and xerothermic-hygrothermic indices to the Indian sub-continent. *Indian For.* 94: 351–362.

MEHER-HOMJI, V. M., – 1968b – Variability, an aspect of bioclimatology with reference to the Indian sub-continent. In 'Proceedings Symposium Recent Advances in Tropical Ecology,' eds. R. Misra & B. Gopal, vol. 1, pp. 144–153. International Society for Tropical Ecology, Varanasi.

MEHER-HOMJI, V. M., – 1971a – Bioclimatic variability. *Tropical Ecology* 12: 155–176.

MEHER-HOMJI, V. M., – 1971b – Analogous bioclimates and introduction of economic exotics. *J. Bombay Nat. Hist. Soc.* 67: 398–413.

MEHER-HOMJI, V. M., – 1973 – A phytoclimatic approach to the problem of Mediterraneity in the Indo-Pakistan sub-continent. *Feddes Repertorium* 83: 757–788.

MEHER-HOMJI, V. M., – 1974a – Variability and the concept of a probable climatic year with particular reference to the Indian sub-continent. *Arch. Met. Geoph. Biokl.*

MEHER-HOMJI, V. M., – 1974b – The climate of Cuddalore: a bio-climatic analysis. *Geogr. Rev. India.*

MEHER-HOMJI, V. M. & R. K. GUPTA, – 1971 – Bibliography on plant-ecology in India. 3. *Excerpta Botanica* B, 11: 161–182.

MEHER-HOMJI, V. M. & K. C. MISRA, – 1973 – Phytogeography of the Indian subcontinent. In 'Progress of Plant Ecology in India,' eds. R. Misra, B. Gopal, K. P. Singh & J. Singh, vol. 1, pp. 9–88, Today and Tomorrow's Printers and Publishers, New Delhi.

MEHER-HOMJI, V. M. & N. P. PERERA, – 1972 – Bibliography on plant-ecology in Ceylon. *Excerpta Botanica* B. 12: 147–157.

MEIGS, P., – 1953 – World distribution of arid and semi-arid homoclimates. In 'Reviews of Research on Arid Zone Hydrology,' pp. 203–210. Arid Zone Research 1. UNESCO, Paris.

MILNER, C. & R. E. HUGHES, – 1968 – Methods for the Measurement of Primary Production of Grassland. IBP Handbook no. 6. Blackwell, Oxford. 70 pp.

MISRA, R. (ed.), – 1974 – Proceedings of IBP Symposium and Synthesis Meetings on Tropical Grassland Biome. Banaras Hindu University, Varanasi. 174 pp.

MISRA, R. & B. GOPAL (eds.), – 1968 – Proceedings of the Symposium on Recent Advances in Tropical Ecology. Parts 1 and 2. International Society for Tropical Ecology, Varanasi. 773 pp.

MOOLEY, D. A. & H. L. CRUTCHER, – 1968 – An Applicaton of the Gamma Distribution Function to Indian Rainfall. ESSA Technical Report, EDS 5. Department of Commerce, Silver Spring. 47 pp.

MUELLER-DOMBOIS, D., – 1968 – Ecogeographic analysis of a climate map of Ceylon with particular reference to vegetation. *Ceylon For.* 8: 1–20 + map.

MUNN, R. E., – 1970 – Biometeorological Methods. Academic Press, New York and London. 336 pp.

MURATA, Y., – 1964 – On the influence of solar radiations and air temperature upon the local differences in the productivity of paddy rice in Japan. *Proc. Crop Sci. Soc. Japan* 33: 59–63.

MURTHY, R. S., – 1971 – Acid sulphate soils in India. In 'Soil Survey and Soil Fertility Research in Asia and the Far East,' pp. 24–29. FAO, Rome.

NEWMAN, J. E. & WANG Jen-Yu, – 1959 – Defining agricultural seasons in the middle latitudes. *Agron. J.* 5: 579–582.

NOY-MEIR, I., – 1973 – Desert ecosystems: environment and producers. *Ann. Rev. Ecol. Systematics* 4: 25–51.

NUTTONSON, M. Y., – 1947 - Ecological Crop Geography of China and its Agro-climatic Analogues in North America. American Institute of Crop Ecology, Washington. Study no. 7. 28 pp.

NUTTONSON, M. Y., – 1949 – Agricultural Climatology of Japan and its Agro-climatic Analogues in North America. American Institute of Crop Ecology, Washington. Study no. 9. 20 pp.

NUTTONSON, M. Y., – 1952 – Ecological Crop Geography and Field Practices of the Ryukyu Islands. Natural Vegetation of the Ryukyus and Agro-climatic Analogues in the Northern Hemisphere. American Institute of Crop Ecology, Washington. 106 pp.

NUTTONSON, M. Y., – 1970 – Some general observations on the physiography and climate of China. In 'The Agricultural Regions of China,' by C. S. Chen, pp. 20–47. American Institute of Crop Ecology, Silver Spring, Md.

OHYA, M., – 1966 – Comparative study of the geomorphology and flooding in the plains of the Cho-Shui-Chi, Chao-Phya, Irrawaddy and Ganges. In 'Scientific Problems of the Humid Tropical Zone Deltas and their Implications,' pp. 23–28. Humid Tropics Research. UNESCO, Paris.

OJO, O., – 1969 – Potential evapotranspiration and the water balance in West Africa. *Arch. Met. Geoph. Biokl. Ser. B* 17: 239–260.

OLIVER, J., – 1969 – Problems of determining evapotranspiration in the semi-arid tropics illustrated with reference to the Sudan. *J. Trop. Geogr.* 28: 64–74.

PALIT, S. E., – 1970 – Forests and land use in India. *Ind. For.* 96: 339–350.

PANDEY, S., – 1969 – Some aspects of arid zone geomorphology. *Ann. Arid Zone* 8: 196–208.

PANDEYA, S. C., S. M. PANDYA, M. S. MURTHY & KURUVILLA, – 1969 – Forest ecosystem – classification of forest vegetation with reference to forests in the River Narmada catchment area. *J. Ind. Bot. Soc.* 46: 412–427.

PANTON, W. P., – 1964 – The 1962 soil map of Malaysia. *J. Trop. Geography* 18:118–124.

169

PAPUA AND NEW GUINEA: ADMINISTRATION OF THE TERRITORY OF/UNESCO, Southeast Asia, – 1960 – Symposium on the Impact of Man on Humid Tropics Vegetation. UNESCO, Southeast Asia. 402 pp.

PARKINSON, D., T. R. G. GRAY & S. T. WILLIAMS, – 1971 – Methods for Studying the Ecology of Soil Microorganisms. IBP Handbook no. 19. Blackwell, Oxford, London, Edinburgh and Melbourne. 128 pp.

PATNAIK, S., – 1971 – Laterite soils in India. In 'Soil Survey and Soil Fertility Research in Asia and the Far East,' pp. 52–58. FAO, Rome.

PENMAN, H. L., – 1948 – Natural evaporation from open water, bare soil and grass. *Proc. Roy. Soc. London, ser. A.* 193: 120–45.

PEREIRA, H. C., – 1970 – Water resources problems: present and future requirements for life. In 'Use and Conservation of the Biosphere,' pp. 71–85. UNESCO, Paris.

PERERA, N. P., – 1968 – Some problems of climate-vegetation correlations with special reference to Ceylon. *Vidyodaya J. Arts. Sci. Lett.* 1: 173–184.

PHILLIPSON, J. (ed.), – 1971 – Methods of Study in Quantitative Soil Ecology (Population, Production and Energy Flow). IBP Handbook no. 18. Blackwell, Oxford, London, Edinburgh and Melbourne. 308 pp.

POISSONET, J. & J. CESAR, – 1972 – Structure spécifique de la strate herbacée dans la savane à palmier Ronier de Lamto (Côte-d'Ivoire). *Ann. Univ. Abidjan sér. E (Ecologie)* 5: 577–601.

POISSONET, P. S., P. M. DAGET, J. A. POISSONET & G. LONG, – 1972 – Rapid point survey by bayonet blade. *J. Range Management* 25: 313.

POISSONET, P. S., J. A. POISSONET, M. P. GODRON & G. LONG, – 1973 – A comparison of sampling methods in dense herbaceous pasture. *J. Range Management* 26: 65–67.

PRAKASH, Mahendra, – 1972 – Forestry in Rajasthan. *Ind. For.* 98: 698–703.

PRIMAULT, B., – 1969a – D'une application pratique des indices bio-météorologiques. *Agricultural Meteorology* 6: 71–96.

PRIMAULT, B., – 1969b – Essai de développement d'une méthode mathématique pour la détermination des indices bio-météorologiques. *Recherche agronomique en Suisse* 8: 380–398.

PRIMAULT, B., – 1970 – Nouvelle conception de la présentation pour l'agriculture, des series climatologiques. *Public. OEPP., sér. A.* no. 57: 41–53.

RAHEJA, P. C., – 1966 – Soil Productivity and Crop Growth. Asia Publishing, New York. 474 pp.

RAMAGE, C. S., – 1971 – Monsoon Meteorology. Academic Press, New York, London. 296 pp.

RAMAMOORTHY, B., V. N. PATHAK & BAJAJ, – 1971 – Soil testing and tissue testing techniques in India. In 'Soil Survey and Soil Fertility Research in Asia and the the Far East,' pp. 85–90. FAO, Rome.

RAMAMOORTHY. B. & M. VELAYUTHAM, – 1971 – Soil test crop response correlation work in India. In 'Soil Survey and Soil Fertility Research in Asia and the Far East,' pp. 96–102. FAO, Rome.

RAMASWAMY, C., – 1967 – The problem of the Indian southwest monsoon. Prince Mukarram Jah Lectures. Indian Geophysical Union, Hyderabad. 100 pp.

RAMASWAMY, C., – 1969 – The basic causes of the large-scale deficiency in the south-west monsoon rainfall over India in 1965 and 1966. *Curr. Sci.* 38: 53–58.

RANDHAWA, N. S., – 1971 – Present status of micronutrient research in India. In 'Soil Survey and Soil Fertility Research in Asia and the Far East,' pp. 91–95. FAO, Rome.

RATHJENS, C., C. TROLL & H. UHLIG (eds.), – 1973 – Vergleichende Kulturgeographie der Hochgebirge des Südlichen Asien. Franz Steiner Verlag, Wiesbaden. 184 pp. + 32 plates.

170

RAWSON, E. R., – 1963 – The Monsoon Lands of Asia. 256 pp.
REED, C. F., – 1969 – Bibliography to floras of southeast Asia. Harrod, Towson, Md. 191 pp.
REGE, N. D. & N. PATNAIK, – 1971 – Country report: India soil surveys. In 'Soil Survey and Soil Fertility Research in Asia and the Far East,' pp. 199–202. FAO, Rome.
RICHARDS, P. W., – 1952 – The Tropical Rain Forest: An Ecological Study. Cambridge University Press, London. 450 pp.
RICHARDS, P. W., – 1973 – The tropical rain forest. *Scientific American* 229: 58–67.
ROBERTSON, V. C., – 1968 – Settlers in the rain forest. *Geogr. Mag.* 40: 1049–1057.
ROBINSON, H., – 1966 – Monsoon Asia. – MacDonald and Evans, London. 559 pp.
ROSSETTI, C., – 1963 – Un dispositif de prises de vues aériennes à basse altitude et ses applications pour l'étude de la physionomie de végétations ouvertes. *Bull. du Service de la Carte Phytogéographique, sér. B.* 7: 211–238.
ROSSETTI, C., – 1967 – Compte Rendu de Missions sur l'Etude des Images photographiques aériennes à grande Echelle de diverses Formations végétales pour le Compte du C.E.P.E. Document no. 38. CNRS, Montpellier. 120 pp.
ROSSETTI, C., P. KOWALISKI & N. HAVE, – 1967 – Relations entre les caractéristiques de réflexion spectrale de quelques espèces végétales et leurs images sur des photographies en couleur, terrestres et aériennes. Actes du IIe Symposium international de Photo-Interprétation. II-27-51. Editions Technip, Paris.
RUMNEY, G. R., – 1968 – Climatology and the World's Climates. Collier-Macmillan, New York, London. 656 pp.
RUPRECHT, E., – 1970 – A quantitative investigation on the aridity of the Desert of Thar. *Bonner Meteorologische Abhandlungen* 14: 83–99.
SAMISH, R. M., – 1971 – Plant Analysis and Fertilizer Problems. Gordon and Breach, London. 660 pp.
SCHULZE, F. E. & N. A. de RIDDER, – 1974 – The rising water table in the West Nubarya area of Egypt. *Nature and Resources* 10: 12–18.
SCHWEINFURTH, U., – 1957 – Die horizontale und vertikale Verbreitung der Vegetation in Himalaya. *Bonner Geographische Abhandlungen* 20: 1–372.
SCOTNEY, D. M. & J. M. de JAGER, – 1971 – The assessment of environmental potential. *Agrochemophysica* 3(4): 71–74.
SCOTT, I. M., – 1974 – Soils of the Central Sarawak Lowlands. Sarawak Department of Agriculture, Soil Survey Memoir no. 2. Kuching.
SEHGAL, J. L. et al., – 1968 – A climatic sequence from the Thar Desert to the Himalayan mountains in Punjab. *Pédologie* 18: 351–373.
SELOD, Y. I., – 1961 – Bioclimats et végétation du Pakistan occidental. Thèses présentées à la Faculté des Sciences de l'Université de Toulouse. 116 pp.
SEN, A. K., – 1972 – Agroclimatic regions of Rajasthan. *Ann. Arid Zone* 11: 31–40.
SHANBAG, G. Y., – 1956 – The climates of India and its vicinity according to a new method of classification. *Indian Geogr. J.* 31: 1–25.
SHAW, R. H., – 1971 – Special meteorological data needs for agriculture. *Iowa State J. Sci.* 45: 529–539.
SHIMWELL, D. W., – 1971 – The Description and Classification of Vegetation. Sidgwick and Jackson, London. 322 pp.
SHYAM SUNDER, S., – 1966 – Exotic forest species – Introduction on basis of study of bioclimate. *Mysore Forest* (Mysore Forest Dept., Bangalore). Part I, January: 5–22. Part II. April: 11–33. Part III. July: 1–16.
SINGH, Pratap, – 1971 – Development of an integrated Indian forest policy. *Ind. For.* 97: 84–88.

SINGH, S., S. PANDEY & B. GHOSE, – 1971 – Geomorphology of the middle Luni Basin of western Rajasthan. *Ann. Arid Zone* 10: 1–14.

SLATYER, R. O., – 1970 – Plant responses to climatic factors. *Nature and Resources* 6(4): 11–15.

SLATYER, R. O. (ed.), ·· 1973 – Plant Response to Climatic Factors. Proceedings of the Uppsala Symposium. UNESCO, Paris. 574 pp.

SMITH, G. D., – 1965 – Lectures on soil classification. *Pédologie,* no. spéc. 4.

SMITH, L. P., – 1961 – Weather and food. *WMO Bull.* 10: 138–144.

SMITH, L. P., – 1968 – Summary of discussions. In 'Agroclimatological Methods: Proceedings of the Reading Symposium,' pp. 379–382. UNESCO, Paris.

SMITH, L. P. (ed.), – 1972 – The Application of Micrometeorology to Agricultural Problems. WMO Technical Note no. 119. WMO, Geneva. 74 pp.

SPATE, O. H. K. & A. T. A. LEARMONTH, – 1967 – India and Pakistan: Land, People and Economy (3rd ed.). Methuen, London. 439 pp.

SPENCER, H. E., – 1964 – The development and spread of agricultural terracing in China. In 'Symposium on Land Use and Mineral Deposits in Hong Kong, Southern China and Southeast Asia,' pp. 105–110. University of Hong Kong.

SPENCER, J. E. & W. L. THOMAS, – 1971 – Asia, East by South: A Cultural Geography (2nd. ed.). Wiley, New York, London, Sydney, Toronto. 669 pp.

STERN, K. & L. ROCHE, – 1974 – Genetics of Forest Ecosystems. Springer Verlag, New York. 350 pp.

STOBBS, A. R., – 1970 – Soil survey procedures for development purposes. In 'New Possibilities and Techniques for Land Use and Related Surveys,' ed. I. H. Cox, pp. 41–62. World Land Use Survey Occasional Paper no. 9. Geographical Publications, Berkhamsted.

SUBRAHMANYAM, V. P., – 1969 – Nature and classification of tropical climates with special reference to India. *Ind. For.* 95: 659–669.

SUKHATME, P. V., N. ERUS & J. M. P. MEMORIA, – 1970 – Need of an assured and controlled supply of water for improving agricultural production. *Rev. Internat. Statistical Institute* 38: 120–139.

TAYLOR, B. W. & R. O. WHYTE, – 1958 – The role of vegetation studies in land classification. In 'Proceedings Symposium on Humid Tropics Vegetation, Tjiawi,' pp. 121–131. UNESCO Science Cooperation Office for South East Asia.

THOM, H. C. S., – 1966 – Some Methods of Climatological Analysis. WMO Technical Note no. 81. WMO, Geneva. 53 pp.

THORNTHWAITE, C. W., – 1948 –, An approach toward a rational classification of climate. *Geogr. Rev.* 38: 85–94.

TOMAR, M. S. & A. R. MASLEKAR, – 1973 – Landuse and forest type classification proposed for aerial photo-interpretation. *Indian For.* 99: 281–295.

TOYODA, H., – 1967 – Report to the Government of Taiwan on Development of Supplemental Irrigation on Slope Lane. UNDP Report TA 2388, FAO, Rome. 29 pp.

TRAPNELL, C. G., – 1953 – The Soils, Vegetation and Agriculture of N.E. Rhodesia. Govt. Printer, Lusaka. 146 pp.

TRAPNELL, C. G. & J. N. CLOTHIER, – 1957 – The Soils, Vegetation and Agricultural Systems of N.W. Rhodesia. Govt. Printer, Lusaka. 87 pp.

TRAPNELL, C. G., J. D. MARTIN & W. ALLAN, – 1950 – Vegetation Soil Map of Northern Rhodesia. Govt. Printer, Lusaka.

TREWARTHA, G. T., – 1961 – The Earth's Problem Climates. Uni. Wisconsin Press/Methuen, Madison and London.

TRIBE, D., – 1970 – Animal ecology, animal husbandry and effective wildlife management. In 'Use and Conservation of the Biosphere,' pp. 123–141. UNESCO, Paris.

TRICART, J., – 1963 – Principes et Méthodes de la Géomorphologie. Masson, Paris.

TRICART, J., – 1966 – La place de la géomorphologie dans l'étude de la mise en valeur des deltas tropicaux. In 'Scientific Problems of the Humid Tropical Zone Deltas and their Implication. Proceedings of the Dacca Symposium,' pp. 15–22. UNESCO, Paris.

TRICART, J., – 1972a – The Landforms of the Humid Tropics. Forests and Savannas. Longmans, London. 306 pp.

TRICART, J., – 1972b – Travaux pratiques de Géomorphologie structurale. Société d'Edition d'Enseignement Supérieur, Paris. 183 pp.

TRICART, J. & A. CAILLEUX, – 1972 – Introduction to Climatic Geomorphology. Longmans, London. 295 pp.

TROCHAIN, J. L., – 1957 – Accord interafricaine sur la définition des types de végétation de l'Afrique tropicale. *Bull. Inst. d'Etudes Centrafricaines n.s.*, nos. 13–14: 55–93.

TROLL, C. (ed.), – 1972a – Geoecology of the High-Mountain Regions of Eurasia. *Erdwissenschaftliche Forschung* IV. Franz Steiner Verlag, Wiesbaden. Individual contributions separately paginated.

TROLL, C., – 1972b – The upper limit of aridity and the arid core of high Asia. *Erdwissenschaftliche Forschung* IV: 237–243.

UCHIJIMA, T. & J. HANYU, – 1967 – On the regional differences in the agro-climatic index on quantity of ripening of rice plant in Japan. *J. Agric. Meteorology*, Tokyo 22: 137–142.

UHLIG, H., – 1969 – Hill-tribes and rice-farmers in the Himalayas and Southeast Asia: problems of the social and ecological differentiation of agricultural landscape types. in *Transactions. Institute of British Geographers.* 47, 1–23.

UHLIG, H., – 1970 – Die Agrarlandschaften des Chenab-Tals in Jammu und Kaschmir. *Tübinger Geographische Studien* 34: 308–323.

UNDERHILL, H. W., – 1971 – Hydrological and hydrometeorological data as essential parameters for design of economic development projects. In 'The Role of Hydrology and Hydrometeorology in the Economic Development of Africa,' Vol. 1, pp. 47–64. WMO, Geneva.

UNITED NATIONS EDUCATIONAL, SCIENTIFIC AND CULTURAL ORGANI-ZATION, – 1964 – International Hydrological Decade, Intergovernmental Meeting of Experts. Final Report. UNESCO, Paris. 17 pp + annexes.

UNESCO, – 1966a – Scientific Problems of the Humid Tropical Zone Deltas and their Implications. Proceedings of the Dacca Symposium. Humid Tropics Research, UNESCO, Paris. 422 pp.

UNESCO, – 1966b – International Hydrological Decade. Co-ordinating Council. Second Session. Final Report. UNESCO, Paris. 40 pp. + annexes.

UNESCO, – 1967a – Methods of Study in Soil Ecology. Proceedings of the Paris Symposium. Ecology and Conservation 2. UNESCO, Paris. 303 pp.

UNESCO, – 1967b – International Hydrological Decade. Co-ordinating Council. Third Session. Final Report. UNESCO, Paris. 46 pp. + annexes.

UNESCO, – 1968a – Regional Seminar on the Ecology of Tropical Highlands. UNESCO Field Science Office for South Asia, New Delhi. 72 pp.

UNESCO, – 1968b – Agroclimatological Methods. Proceedings of the Reading Symposium. Natural Resources Research 7. UNESCO, Paris. 392 pp.

UNESCO, – 1968c – International Hydrological Decade. Co-ordinating Council. Fourth Session. UNESCO, Paris. 32 pp. + annexes.

UNESCO, – 1970 – Vegetation Map of the Mediterranean Zone. Explanatory Notes. Arid Zone Research 30. UNESCO, Paris. 90 pp.

UNESCO, – 1972a – Influence of man on the hydrological cycle: guidelines to policies for the safe development of land and water resources. In 'Status and Trends of Research in Hydrology, 1965–1974,' pp. 31–70. UNESCO, Paris.

UNESCO, – 1972b – Ground-Water Studies. An International Guide for Research and Practice. 18 chapters, loose-leaf with binder. UNESCO, Paris.

UNESCO, – 1972c – Status and Trends of Research in Hydrology, 1965–1974. UNESCO, Paris. 148 pp.

UNESCO, – 1972d – Teaching Aids in Hydrology. UNESCO, Paris. 64 pp.

UNESCO, – 1973 – International Classification and Mapping of Vegetation. Applicable to vegetation maps at scales of 1:1,000,000 or less. UNESCO, Paris. 93 pp. + 1 sheet.

UNESCO, – 1974 – Natural Resources of Humid Tropical Asia. Natural Resources Research 12. UNESCO, Paris. 456 pp.

UNESCO/WMO/IASH – 1974a – Design of Water Resources Projects with Inadequate Data: Proceedings of the Madrid Symposium. UNESCO, Paris.

UNESCO/WMO/IASH – 1974b – Mathematical Models in Hydrology. Proceedings of the Warsaw Symposium. UNESCO, Paris. 1360 pp. (approx.)

UNITED STATES AGENCY FOR INTERNATIONAL DEVELOPMENT, – 1962 – Techniques and Procedures for Cambodia: Forest Inventory Manual. Paginated under five chapters.

U.S. SALINITY LABORATORY, – 1954 – Diagnosis and Improvement of Saline and Alkaline Soils. U.S. Dept. Agric. Handbook no. 60.

VERSTAPPEN, H. T., – 1969 – Aerial survey and rural development in South Asia. in 'Problems of Land Use in South Asia,' Yearbook of the South Asia Institute, Heidelberg University, 1968–69, pp. 1–6. Otto Harrassowitz, Wiesbaden.

VERSTAPPEN, H. T., – 1973 – A Geomorphological Reconnaissance of Sumatra and Adjacent Islands (Indonesia). Wolters-Noordhoff, Groningen. 182 pp.

VIDAL, J. E., – 1972 – Bibliographie botanique Indochinoise. *Bull. Soc. Etudes Indochinoises, n.s.* 47: 657–749

VINK, A. P. A., – 1960 – Quantitative aspects of land classification. Paper v.52. In 'Trans. Seventh International Congress of Soil,' *Science IV, Madison.* 371–378 pp.

WALL, J. R. D., – 1967 – The Quaternary geomorphological history of north Sarawak with special reference to the Subis karst, Niah. *Sarawak Museum J.*: 15: 97–125.

WALLACE, B. J., – 1970 – Hill and Valley Farmers: Socio-economic Change among a Philippine People. Schenkman, Cambridge, Mass. 137 pp.

WALLEN, C. C., – 1963 – Aims and methods in studies of climatic fluctuations. In 'Changes of Climate,' pp. 467–473. UNESCO, Paris.

WALLEN, C. C., – 1966 – Arid zone meteorology. In 'Arid Lands,' ed. E. S. Hills, pp. 31–51. Methuen/UNESCO, London and Paris.

WALLEN, C. C., – 1968a – Agroclimatological studies in the Levant. In 'Agroclimato logical Methods. Proceedings of the Reading Symposium,' pp. 225–232. UNESCO, Paris.

WALLEN, C. C., – 1968b – Needs for application of climatology to agriculture and how to meet them. In 'Agroclimatological Methods: Proceedings of the Reading Symposium,' pp. 375–377. UNESCO, Paris.

WALTER, H., – 1973 – Vegetation of the Earth. English Universities Press. London/ Springer-Verlag, New York. Heidelberg. 237 pp.

WANG Chi-wu, – 1961 – The Forests of China. Maria Moors Cabot Foundation Publication Ser. no. 5. Harvard, Mass. 313 pp.

WATER RESOURCE PLANNING COMMISSION, MINISTRY OF ECONOMIC AFFAIRS, TAIWAN, – 1968 – First Stage Planning Report in Kaoping Chi Basin Water Resources Development. Taipei, 181 pp.

WATSON, A. (ed.), – 1970 – Animal Populations in Relation to their Food Resources: A Symposium, Aberdeen. Blackwell, Oxford. 477 pp.

WEBB, L. J., J. G. TRACEY, W. T. WILLIAMS & G. N. LANCE, – 1971 – Prediction of agricultural potential from intact forest vegetation. *J. Appl. Ecol.* 8: 99–121.

WHITTAKER, R. H. (ed.), – 1973 – Ordination and Classification of Communities. Part 5. Handbook of Vegetation Science. Junk, The Hague. 838 pp.

WHYTE, R. O., – 1958 – The classification and utilization of grazing lands in the tropics. In 'Humid Tropics Symposium on Climate, Vegetation and Land Utilization,' pp. 9. UNESCO, Paris.

WHYTE, R. O., – 1964 – The Grassland and Fodder Resources of India (2nd ed.). ICAR Scientific Monograph no. 22. ICAR, New Delhi. 553 pp.

WHYTE, R. O., – 1968 – Grasslands of the Monsoon. Faber, London/Praeger. New York. 325 pp.

WHYTE, R. O., – 1972a – Rural Nutrition in China. Oxford University Press, Hong Kong, London, New York. 54 pp.

WHYTE, R. O., – 1972b – The Gramineae, wild and cultivated, of monsoonal and equatorial Asia. 1. Southeast Asia. *Asian Perspectives* 15: 127–151.

WHYTE, R. O., – 1974a – Tropical Grazing Lands: Communities and Constituent Species. Junk, The Hague. 222 pp.

WHYTE, R. O., – 1974b – Rural Nutrition in Monsoon Asia. Oxford University Press, Kuala Lumpur, London, New York. Melbourne. 296 pp.

WHYTE, R. O., – 1974c – Grasses and grasslands. In 'The Natural Resources of Humid Tropical Asia,' pp. 239–262. Natural Resources Research 12. UNESCO, Paris.

WHYTE, R. O., – 1975a – The genetic resources in Asian ecosystems containing perennial species of the Gramineae and Leguminosae. Paper to Symposium on South East Asian Plant Genetic Resources, Bogor, Indonesia.

WHYTE, R. O., – 1975b, – The Gramineae of western monsoon Asia: their antiquity and evolution. *Biotropica* 7:

WIKKRAMATILEKE, R., – 1956 – Climate in the south-east quadrant of Ceylon. *J. Trop. Geogr.* 8: 55–72.

WILFORD, G. E. & J. R. D. WALL, – 1965 – Karst topography in Sarawak. *J. Trop. Geogr.* 21: 44–70.

WILLIAMS, C. N. & K. T. JOSEPH, – 1970 – Climate, Soil and Crop Production in the Humid Tropics. Oxford University Press, Kuala Lumpur and Singapore. 177 pp.

WILLSON, M. F., – 1973 – Tropical plant production and animal species diversity. *Tropical Ecol.* 14: 62–65.

WINSTANLEY, D., B. EMMETT & G. WINSTANLEY, – 1974 – Climatic changes and the world food supply. (Paper to) Conference of American Society of Civil Engineers, Biloxi, Miss. (mimeogr.). 38 pp.

WITHINGTON, W. A., – 1960 – Amount and variability of tropical rainfall in north Sumatra. Communications of Research Institute of Sumatra Planters' Assoc. General Ser. no. 75. 14 pp.

WOOLDRIDGE, S. W. & R. S. MORGAN, – 1966 – An Outline of Geomorphology. Longmans, London.

WORKING GROUP ON THE INFLUENCE OF MAN ON THE HYDROLOGICAL CYCLE, – 1972 – Influence of man on the hydrological cycle: guidelines to policies for the safe development of land and water resources. In 'Status and Trends of Research in Hydrology, 1965–74,' pp. 31–70. UNESCO, Paris.

WORLD LAND USE SURVEY, – 1970 – New Possibilities and Techniques for Land Use and Related Surveys, with Special Reference to the Developing Countries. Geographical Publications Ltd., Berkhamsted, Herts. 138 pp.

WORLD METEOROLOGICAL ORGANIZATION, – 1963 – Guide to Agricultural Meteorological Practices. WMO No. 134 TP 61. Pagination in sections. With supplements 1965 and 1968. WMO, Geneva.

WMO, – 1966 – Some Methods of Climatological Analysis. WMO Technical Note no. 81, WMO, Geneva. 53 pp.

WMO, – 1969 – Hydrological Forecasting. Proceedings WMO/UNESCO Symposium. WMO Technical Note no. 92. WMO, Geneva. 328 pp.

WMO, – 1970 – The Planning of Meteorological Station Networks. Technical Note no. 111. WMO, Geneva. 35 pp.

WMO, – 1972 – The Application of Micrometeorology to Agricultural Problems. Technical Note no. 119. WMO, Geneva. 74 pp.

YAO, A. Y. M., – 1969a – Characteristics and Probabilities of Precipitation in China. ESSA Technical Report EDS 8. U.S. Department of Commerce. Silver Spring, Md. 35 pp.

YAO, A. Y. M., – 1969b – Climatic hazards to the agricultural potential in the north China plain. *Agric. Meteorology* 6: 33–48.

YAO, A. Y. M., G. L. BARGER & H. L. CRUTCHER, – 1971 – Precipitation Probability for Eastern Asia. U.S. Environmental Data Service. Silver Spring, Md. 71 pp.

YOSHINO, M. M. (ed.), – 1971 – Water Balance of Monsoon Asia: A Climatological Review. Uni. Hawaii Press, Honolulu. 308 pp.

ZONNEVELD, I. S., – 1970 – The contribution of vegetation science to the exploration of natural resources. Misc. Papers 5, Landbouwhogeschool, Wageningen, 31–44.

ZONNEVELD, I. S., – 1972 – ITC Textbook of Photo-Interpretation. Vol. 7. Use of Aerial Photographs in Geography and Geomorphology. Chapter 7, 4: Land Evaluation and Land(scape) Science. ITC, Enschede, 106 pp.

4 EVALUATION OF ECONOMIC CONSTITUENTS

4.1 Crops

4.1.1 Origin and conservation of genetic resources

The Technical Advisory Committee of the FAO Consultative Group on International Agricultural Research, at its fourth meeting in Washington D.C., August, 1972, considered a proposal for the establishment of an international network of centres for the conservation and documentation of plant genetic resources. The preamble reads as follows: The sources of varieties (cultivars) of crop plants are not distributed equally all over the world, but concentrated, due to biological and/or historical factors, in limited regions called 'centres of crop diversity', or of 'crop variability'. All of them are situated in the developing countries. In several of them the richness in primitive types is fast disappearing, as new and superior man-made cultivars replace the old varieties. However, it is to these old varieties that the breeders in all countries have to turn to obtain the factors of resistance, quality and other characteristics for the further improvement of the current cultivars. Thus, the conservation of the primitive types of crop plants and the related wild species is an international task which requires the exploration of these resources in the field; their conservation on a long-range basis, and their evaluation and utilization in crop improvement programmes.

The world network is to be based on a. existing centres in the developed countries, b. international centres working on specific crops, c. new regional centres in areas of crop diversity, and d. a co-ordinating centre. Major documentation centres are essential for the functioning of such a global network, e.g. the data banks at Beltsville, Maryland, and at five regional centres in the U.S.A.; also those at Volkenrode, Bari, Izmir (proposed), Leningrad, Canberra, New Delhi and the International Rice Research Institute, Los Baños.

In the assessment of the present crops with a view to their improvement, it is first necessary to consider their origin from wild species or primitive cultivated forms, and to ensure that the basic genetic resource, whatever and wherever it may be, is conserved for future use. Proposals recently put forward regarding the origin of Asian cereals in general (Whyte, 1973, 1974c), of rice in particular (Whyte, 1975a), and also of the grain legumes of Near Eastern and Asian agriculture (1975b) will probably and in due course lead to a marked change of emphasis in the general approach. It may come to be accepted that the annual cereals and grain legumes arose from wild perennial species in specific zones and at particular periods in the palaeoclimatic history of Asia.

Then field collections will have to be extended into the vegetation communities which contain these perennial ancestral forms (if they still exist). Here one may expect to find, as a sequel to a series of years of drought combined with intense heat, recently evolved annual forms with a wider spectrum of variability than those annual types of modern agriculture which have been exposed for millenia to climatic and human selection. In these latter types, the original variability has been eroded away, and the process is being accelerated by modern plant breeding, concerned with the production of a few high-yielding varieties in place of the many.

The proposal that annual cereals and grain legumes became available to primitive man quite suddenly in his long history and over quite a short period of time can also be applied to all other annual and biennial crops and vegetables. This provides a vast new field of research for plant collectors, taxonomists and geneticists. Those concerned with rice in particular may refer to the manual of IRRI (International Rice Research Institute, Los Baños, Philippines), designed to help field workers to collect and conserve indigenous cultivars, primitive forms and wild species of the genus *Oryza* (Chang et al., 1972). Aspects covered relate to the planning of field collections, operations related thereto, and procedure for preliminary screening and seed increase (see Tables 4.1 and 4.2 for a statement of the morpho-agronomic features for identifying and distinguishing between rice varieties, and a specimen field record form).

The collection and conservation of genetic resources have also represented an essential initial step in the research programme of the Asian Vegetable Research and Development Center, established at Shanhua, Tainan, Taiwan, under the auspices of the FAO Consultative Group on International Agricultural Research (P.A. Oram, Secretary of Technical Advisory Committee). Regional studies on Asian vegetables will make a major contribution to raising the present low proportion of protective foods in the average diet. The Asian Development Bank has provided direct grants to the Center, and also technical assistance grants for outreach programmes in the Republic of Korea, the Philippines and Thailand.

4.1.2 Assessment of comparative protein quantity and quality

The FAO/WHO/UNICEF Protein Advisory Group (PAG)* recognized at its meetings in 1969 and 1970 that for populations in developing countries the foodgrain cereals and the grain legumes provide a significant if not the major proportion of the daily protein intake. In developed countries the main emphasis of plant breeders is on yield, response to fertilizers, and the minimizing of crop losses. In Asia it is equally or even more important to find and to develop the use of crop varieties (cultivars) having the potential to produce larger amounts of protein of a desired amino-acid balance for human nutrition, pro-

*now renamed the Protein-Calorie Advisory Group, but still retaining the short title: PAG.

Table 4.1. IRRI – morpho-agronomic features for identifying and distinguishing among rice varieties (Chang and others, 1972).

Pigmentation of plant parts (basal leaf sheath, leaf blade, internodes): green or shades of purple.

Pubescence of leaves and glumes: hairy vs. glabrous (smooth).

Leaf characteristics: dimensions and shape; angle of attachment; angle of openness (erect or drooping); degree of greenness; rate of senescence.

Culm characteristics: angle (erect vs. spreading); outer diameter; length; wrapping of internodes by sheath; number of tillers per plant.

Flagleaf features: dimensions and shape: angle.

Panicle features: degree of exsertion; number per plant; length; pattern of branching; clustering on secondary branches; number of spikelets on the panicle; threshability.

Grain (spikelet) features: length, width, shape, thickness, colour of glumes; presence, or absence of brown furrows; presence or absence of awns; length of the sterile lemmas ("outer glumes"); grain weight.

Caryopsis (brown rice): presence or absence of red pigments in the seed-coats; dimensions and shape; translucency of starchy endosperm; chalkiness in the endosperm; hardness.

vided that this can be done without reducing yield (FAO, 1970a, on amino acid content of foods and biological data on proteins). The specific nutritional emphasis to be given to a plant breeding programme will be determined by the crop under consideration (Young, 1970, Austin et al., 1969, Ramasastri & Srinavasa Rao, 1969; Mohan & Deosthale, 1969; Kaul and others, 1969). Cereals are generally low in protein concentration and unbalanced in amino-acid composition:

maize – lysine and tryptophan limiting, protein content about 8 to 10 per cent; introduction of mutant gene, opaque–2, into common hybrid varieties has increased lysine and tryptophan concentrations, created a better isoleucine: leucine balance and thus improved protein quality in general (Deosthale & Pant, 1971, on nutrient composition and amino acid pattern of some high-yielding varieties).

rice – quality of protein relatively good, although lysine and then threonine are limiting amino acids; IRRI have detected or reconfirmed (1972 Report) high levels of protein in a few breeding lines that have improved plant type, and are attempting to combine these characters with disease and insect resistance; a modified recurrent selection scheme is designed to reinforce genetic characters for high protein content (also Deosthale & Pant, 1970, on protein, thiamine, riboflavin and niacin in red rice, preferred in some parts of India and Sri Lanka for its taste).

wheat – the nutritive value of wheat would be greatly improved by an increased lysine concentration, plus a protein content above the present average of

179

```
┌─────────────────────────────────────────────┐
│        Rice Collection No.  _____  ┌──────┤
│                            (District)  (Sample)│
│  Variety name(s) _____    │
│                  _____    │
│                  (give species name, if needed) │
│  Maturity:_____ days,_____ season   │
│                                               │
│  Type:_____ lowland, _____ rainfed _____ upland│
│                                               │
│        deep water, _____ water logged       │
│                                               │
│  Altitude: _____ metres                │
│                                               │
│  Locality: _____ village      │
│                                               │
│             _____ town/city   │
│                                               │
│  Grower's name:_____        │
│                                               │
│  Team or collector:_____ Date:_____   │
│                                               │
│                                               │
│  NOTES: (on soils, topography, diseases and in-│
│          sects, special plant characteristics, etc.,│
│          if warranted).                       │
│                                               │
│         _____            │
│                                               │
│         _____            │
│                                               │
│         _____            │
│                                               │
│  Accession no.:                               │
│                  (given by national centre)   │
└─────────────────────────────────────────────┘
```

Table 4.2. IRRI – sample of record form of
use in field collections of rice varieties
(Chang and others, 1972).

12 per cent (Austin and others, 1970; Mattern, Schmidt & Johnson, 1970, on
the techniques adopted at the University of Nebraska in the screening for lysine
content of the softer endosperm types of common wheats in the USDA World
Wheat Collection – 16,000 entries in all).
millets and sorghums – deficient in lysine and next in threonine; high leucine
probably disadvantage; protein content about 10 per cent; with essentially
equal protein and lysine contents, protein digestibility (availability) of a num-

180

ber of new cultivars varied in one study from 30 to 70 per cent and in another from 49 to 77 per cent (see also Purdue-AID booklet on protein content and amino acid composition for 832 lines of the World Sorghum Collection; also Deosthale & Mohan, 1970, on locational differences in protein, lysine and leucine content of varieties; Deosthale, Nagarajan & Pant, 1970, on protein and amino acid content of *Eleusine coracana*).

The PAG has noted that serious problems have emerged in the improvement of protein quality and quantity in staple food crops. These relate to the availability and adequacy of rapid and reliable techniques for screening new cultivars, or large numbers of mutants, not only for potential protein quality but also for factors which influence the nutritional availability of protein. A new screening model is needed for protein quality and quantity, taking into account not only essential amino acids, but also the ratios between types of protein constituents (albumins, globulins, glutelins, prolamins), modification of which is the primary cause of change in several amino acid patterns. Progress in breeding is limited by lack of knowledge of the fundamental metabolic mechanisms which control the synthesis of amino acids and protein in plants. In cereal breeding programmes, mutants may be discarded after an assessment for protein content and essential amino acids without evaluation of other important properties such as better processing quality and palatability.

In the absence of adequate supplies of cheap protein from animal sources in Asia, the grain legumes (or pulses) might be expected to play an important role in supplementing the supplies of vegetable protein provided by the cereals. Grain legumes have been popular in the diets of people in Asia, they make a substantial contribution to energy needs, they contain two to three times as much protein as the cereals, and the amino acid profile of grain legumes complements that of cereal proteins (Krober, 1969, Whyte, 1974b). But any attempt at increasing legume production can succeed only if their productivity becomes more competitive with the high-yielding varieties of cereals (PAG Bulletin 3, no. 4, 1973), the cultivation of which is actually causing a fall in the area under grain legumes.

Rockefeller Foundation has proposed that preliminary attention be given to only six species: low humid tropics — *Cajanus cajan* and *Vigna sinensis;* semi-dry or seasonal tropics — *Arachis hypogaea;* tropical intermediate elevations to temperate zones — *Glycine max* and *Phaseolus vulgaris;* cool weather, high elevation zone — *Cicer arietinum;* (see also PAG Bulletin vol. 3, no. 2, 1973, for more extensive list).

Characters which require analysis in an assessment of potential include the inherent genetic potential to capture and convert radiant energy; photosynthetic efficiency; the harvesting in the seeds of a greater portion of the energy fixed by the plant; response to fertilizers and efficiency of symbiotic nitrogen fixation; the development of reliable and simple screening techniques for monitoring nutritive value, especially protein content and amino acid balance (PAG Bulletin vol. 3, no. 2, 1973); analyses for toxicity and digestibility (PAG

181

Bulletin, ibid.). If these approaches combined with the collection, assessment and maintenance of germ plasm result in genotypes with a higher innate potential, agronomic methods and inputs may progress from the subsistence to the intensive level (PAG Bulletin vol. 3, no. 4, 1973).

PAG notes (Bulletin vol. 2, no. 1, 1972) that the Federal Experiment Station, Mayaguez, Puerto Rico, has studied the tubers of 40 varieties of yams and finds a protein content varying from 6.3 to 13.4 per cent, comparing favourably with cassava 1 to 6 per cent, sweet potato 4 to 6 per cent, and taro 3 to 5 per cent.

4.1.3 Techniques for field experiments

IRRI has published a handbook (Gomez, 1972) on the techniques that they recommend for field experiments with rice; the text is presented under two main heads, and 23 subheads, three of which are given below in full:
field technique — plot size, shape and orientation; number of replications; experimental design; blocking; randomization; soil heterogeneity; border effect; missing hills; replanting dead hills; residual effect of fertilizers; residual effect of unplanted alleys; number of seedlings per hill; off-types; pest and disease damage; minor sources of variation
collection of data — plot sampling; measuring grain yields, plant height and tiller number, yield components, leaf area index, incidence of stem borer; sampling in broadcast rice, sampling for protein determination

Measuring leaf area index:
The leaf area index (LAI) is the area of the leaf surface per unit area of land surface. Methods of measuring LAI are given for two conditions, one in which leaves are not removed from plants, and another in which leaves are removed.

Measuring LAI with leaves not removed from plants: Select at random n hills from each plot, making sure that each hill is surrounded by living hills. The values of n for any specific degree of precision are given in Table 4.3. In general, n = 10 is sufficient. Count the tillers for each sample hill in each plot.

Measure the length and maximum width of each leaf on the middle tiller and compute the area of each leaf based on the length–width method: leaf area = K x l x w where K is the 'adjustment' factor. l is the length, and w is the width. The value of K varies with the shape of the leaf which in turn is affected by the variety, nutritional status, and growth stage of the leaf . Under most conditions, however, the value of 0.75 can be used for all stages of growth except the seedling stage and maturity for which the value of 0.67 should be used.

Compute the leaf area per hill and leaf area index: Leaf area/hill = total leaf area of middle tiller x total number of tillers

$$LAI = \frac{\text{Sum of leaf area/hill of } n \text{ sample hills (sq. cm.)}}{\text{area of land covered by } n \text{ hills (sq. cm.)}}$$

Measuring LAI with leaves removed from plants: Select at random n hills from each plot, making sure that the hills are surrounded by living hills. The values of n for any specific degree of precision are given in Table 4.3. In general, n = 8 is sufficient.

For each plot remove the sample hills from the soil. From each sample hill, separate the middle tiller from the rest of the tillers. Remove all green leaves from the selected tiller. Make sure the leaves do not dry and curl before their leaf areas are measured. To avoid drying and curling, place the sample leaves in a test tube containing a small amount of water. Measure the area of the leaves. With an automatic area meter, all leaves from each sample tiller may be read together. With the length–width method, however, the area for each leaf separately should be obtained. Dry the leaves and weigh.

Remove the leaves from the rest of the tillers in the sample hill and obtain their dry weight. Compute leaf area per hill and leaf area index:

$$\text{Leaf area/hill} = \frac{aW}{w}$$

$$\text{LAI} = \frac{\text{Sum of leaf area/hill of } n \text{ sample hills (sq. cm.)}}{\text{area of land covered by } n \text{ hills (sq. cm.)}}$$

where a is the total leaf area of sample tiller, w the dry weight of leaves from sample tiller, and W the dry weight of all leaves in the hill (including those from the sample tiller).

Sampling in broadcast rice

The greatest difficulty in sampling in broadcast rice plots lies in the identification of sampling units. In transplanted rice, the sampling unit used is based on hills, which is not applicable to broadcast rice. In broadcast rice, the sampling unit must be identified in terms of area. But demarcating sampling areas for measuring various rice characters is an additional problem.

The non-uniform plant density in broadcast rice plots also causes higher variability than in transplanted rice. Thus, for most characters the sample size in broadcast rice should be larger than in transplanted rice.

How to sample broadcast rice: For grain yield in a broadcast plot, just before seeding outline the desired harvest area in the centre of the plot with four small stakes connected with string. The string should be tied flush with the soil to avoid any effect on the seeding operation. Thus the string should be durable enough to withstand being covered with mud for the entire crop season. At maturity, harvest all plants in the demarcated area of each plot.

For other plant characters, construct rectangular wire frames, each 50 × 30 sq. cm. Place three to four frames at random in each plot just before seeding. Various characters, such as tiller count and plant height, can be measured from plants within these units. If measurements are to be made at early growth stages, place these frames at random just outside the harvest area to prevent trampling of the harvest area.

Sampling for protein determination:

The measurement of protein content consists of the sampling procedure for selecting the sample grains and the chemical analysis of the selected grains. Since only a small sample is used in the chemical analysis, the sample must be properly selected.

Selection of sample grains. For protein analysis of individual plants, take the sample after all grains have been threshed and bulked. Do not sample panicles. For protein analysis of whole plots, take sample grains from the bulk harvest used for plot yield determination. Do not take hill samples.

Size of sampling unit. Use 100-grain (or 2-grain) samples and make two determinations per sample. When only a grinder with small capacity is available, 10-grain samples can be used. If a 10-grain sample is used analyse at least two separate 10- grain samples, without duplicate analysis. A sample size smaller than 10 grains should not under any circumstances be used.

Table 4.3. IRRI: Estimated standard error of a plot mean (not including measurement error) for measuring leaf area index for varying sample size (Gomez, 1972).

Sample size *	Standard error (per cent)	
	Unfertilized	Fertilized
2	20.2	30.0
3	16.4	24.5
4	14.2	21.2
5	12.7	19.0
6	11.6	17.3
7	10.8	16.1
8	10.1	15.0
9	9.5	14.2
10	9.0	13.4
12	8.2	12.3
14	7.6	11.4

* No. of hills per plot. To take care of a higher measurement error when leaves are not removed a slightly larger sample size is required.

See also Yoshida, Forno, Cock & Gomez, 1972, for a laboratory manual for physiological studies of rice. IRRI (1974) has also published a Handbook describing methodology for transplanting rice in straight rows, instructions for laying out experimental plots, and the direct seeding of rice.

4.1.4 Assessment of drought injury and tolerance

The first objective of the symposium sponsored by the Crop Science Society of America on drought injury and resistance (Larson & Eastin, 1971) related to the techniques and methods plant breeders may adopt in breeding for drought tolerance and avoidance. The contribution of C. Y. Sullivan of the University of Nebraska was concerned with techniques for measuring plant stress under drought conditions (extensive bibliography). Three types of measurements are suggested: a. desiccation tolerance tests or related heat tolerance tests give information on how much tissue drying can be tolerated before severe injury occurs; b. field measurements of water potential (or relative water content) show how far the internal water status is kept above the critical point during the drought; c. diffusive resistance or stomatal observations indicate if the internal water potential is kept up by retarded transpiration, or whether an efficient root and conducting system is keeping the plant shoot supplied with water.

In answer to the question: can we breed for drought resistance, E. A. Hurd of Canada Department of Agriculture, Swift Current, Saskatchewan, gives a confident and optimistic affirmative.

'The production of cultivars having a large number of favourable genes for yield in any environment has been more talked about than worked on by plant breeders. I do not believe we have reached a barrier in semi-arid climates . . . we are on a plateau of our own making . . . I am not referring to complex theories and formulae, but to a very simple application of the normal curve to segregating populations. . . In breeding for drought resistance it is more important to breed for maximum yield in a good year. . . one method is to grow large populations in early generation yield trials (F_3 on) under typically dry growing conditions . . . more information is required from basic work by the physiologist and others.'

The International Rice Research Institute, Los Baños, Philippines, uses the following criteria in rating cultivars and breeding lines for reaction to drought (Chang, Loresto & Tagumpay, 1974): a. Vegetative phase (from 40 days after seeding to booting), 1. Plasticity in leaf rolling and unfolding. Leaves of susceptible varieties roll tightly and do not unfold completely the next morning; in extremely susceptible plants, the new leaves that later grow out are much smaller and the plants are stunted. Resistant genotypes roll gently and unfold at night. Sustained leaf rolling occurs at a later date. 2. Extent of leaf death. The lower leaves of susceptible genotypes die soon after water stress has set in; the tips of the young leaves also show extensive die-back, resembling sunscalding. Few of the lower leaves of tolerant genotypes die and those that do die slowly. The upper leaves may turn yellowish 3. Degree of stunted growth. After severe water stress, susceptible genotypes produce dark leaves that are distinctly stunted. Such new leaves fold early when plants are subjected to another period of water stress. Plant height is reduced and the adult plants are stunted at flowering. b. Reproductive phase (heading to maturity), 1. Effect on heading date and panicle exsertion. Heading of susceptible varieties is delayed and/or panicle exsertion is incomplete. 2. Effect on panicle size, shape and colour. Susceptible genotypes produce shorter panicles, some with deformed rachises or aborted terminal spikelets. White glumes are another symptom of severe drought injury. 3. Effect on spikelet development. Drought injury markedly reduces spikelet fertility of susceptible varieties. Light and poorly developed grains are another distinct sign of drought damage. 4. Reduction in grain yield. Reduction in grain yield as related to drought is another useful index.

'On the basis of repeated observations and recordings of the above criteria during the crop season, the test varieties are rated by comparing the sequence and extent of stress symptoms with the check varieties (TN_1, IR8, IR5, Rikuto Norin 21 and OS4). Scores taken during both the vegetative and the reproductive stages are combined in the final rating. The classes are susceptible (S), moderately susceptible (MS),, intermediate (I), moderately resistant (MR), and resistant (R); the code numbers, 1 to 9. Varieties sometimes fall between two classes; e.g. I-MR or the code number 6.'

4.1.5 Assessment of crop losses from pests and diseases

Over a period of five years physiologists from IRRI collected information on
nutritional disorders of rice as reported in the scientific literature. They also
travelled to countries which had reported these disorders, examined the symptoms in the field, analysed the nutrient content of the soils and the plants,
and brought samples of the soil to Los Baños in order that plants might be
grown under controlled conditions. In Table 4.4 are given the countries reporting rice diseases, the name of the disease and its probable cause. In Table
4.5 the nutritional disorders of rice are classified in relation to soil pH and
other characteristics of the soils from which disorders have been reported.

'There should be a general awareness of the necessity to appraise crop
losses before more effective and safer measures for pest and disease control
can be developed. Only by uncovering real losses can we focus on absolute
opportunities for gain, through plant protection resources (including opportunities for avoiding expensive research on unimportant pest and disease problems). One of the major steps in the right direction has been the establishment of the FAO international collaborative programme for the development
of reproducible methods for the assessment of crop losses' (Chiarappa, Chiang
& Smith, 1972).

An outcome of this programme has been the publication by the Commonwealth Agricultural Bureaux, by arrangement with FAO, of the loose-leaf
manual on crop-loss assessment methods (FAO, 1970b), on the evaluation
and prevention of losses by pests, diseases and weeds. The five entries on
pp. 189–194 are extracts from the manual relating to crops important in
Asian agriculture (see also Ou of IRRI, 1972, on diseases of rice caused by
fungi, bacteria, viruses and mycoplasmas, nematodes and physiological
factors; also a shorter version by the same author, 1973, entitled *A handbook
of rice diseases in the tropics*; also K. C. Ling of IRRI (1972) on the diseases
of rice caused by viruses and mycoplasma, with particulars of insect vectors).

In any study of the estimation of incidence of diseases and pests under cultivators' conditions over a wide area, it is necessary to know whether the required data can be collected on a sample basis. A sampling technique for the
estimation of incidence of gall-fly, stem borer and blast in field experiments
on padi in India has been discussed by Abraham et al. (1963) and Abraham
(1966).

The Institute of Agricultural Research Statistics, New Delhi, made pilot
sample surveys (Sardana, Khosla & Rao, 1971) in the Districts of Cuttack,
Thanjavur and West Godavari to evolve suitable techniques for sampling and
measurement of the incidence of pests and diseases in the padi crop, and so
for the assessment of consequent reduction in yield. On a district level, it may
be possible to estimate the incidence of most of the major pests and diseases
of padi with a standard error of not more than 10 per cent, by taking a sample of 100 villages, four fields per selected village, and two plots of 1 sq. m.

Table 4.4. Physiological disorders of rice reported from various rice-growing countries (from *The IRRI Reporter* vol. 6, no. 5, 1970)

Country	Physiological disease	Possible cause
Burma	Amiyi-Po	K deficiency
	Myit-Po	P deficiency
	Yellow leaf	S deficiency
Ceylon	Bronzing	Fe toxicity
Colombia	Espiga erecta	?
India	Khaira disease	Zn deficiency
	Bronzing	Fe, Mn, H_2S toxicities
	Yellowing	?
Indonesia	Mentek	Virus disease ?
Japan	Akiochi	H_2S toxicity, K, Mg, Ca, Si deficiencies
	Akagare I	K deficiency (Fe toxicity)
	Akagare II	Zn deficiency
	Akagare III	I toxicity
	Aodachi	?
	Hideri-Aodachi	?
	Straighthead	?
Korea	Akiochi	H_2S toxicity, K, Mg, Ca, Si deficiencies ?
Malaysia	Penyakit Merah (Yellow type)	Virus disease
	Penyakit Merah (Brown spot type)	Fe toxicity
Pakistan	Pansukh	?
	Bronzing (Hadda)	Zn deficiency
Portugal	Branca	Cu deficiency ?
Taiwan (southern)	Suffocating disease	Mainly virus disease
U.S.A.	Straighthead	?
	Alkali disease	Fe deficiency

each per selected field. To estimate the intensity of incidence of *Helminthosporium oryzae* in terms of score per plant, two plants per selected plot in the above scheme of sampling may give an estimate of the desired precision. Subdivision of the field into peripheral, interior and central sectors for location of sampling plots is not likely to increase the efficiency of sampling. If sampling plots within fields are kept fixed for recording observations throughout a season, this may not result in any loss of efficiency in sampling, as compared with selection of new plots on each occasion of sampling.

Many workers have proposed methods of sampling for the estimation of populations of parasitic nematodes in soils. Sampling techniques for this

Table 4.5. Classification of nutritional disorders in rice in Asia (from *The IRRI Reporter* vol. 6, no. 5, 1970)

Soil		Disorder	Local name
Very low pH	(Acid sulfate soil)	Iron toxicity	"Bronzing"
High in active iron	Low in organic matter	Phosphorus deficiency	
	High in organic matter	Phosphorus deficiency combined with iron toxicity	
	High in iodine	Iodine toxicity combined with phosphorus deficiency	"Akagare Type III"
	High in manganese	Manganese toxicity*	
Low pH — Low in active iron and exchangeable cations.	Low in potassium	Iron toxicity interacted with potassium deficiency	{ "Bronzing" "Akagare Type I"
	Low in bases and silica, with sulfate application	Imbalance of nutrients associated with hydrogen sulfide toxicity	"Akiochi"
High pH	High in calcium	Phosphorus deficiency / Iron deficiency / Zinc deficiency	{ "Khaira" "Hadda"
	High in calcium and low in potassium	Potassium deficiency associated with high calcium	"Taya—Taya"
	High in sodium	Salinity problem / Iron deficiency / Boron toxicity*	"Akagare Type II"

* Probably rare.

purpose have been evolved for estimation of such populations in rice soils. At the Central Rice Research Institute, Cuttack (Rao et al., 1971) in the nursery area, one aliquot sample from five soil cores taken at random in an area of 50 sq. m. is sufficient. Under field conditions, strata are marked to comprise 5m. x 5m. squares and one sample for every 25 sq. m. area is necessary. In the extraction of migratory endoparasitic stages from the roots of rice, the cutting of roots into pieces 3 to 5 mm. in length and processing them through an improved Baerman funnel is most effective. In sampling rice roots for endoparasitic stages a minimum of 10 to 20 per cent wet weight of roots is necessary.

The concept of integration has been introduced into programmes concerned with pest control. Integration in this context refers to the rational use of insecticides combined with techniques of crop husbandry, the use of resistant or tolerant varieties, and other methods of control by cultural means. FAO sponsored a conference in September, 1965 on integrated pest control (FAO, 1966). A Conference of (British) Advisory Entomologists considered the subject in April, 1967 (Empson, 1968), in relation to plant breeding, husbandry, predation and parasitism, effects of pesticides, selectivity of pesticides, direct population management, shelter, birds, herbicides and pollination, and the Californian experience of integrated control. The International Organization for Biological Control of Noxious Animals and Plants (O.I.L.B.) in 1970 formed a sub-commission on integrated control with working groups on annual crops, and proposed to follow this with other groups responsible for pests of cotton, and the soil pest complex (*Bull. British Ecological Society*, December, 1970, p. 19).

Representative extracts from the loose-leaf manual on crop loss assessment methods, prepared by Plant Production and Protection Division, FAO, and published by Commonwealth Agricultural Bureaux (FAO, 1970b).

Crop	Pest or Weed	Summarized by:
cotton	American bollworm	P. T. Walker
sugarcane	sugarcane froghopper	D. W. Fewkes
rice	black rice bug	Eiji Kawase, Hisatsugu Ishisaki, Ichitaro Tamura
grain sorghum	pigweed	A. F. Wiese
maize/soybean	smooth pigweed	E. L. Knake

Host: **Gossypium** spp. (cotton) **Organism: Heliothis armigera** (American bollworm)

Method developed in: Rhodesia. *Field symptoms:* Buds, flowers and green bolls show holes, darkening and discolouring due to feeding of larvae. Bracts spread out and curl downwards. Frass (excreta) may be seen outside the bolls. Larvae are also visible as they often feed with much of their body exposed on the outside of the boll. *Effect on crops:* Loss of seed cotton by destruction of flowers and bolls. Damage to cotton lint by discoloration and staining by moulds.

Procedures: The number of larvae/plant on at least 12 plants at random along the field diagonals are counted/sample, twice a week from the fourth leaf stage to boll opening. The average number of eggs/plant is also determined at each count. Examine different plants each time to avoid damaging them. Sprays are applied according to the number of *Heliothis* eggs, in relation to other insect pests of cotton, to obtain different yield levels.

Intensity/loss relation: When insecticide sprays were applied at different infestation levels in two consecutive years, the following relations were found between number of *Heliothis* eggs and yield:

Year	Average no. of eggs/plant	Total seed cotton lost kg/ha
1963–64	0·5	138
	1.0	508
1964–65	0·25	112
	0·50	198
	1·00	820

Similar relations, although less reliable, were obtained when numbers of larvae per plant were used in place of egg numbers.

Limitation or prevention of damage: Infestations can be reduced by early uprooting and burning crop residues, clean weeding, early thinning, avoidance of maize nearby, and by application of a recommended insecticide at times advised locally. When using insecticides, good coverage of the upper half of the plant is essential, as this is where most eggs are found.

Source: Matthews, G. A. & Tunstall, J. P. 1968. Field trials comparing carbaryl, DDT and endosulfan on cotton in Central Africa. *Cott. Grow. Rev.* 45: 115–127.

Cross references: The whole insect pest complex on cotton must be considered in each area. *Diparopsis* spp., *Earias* spp, *Empoasca, Aphis, Tetranychus* and various leaf-eating caterpillars, *Cosmophila, Acontia, Plusia,* etc., all contribute to crop loss.

Host: Saccharum officinarum (sugarcane)
Organism: Aeneolamia varia saccharina (sugarcane froghopper)

Method developed in: Trinidad, West Indies. *Field symptoms:* Nymph may cause slight yellowing of leaves and check growth. Main damage due to adults feeding on leaves. Necrotic streaks develop from feeding punctures, spreading longitudinally to form brown streaks of dead tissue. In heavy attacks the 'blight' streaks coalesce and the whole leaf becomes brown and dies. Blighted fields have scorched appearance. *Effect on crops:* Yields of cane and sugar reduced; in heavy attacks cane/sugar ratio may be adversely affected.

Procedures: Although main damage is by adults, the following is based on nymphs, partly because of the difficulty of assessing adults in small plots. All nymphs above ground in 2 stools per one-eightieth of an acre ($50 \cdot 6m^2$) plot, 3 or 6 replicates, are counted when the crop is 8 months old and about 2m high. Number of stalks per stool is recorded and counts are adjusted to standard sampling unit of 10 stalks. Adjusted samples (*n*) are transformed to $\log_{10}(n+1)$ before analysis. Whole plots harvested at 12 months and weighed to estimate cane yields. Cane/sugar ratio is determined by crushing stalks and juice analysis. Differences in infestation level artificially produced by insecticide treatment when crop is 6 months old.

Intensity/loss relations: Yield data from 3 insecticide trials for the control of nymphs on variety B41227 in 1964 shows a reduction in yield of cane/ha (y) with increase in maximum number of nymphs/10 stalks (n), 6 to 10 weeks after application of insecticides. This relation is given approximately by: $y = a - 23x$, where $x = \log_{10} (n + 1)$. This means very roughly a loss of 6·5 tonnes cane/ha for every x 2 increase in infestation, and 23 tonnes for every x 10 increase. Equivalent loss of sugar is found from a cane/sugar ratio of about 10/1, but severe damage may increase the ratio. For the highest yielding trial the relation is best for more than 5 nymphs/10 stalks and may be sigmoid, with minor losses at low levels of infestation.

Limitation or prevention of damage: Control by insecticides:
(a) Application of dusts to the base of cane stools to control nymphs (before (1965).
(b) Aircraft spraying of foliage to control adults (main method 1967 onwards). No cultural or biological control is practised.

Source: Fewkes, D. W. & Buxo, D. A. 1966. Yield losses in sugarcane due to froghopper infestations. *Ann. Rep. Tate & Lyle Cent. agric. Res. Stn.* 1965. Trinidad. 364–372.

Host: Oryza sativa (rice) **Organism: Scotinophara lurida** (black rice bug)

Method developed in: Japan. *Field symptoms:* Brownish, irregular specks appear on the sheath while yellowish spots are observed on the tip and middle part of the leaves emerging from these sheaths. Heavily injured leaves are twisted and yellowish while the sheath becomes a 'dead heart'. Injured nodes turn brown and brown spots occur on the surface of injured grains. *Effect on crops:* The numbers of 'dead hearts', 'white heads' and empty ears increase on heavily infested plants, and the growth of tillers becomes disorderly. There are more panicles/hill of injured plants than of healthy plants, but less ripe grains/hill because of the emergence of weak tillers. 'Brown rice' of injured plants is of lower quality as it contains numerous green and chalky kernels, and cracked and darkened rice.

Procedures
(1) The effect of insect numbers was measured by sowing four hills/concrete pot, 36 × 36 cm, kept under paddy field conditions and covered with wire netting (two replicates). There were no insects on one treatment (a), two overwintering imagines/hill were put about four weeks after sowing in another treatment (b), and 10 first generation larvae were placed/hill about eight weeks after sowing in a third treatment (c). The growth of tillers and yield were measured about four months after sowing (1).

Intensity/loss relations: Example 1.

	No. of stems	No. of panicles	No. of grains
(a) Unattacked	26	25	3,633
(b) Two imagines	29	27	3,272
(c) 10 larvae	32	26	3,299
		(per hill)	

191

	Weight of 'brown rice' (g/hill) per developed tiller								
No. of tillers	0	4	5	6	7	8	9	10	Total
(a) Unattacked	2·0	16·0	11·0	10·4	7·0	3·7	1·7	1·7	54·3
(b) Two imagines	0·3	12·7	12·6	10·7	7·3	2·8	1·1	0	47·5
(c) 10 larvae	1·8	5·7	13·6	11·9	7·3	3·8	1·4	0	45·5

(Var. Norin 1)

There is thus a reduction of 13% in 'brown rice' in (b) and 15% in (c) over (a).

Example (2)

	Percentage loss in yield over unattacked hill
4 imagines	60
8 imagines	80
30 larvae	40
60 larvae	80

(2) A comparison of yield from damaged and undamaged hills in the field was carried out by selecting several plots, dividing into grades of yield and multiplying by the number of plots in each grade (2).

Limitation or prevention of damage: Late varieties are less damaged than early varieties. The imagines of the over-wintering generation can be controlled by insecticide dust. High population levels need additional application of dusts at the stage of first generation larvae. For the control of the following generation images, organophosphorus dust is most effective.

Sources: 1. Kawase, E. & Ishisaki, H. 1956. *Proc. Assoc. Pl. Prot. Hokuriku,* No. 4.
2. Tamura, I. 1957. Diagnosis of rice damaged by injurious insects. *Plant Protection* 11, 2. 1957. (In Japanese.) 3. Kawase, E., Katsumoto, K., & Ishisaki, H. *Bull. Ishikawa Agr. Exp. Sta.*

Cross-references: Symptoms are often confused with those of the rice stem maggot, *Chlorops oryzae,* and nematodes, *Aphelenchoides besseyi.*

Host: **Sorghum bicolor** (grain sorghum) **Organism: Amaranthus** sp. (pigweed)

Method developed in: U.S.A, Texas. *Field symptoms:* At harvest time infested fields contain various numbers of pigweed which protrude above the crop. *Effect on crops:* Competition for soil moisture, nutrients and light causes reduction in crop yield.

Procedures: Sorghum was planted by conventional methods on soil beds prepared for furrow irrigation. Pigweed seed was spread with a whirlwind seeder over the entire test area. This was then furrow-irrigated to germinate the sorghum and weed seed. At approximately one week after the weeds and sorghum had emerged the weeds were hand thinned to populations ranging from 1 per 8 ft of row(2.4 m) to 4/ft of row (30·5 cm.). This population was maintained throughout the year by further hand weeding.

Intensity/loss relation: There was a curvilineal relation between pigweed population and sorghum yield which could be expressed by the equation: $Y = 4924 + 564X - 2933\sqrt{X}$, in which Y = expected yield of grain sorghum in lb/acre, and X = number of pigweed plants per ft of row.

Using this equation it was calculated that 1 weed in 8, 4, 2 and 1 ft of row caused yield reductions of 20, 27, 36 and 48%, respectively.

Limitation or prevention of damage: Under dryland farming conditions, pigweed growth is reduced by cropping sequences, involving rotation with wheat, and by the use of suitable post-emergence herbicides. Under irrigated conditions, certain pre- and post-emergence herbicides give satisfactory control.

Sources: 1. Shipley, J. L. & Wiese, A. F. 1969. Economics of weed control in sorghum and wheat. *Texas Agr. Exp. Sta.* MP-909. 2. 1969 Suggestions—Weed control with chemicals. *Texas Agr. Exp. Sta.* B-1029, p. 7.

Host: Zea mays (maize, corn) **Glycine max** (soybean) **Organism: Amaranthus hybridus** (smooth pigweed)

Method developed in: USA, Illinois. *Field symptoms:* Fields with unkempt appearance. *Amaranthus* protruding above crop plants. *Effect on crops:* Weed competition for nutrients, moisture and light results in reduction of growth.

Procedures: Experiments were conducted over a three-year period. A randomized complete block design with five replications were used. For corn, individual plots were 2 × 12 hills. Hills were 20 inches (51 cm) apart and rows 40 inches (102 cm). For soybean plots consisted of 2 rows, 40 inches apart and 16·5 ft (5 cm) long.

Pigweed infestations were obtained by seeding immediately after planting of both crops. Seven intensities were established by hand thinning as follows: pigweeds in a band 4 to 6 inches wide (10·2—15·2 cm); pigweed spaced 1, 5, 10, 20 and 40 inches apart (2·5, 12·7, 15·4, 30·8 and 61·6 cm); weed-free check.

At maturity, the corn was harvested by hand, shelled, weighed and tested for moisture. Dry weight of stalks and cobs was also determined. Pigweed plants were separated by hand from soybeans at harvest and dry weight determined for beans and straw. Dry weight of pigweed was also determined in both the corn and soybean plots.

Intensity/loss relation: The following relations were observed between weed density and losses in yield of the two crops studied:

(a) *Corn*

Pigweed spacing (inches)	Weed density kg/ha (dry weight)	Percentage loss in yield of shelled corn when average yield (dry weight) in absence of weeds was 5900 kg/ha
Band	3,192	38·8
1	3,091	36·2
5	2,274	27·1
10	1,512	15·1
20	1,042	11·9
40	515	4·9

Yield reductions of corn were greatest in 1960 when higher rainfall favoured weed growth.

193

Pigweed spacing (inches)	Weed density kg/ha (dry weight)	Percentage loss in yield of soybean grain when average yield (dry weight) in absence of weeds was 2,340 kg/ha
Band	3,752	56·4
1	3,125	52·6
5	2,811	45·9
10	2,363	32·5
20	1,691	27·7
40	1,120	17·7

Limitation or prevention of damage: Pre-emergence herbicides, rotary hoeing and row cultivations.

Source: Moolani, M. K., Knake, E. L. & Slife, F. W. 1964. Competition of smooth pigweed with corn and soybeans. *Weeds* 12: 126–128.

The first important reference to the use of aerial photography for the field study of plant diseases was made by Colwell (1956). That account deals almost entirely with the cereal rusts; the author examines the technical requirements, and indicates the difficulties, limitations and undoubted benefits of the method. Brenchley (1964, 1968a and b) has reviewed the method from the point of view of an officer in the (British) Agricultural Development and Advisory Service based in Cambridge, who used aerial photography during the preceding period of eight years to investigate difficult problems of plant pathology in the field.

The most important characteristic is that aerial photography often reveals patterns of incidence, intensity and development of a disease which cannot be seen as well, if at all, from the ground. It also makes possible the study, on an extensive scale, of diseases where they actually occur. Aerial photography is often an ideal complement to ground studies in the complete investigation of a disease, and leads to collaboration with specialists in other relevant disciplines, such as soil physics, soil chemistry, agronomy and mechanization, in the interpretation of the disease patterns revealed in the photographs (Fig. 4.1a to c).

The key to the distinction between the diseased and healthy parts of a crop lies in the use of the appropriate film/filter combination (Brenchley, op. cit.). In some cases, it may be necessary to use two or more such combinations simultaneously — multiband photography or multispectral remote sensing. Four main types of films have been used by workers in this field; panchromatic, infrared, normal colour, and colour infrared.

The Science Arm of the ADAS report for 1972 refers to the continued use of aerial photography in Britain to detect outbreaks of crop diseases at a very early stage. The Service's unit is still very small, but it has been re-equipped to speed interpretation of results, and given three years for further experimentation.

194

Fig. 4.1a. England/East Anglia. Crop failure of the 'Docking Disorder' type in barley near Holt, Norfolk – infrared – poor patches dark (Photo G. H. Brenchley).

Fig. 4.1b. England/East Anglia. Patterned failures in sugar beet and other crops – panchromatic – poorer parts of crops light (Photo G. H. Brenchley).

195

Fig. 4.1c. England/East Anglia. Blight developing on a potato field (Photo G. H. Brenchley).

4.2 Domestic livestock

In the language of ecosystematics, we now come to consider the major group of consumers of the primary products of the food and fodder chain, after man himself. In assessing the potential contribution of the diverse types of livestock and of animal husbandry in monsoonal and equatorial Asia, one has to accept the fact that the full potentials, as defined by specialists in the breeding, nutrition, husbandry and health of domestic animals can rarely be achieved. The animal industries concerned with the production of protein-rich food, raw materials for clothing and draft for field work and transport all demand ample feed and fodder from natural grazing lands or from croplands. Those animal industries producing animal protein for human consumption are primarily dependent on feeds, fodders and concentrates from cultivated land.

196

The human population of Asia is increasing at something approaching 75 million per year, without a compensatory increase in the total and locally available areas of cultivated land. The International Rice Research Institute estimates that the demand for rice will increase by 30 per cent over the next ten years. The world reserves of foodgrains from which supplies may be purchased are steadily dwindling and becoming more costly. To an ever-increasing extent, the countries of Asia will have to rely on the produce from their own land, and that produce will increasingly be foods for direct human consumption: cereal foodgrains, grain legumes or pulses, vegetables and fruits. Whatever land can be spared from this production will be devoted to cash crops for sale within the country or for export. In such a developing situation of intense competition between the two major groups of consumers in the economic ecosystems, man and domestic livestock, there will be little, and progressively less and less land available for the feed and fodder crops and concentrates required by food-producing animal industries.

It will become increasingly unrealistic to talk about targets for minimum effective intake of animal protein per head of human population, or even per head of those women and children within the vulnerable groups whose need is greatest. The potential animal industries of the future will continue to produce costly foods to the economic limits imposed by the demand from urban and industrial communities with the necessary purchasing power, centres of tourism and the like.

In such a situation, it is necessary to resist the persuasive arguments of the protagonists of one or other of the major types of food-producing animal industries, or of specific types and breeds of animals. It is probably correct to say that the national, international and bilateral agencies which have been engaged over the past fifteen years in promoting and supporting the dairy industry in India, for example, should be brought to realize that they have reached a plateau of national and regional production in terms of cow and buffalo milk, as governed by actual and potential future availability of feeds and fodders and by the market for a costly commodity. The objective observer would draw attention to the low efficiency of conversion of feeds by cow and buffalo as compared with the chicken or the duck, and would advise more attention to the latter, more efficient, less capital-demanding industries. One may query whether FAO and UNICEF were correct in giving such strong support to animal industries based on feeds and fodders from cultivated land, without first surveying the actual and potential fodder production within milk procurement areas, rather than to the production of goat and sheep meat from uncultivated land, provided their grazing grounds could be closed to ingress from the politically-supported surplus cattle (Table 4.6).

Table 4.6. Gradients of increasing availability of soil moisture, soil fertility and intensification of feed and livestock production (Whyte, 1974a). (see also annex 1 in Tribe, 1970, giving characteristics and requirements of stages in livestock production).

Resource	System of use	Type of livestock	End product
1. Unimproved range, mostly arid and semi-arid	Free range grazing, mostly nomadic and/or migratory	Cattle, sheep, goats, camels, wild animals, adapted to wide seasonal fluctuations in supply of feed and water	Beef, meat, wool, hair, hides, skins, wildlife products, tourism
2. Improved range, semi-arid or savannas in humid tropics	Numbers and movement of stock controlled by herding, or by closure of watering points	Similar	Similar
3. Dryland fodder production, crop residues and stubble grazing	Stubble grazing in dry season in a monsoonal environment, with yard feeding or free-range grazing in the rainy (cropping) season	Slight improvement in quality of livestock, in line with slight improvement in total and seasonal availability of feed	Specially relevant to cattle populations in villages of India, (25% of world's total) giving minute amounts of milk in a four-month lactation
4. Sown or planted synthetic stands for grazing	Rotational or fixed grazing during growing season in fenced pastures, other arrangements for dry season	Change of type of livestock and introduction of superior breeds justified, especially milch animals or cattle for fattening.	Milk, meat or beef according to social acceptability
5. Semi-intensive fodder production on cultivated land	Cut and carry, zero-grazing system for crops such as lucerne, berseem or grasses growing alone or with legumes	Production from superior animals ensured by availability of high-protein cut green fodder, or hay or silage	Similar
6. Intensive fodder production	Gross-feeding African grasses grown with irrigation water plus heavy dressings of nitrogen or with cowshed wash or with treated city sewage	Ample green feed, requiring supplementation by concentrates, for productive milch animals in specialist dairy units or milk colonies	Milk
7. Import of feeds, fodders, concentrates or coarse grains from elsewhere in same country or from abroad	Feeding as supplements to roughages and cut green fodder, in stalls, yards, or feedlots	(a) Productivity of high quality milch cows and buffaloes assured. (b) Cattle brought in from range or mountain pastures for fattening	(a) Milk (b) Beef

In spite of the advice given in 1974 by an FAO buffalo enthusiast to the animal husbandry specialists in the People's Republic of China, that this animal, now kept for draft, should be grown for beef, the Chinese planners are correct in the action they propose to take. This is to phase out the buffalo as a draft animal, to replace it with mechanized cultivation, and so release more feed from cultivated land for the production of meat from the pig, which is a more efficient converter of feed and the traditional form of animal protein in the Chinese diet. It would be unrealistic of the Chinese to attempt to increase the production of cow milk from cultivated land, when it is already necessary to import 7 million tons of foodgrains per annum, other than to meet the demand from such hospitals and child-rearing centres and clinics as use it, for export to Hong Kong, and as a component in the customary diet of the increasing number of foreigners now visiting the People's Republic.

4.2.1 The mathematics of livestock planning

Officers trained in animal husbandry and its ancillary industries demonstrate their professional enthusiasm by persuading governments to devote large sums to development projects, usually of a costly western type. These may be planned without reference to the high cost of the final product in relation to the purchasing power of most of the population, and without consideration of the targets for the daily or annual consumption of different forms of animal protein. Above all, planners of livestock development rarely consider in sufficient detail whether resources can be provided for the efficient feeding of greatly increased livestock populations, of an improved genotypical composition, and therefore requiring higher and less fluctuating planes of nutrition.

Targets for human nutrition should be defined in terms of the total number of animals needed to provide specified amounts of animal protein. Animal specialists have not yet become accustomed to expressing, say, one million litres of milk a day or one million eggs, or kilograms of beef or pork or poultry meat a year, in terms of the total numbers of livestock required to produce these quantities from the productive part of that population.

Monsoon Asia has several types of animals and forms of husbandry for its wide range of environments. Examples include sacred milch cows in India, Sri Lanka, Nepal and Burma, and profane milch cows elsewhere in Asia; buffaloes for milk; cattle or buffaloes for beef; cattle or buffaloes for draft; camels for draft and milk; sheep and goats for wool, mohair, meat and draft; pigs for meat and manure; and poultry (hens and ducks for eggs and meat) (see Table 4.6).

Before making calculations for input and output of relatively efficient forms of animal production in Asia, it should be remembered that millions of animals of all types are maintained under scavenging conditions in and around the villages. No report has been found giving their actual and potential contribution to the food resources as a whole, and this would be a most difficult undertaking. Many local studies refer incidentally to this type of animal husbandry. In Java and Madura, more than half the families out of a sample of 900 farms studied in six counties (Hill, 1973) kept ruminant animals (bovines for draft, goats secondary in importance). Lower income families with small farms or unfavourable environment kept animals to augment their gross earnings to an average annual figure per caput of U.S.$21, compared with the overall average of U.S.$27. The cash cost of raising these animals is practically nil, beyond the initial investment for the purchase. There is no significant difference in the consumption of animal products between families raising animals and those without animals at equal income levels.

The efficiency of animal production under scavenger and extensive systems of management has been studied. Odend'hal (1972) has calculated the energetics of Indian cattle on the basis of a study of cattle and other domestic animals in an area of 15 sq. km. in rural West Bengal. Feed consumption and measurements of productivity are extrapolated to the entire cattle population and an energetic balance sheet presented for a one-year period. The cattle population appears to be fairly stable, despite its high density. The age distribution reveals a demographic imbalance in both the younger and older age groups which contributes towards a more efficient utilization of available feed supplies. A major role of the cattle is to convert items of little human value into products of direct human utility. The gross energetic efficiency of the entire cattle population is approximately 17 per cent. The current management of cattle is considered to be appropriate for the ecosystem in which they exist.

Animals maintained on grazing ecosystems, particularly in arid and semiarid regions (Whyte, 1971), experience periods of undernutrition at some time each year as a result of both qualitative and quantitative deficiencies in the diet (Allden, 1970 – about 170 references). A knowledge of the effects on livestock of feed deprivation becomes of economic importance during times of feed scarcity, because the producer needs to keep his animals alive at the lowest cost possible. This is usually achieved by allowing animals to use their own energy conservation mechanisms, a normal physiological response in both wild and domesticated herds and flocks, namely, the accretion of body tissue during periods of feed surplus and the subsequent depletion of these reserves during periods of scarcity. This is an inefficient way of using feed, but the alternative of controlling seasonal deficiencies by the feeding of conserved fodder is not generally economic. Allden (op. cit.) finds that in field situations there is little likelihood that the capacity of animals to produce and reproduce in later life will be affected by nutritional deprivation in early life. Under the changing climatic environment of the field, it is improbable that cattle and sheep could survive diets which might influence adversely their future productivity. Thus feeding for survival would in itself ensure the stability of future productivity.

It is now accepted that, in the breeding of dairy cows under tropical conditions, crossbreds between local breeds and European dairy breeds not only produce more milk than the local breed but also more than pure European maintained under the same conditions (I. L. Mason, personal communication, 1974). Optimal performance may be obtained by some intermediate between the two breeds, maintained as a crossbred population in three different ways. This raises the question of the creation of new breeds, by a. introducing European bulls into a zebu population, b. introducing zebu bulls into a European breed population, c. using three or more breeds to produce breeds of multiple racial origin, and d. crisscrossing between local and improved breed.

In their study of the developing dairy industry of India from the points of view of correct animal nutrition and the provision of feeds and fodders for this relatively intensive form of animal husbandry, Whyte & Mathur (1968) have expressed targets of milk production per day or per year in terms of total bovine population and bovines in milk needed to provide these daily levels of production, the amounts of concentrate feeds and green and dry fodders that these bovine populations need if they are to be well fed (grazing pastures are not a practicable proposition in this environment), and the hectares of land that are needed to produce these gross requirements.* This was done in respect of national targets relating to the total population of India in 1965 of 480 million, a total which has now reached almost 600 million. Calculations were also made on the basis of individual milk processing plants with planned throughputs ranging from 2,000 to 250,000 litres per day.

To obtain an expression of litres input of milk plant in terms of the number of bovines and of hectares of land needed to maintain their optimal level of milk production, the agricultural and livestock targets may be stated as follows: 1. the total productive bovine population (in milk, dry and growing) that is required to provide a sufficient percentage of animals in milk at any one time to produce a given quantity of milk per day; 2. the total amounts of the locally available or potential green and dry fodders, concentrate feeds and grazing needed to feed this total livestock population at all periods of the year; and 3. the area of farm or other land needed to produce these feeds, fodders and grazing, taking into consideration the average local standards of farming, and the average yields of the fodder or cereal crops or the carrying capacity of the grazing land.

The amount needed to meet the Government of India objective of 0.17 litre or 6 oz. per head per day for a human population of 600 million is 100 million litres of milk per day.

European x Zebu crossbred cows or milch buffaloes with a performance

---

* In the calculations made in this section with regard to dairy cattle, pig and poultry production, the final figures indicate only the livestock population that would be needed to yield the products in question, not the *total* livestock population from which allowance should be made for 30 to 50 per cent losses through disease (see 4.2.4).

of 2,000 litres per lactation (two to four times the all-Indian average) may be taken as an example. Bovines giving 2,000 litres or above per lactation may be assumed to demonstrate relative efficiency of feed conversion in relation to average Indian animals. Bovines with this high productivity also make it possible to argue that the money return per hectare from fodder crops in the ultimate form of milk may be as high or even greater than that obtained from the usual arable crops of the region. Allowance has also to be made for the relative efficiency of artificial insemination, loss through disease, and the need for a high culling rate in European x Zebu crossbreds (which cannot be applied in India).

To provide 100 million litres per day, it would be necessary to have a total productive bovine population of 80 million, of which 20 million would be in milk at any one time. This population would require 48 million tons of concentrates, plus 360 million tons of green fodder (produced on 15 million hectares of cultivated land) plus 250 million tons of dry fodder (cereal straws from 70 million hectares).

The actual production of milk in India in 1965 was about 3 million litres per day. It was difficult to visualize a total daily milk production of much more than 10 million litres, or 0.02 litre per head of human population per day at that time, because of: a. the insoluble cattle problem involving the need to feed a co-existent population of uneconomic animals at least with straw and rough grazing; b. the lack of sufficient cultivated land to produce the required amounts of concentrates and green and dry feeds and fodders to meet the special needs of productive bovines; and above all d. the high cost of the final product in relation to average purchasing power.

Thomas Yu of the Animal Industry Division, Joint Commission for Rural Reconstruction, Taipei, has applied the same technique to the planning of pig production in Taiwan, to meet specified targets of a population which is 80 per cent rural (see Table 4.7). Calculations for the supply and demand of pig feeds in Taiwan indicate sufficiency in respect of sweet potato, sorghum and rape seed oil meal, and deficit in respect of wheat bran, rice bran, maize, soybean and peanut oil meal. It is stated that, for the total pig population in Taiwan in 1968 of about 6.2 million, the total feed requirement was about 4 million tons, including 3.2 million tons of sweet potato.

From 1949 to 1959, the pig population in the People's Republic of China varied from 57 million to 180 million, and has subsequently been said to have increased considerably. For ease of calculation, we may take the figure of 240 million. The annual slaughter from such a population, taking the Taiwan figures as the basis, would be 126 million pigs. The average dressed weight at slaughter, based on the average of 2 million pigs imported per year into Hong Kong from China (a major factor limiting pig production in Hong Kong itself – Mason, Mack & Tsui, 1968), is 50 kg. Thus the amount of pig meat available per annum for the total population of China, urban and rural, is 8 kg. per head.

Most of the pigs produced in mainland China are purchased for the urban people, confirming the vegetarian nature of the rural diet. There is no uniform system for rationing of pork in Chinese cities. When supplies are plentiful, pork is freely sold, but in times of scarcity, it may be rationed for several months at a time.

If sweet potato vines and tubers are used in pig rations in China to the extent reported for Taiwan, one million pigs will consume 0.5 million tons of this feed. If the 8 million hectares under sweet potato in China give the average Taiwan yield of 16 tons per hectare, the Chinese production is 128 million tons per annum. A population of 240 million pigs in China consuming sweet potato at the Taiwan rate would require 120 million tons. In order to raise per caput consumption of pig meat annually to 25 kg., 800 million pigs would be required, calling for a production of 400 million tons of sweet potato from 25 million hectares of land, plus the other essential grain and concentrate feeds already in short supply, with perhaps saccharified crop residues.

In most of rural Asia, pig husbandry is at the scavenger level characteristic of subsistence economies. 'Unless the whole of traditional subsistence agriculture is changed, there is no scope for commercial pig production' and the upgrading of stock will fail because the improved animals would not be adapted to the health and nutritional rigours of a scavenging ecosystem. The great increase in the production of pig meat visualized in the FAO Indicative World Plan is dependent upon better use being made of by-products. There are huge amounts of rice bran available (15 kg. of processed product for human use corresponds with 5 kg. of useful byproducts, from which 1 kg. of pork can be produced if fed with suitable concentrates to highly productive pigs); also many other vegetable wastes, the value of which is often underestimated as a result of poor processing and inadequate compounding. Research is necessary on optimal utilization of agricultural and industrial byproducts in animal nutrition. Small-scale processing plants may be established within the framework of rural development, integrating pig and poultry production in small and medium-scale commercial units (D. Kroeske, personal communication, 1974). All this depends ultimately on the dietary preferences and purchasing power of the consumer.

Calculations can be made for poultry (hens and ducks) in order to raise the egg consumption in India and China, for example, from twelve to 25 or more per head per year. Again these forms of production of animal protein are dependent primarily on the produce from cultivated land, and therefore come into competition with both the human population and the bovine population being maintained for milk production. The situation in India is such that, in those parts of the country where dairy development has been vigorously promoted, it is accepted that only some 30 per cent of the hen eggs will be produced in commercial units from hens giving an average of 150 eggs per year, and 70 per cent from birds maintained on a scavenger basis and giving only 60 eggs per year in a seven-month production period.

Table 4.7. Taiwan: Planning pig production to meet targets in human nutrition (Thomas Yu, personal communication, 1968).

A. Requirement for pig meat

Year	Human population (million)	Actual and anticipated consumption per head per annum (kg.)	Total annual production to meet these targets (metric tons)
1965	13.0	14.48	188,709
1966	13.3	14.76	196,973
1967	13.6	15.08	205,223
1968	13.9	15.45	214,523
1969	14.1	15.85	224,119
1970	14.4	16.29	234,446

B. Livestock population required to achieve these targets

Year	Total number of pigs to be slaughtered a year[1] (million)	Sows required to maintain this production	Boars per year	Total pig population (million)	Growing animals (million)
1965	2.7	206,200	2800	5.5	2.8
1966	2.9	217,250	2750	5.7	2.9
1967	3.0	229,300	2700	6.0	3.1
1968	3.2	242,350	2650	6.2	3.3
1969	3.4	256,400	2600	6.5	3.5
1970	3.6	270,450	2550	6.8	3.7

[1] Average slaughter weight, six to ten months old: total liveweight 90 kg.; meat 63 kg.

| 1965 | 2.7 | 206,200 | 2800 | 5.5 | 2.8 |
| 1966 | 2.9 | | | | |

C. Supply and Demand of Livestock Feeds in Taiwan in 1968 (metric tons)

Feedstuff	Local production	Available for Feed	Requirement for Feed	Shortage
Rice bran	244,960	244,960	256,302	11,342
Sweet potato	3,795,000	3,219,309	3,219,309	–
Wheat bran	8,450	8,450	89,013	80,563
Maize	96,000	86,700	207,356	120,656
Sorghum	20,400	18,400	18,400	–
Soybean	91,800	–	211,300	211,300
Peanut oil meal	45,500	45,500	69,778	24,278
Rapeseed oil meal	27,300	27,300	20,220	–

D. Areas of Cropland Required

Crop	1965 Plant Area	1965 Harvested Area Ha.	1966 Plant Area	1966 Harvested Area Ha.
Sweet potato	234,145	234,060	235,567	235,443
Wheat	11,119	11,119	14,507	14,356
Maize	18,704	18,615	22,328	22,220
Soybean	53,176	53,156	51,326	51,323
Peanut	103,642	103,621	98,244	98,026
Rapeseed	17,593	17,593	17,783	17,679
Sorghum	4,101	4,090	5,102	5,100

In assessing the overall potential for the production of human food from the animal industries of Asia, we should perhaps consider production at only two contrasting economic levels (the polarization discussed by Whyte, 1971, in another connection), the highly efficient units ensuring maximum utilization of feeds for production of eggs for urban markets, and at the other extreme, the scavenger birds around every village in Asia, which provide sporadic amounts of animal protein in the rural diet.

For village conditions, Allan McArdle (1965) has proposed a simple approach to the upgrading of Indian village desi stock with crossbred males (Fig. 4.2). If this programme is continued over two years up to the standards indicated, the village should then have birds capable of laying 90 or even 100 eggs per year, under reasonably good scavenging and range conditions, provided full expression of the increased but intermediate genetic potential may be obtained by increasing somewhat the intake of locally available feeds (McArdle, 1972).

As supplies of appropriate poultry feeds become increasingly available, it becomes possible to consider further change from a free-range scavenger system to a deep litter enterprise. This leads to the evolution of quite a new productive ecosystem, in which the poultry droppings which were formerly lost now become available as a rich deep litter organic manure. A. A. McArdle & J. N. Panda have shown (personal communication, 1965) that 33 to 40 laying birds can produce in one year a no-cost byproduct of one ton of deep litter manure which can make possible the production of an extra ton of grain from one acre (0.4 ha.) of irrigated land.

4.2.2 Origin and conservation of genetic resources

The modern breeder of the different types of domestic livestock on the basis of resources at present available to him should be informed on their origin and evolution. This is the field of the palaeontologist and the archaeologist working in association with specialists in the recent and modern distribution

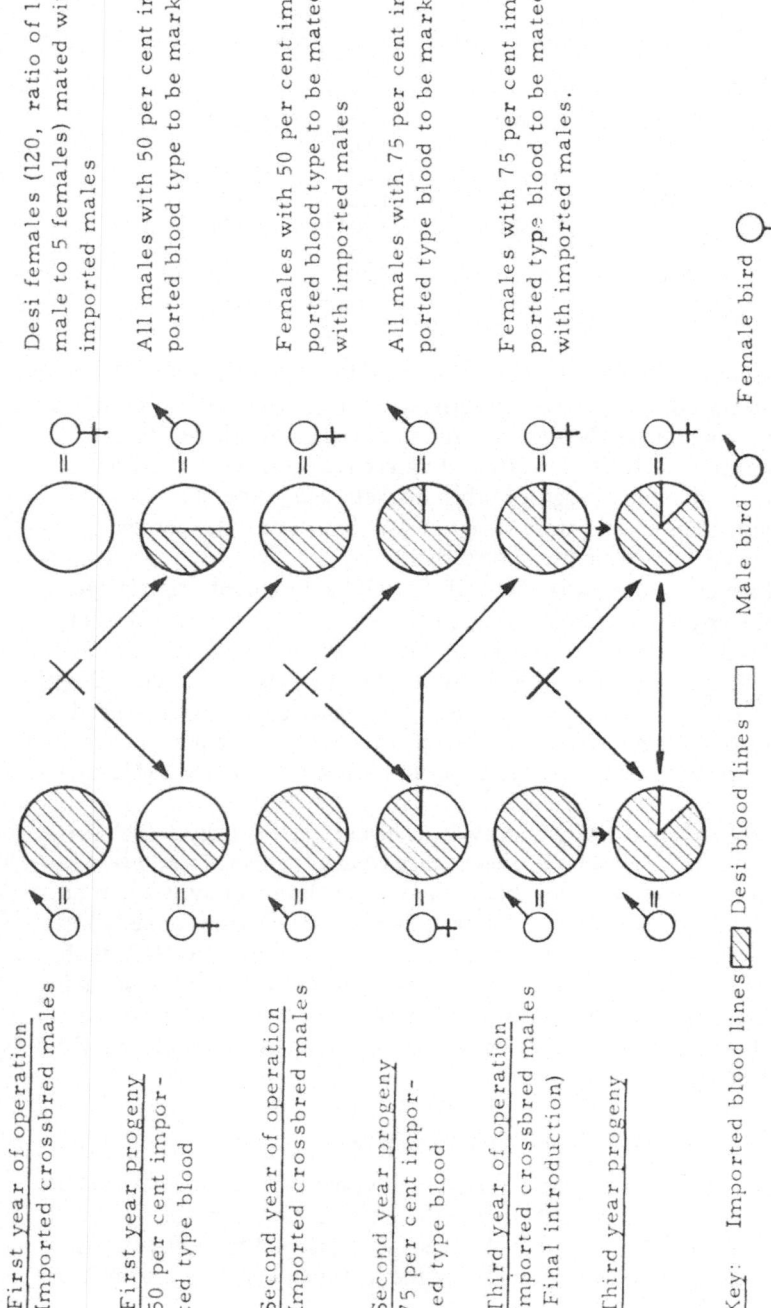

First year of operation
Imported crossbred males

Desi females (120, ratio of 1 male to 5 females) mated with imported males

First year progeny
50 per cent imported type blood

All males with 50 per cent imported blood type to be marketed

Second year of operation
Imported crossbred males

Females with 50 per cent imported blood type to be mated with imported males

Second year progeny
75 per cent imported type blood

All males with 75 per cent imported type blood to be marketed

Third year of operation
Imported crossbred males
(Final introduction)

Females with 75 per cent imported type blood to be mated with imported males.

Third year progeny

Key: Imported blood lines   Desi blood lines   Male bird   Female bird

Fig. 4.2. Diagram of a simple approach to upgrading village desi poultry stock with cross-bred males (McArdle, 1965).

and migration of the wild ancestors of the present types; hopefully also with the palaeobotanists, to relate the ancient and recent vegetation to the fauna for which it provided food and shelter in the evolving biological ecosystems. An introduction to the subject and to the extensive literature is provided by Zeuner (1963) on the history of domesticated animals, by the proceedings of a research seminar on the domestication and exploitation of plants and animals held in the University of London (Ucko & Dimbleby, 1969), and, with particular reference to the goat, by the study of French (1970) which commences with the evidence for the existence of goat-like animals in Miocene and Pliocene times, and of goat remains found in Europe and Asia in the Pleistocene. The early history of the buffalo in Asia is discussed by Cockrill (1974), and in India by Whyte & Mathur (1966).

In 1965/1966, FAO noted the existence and the need to conserve the many types and breeds of animals which are little known outside their relative habitats in developing countries. Some of these have obvious merits, including especially that of adaptability to the environmental conditions which often left much to be desired with regard to climate, nutrition, husbandry, and disease or parasite control. Many of these local types, of potential value in breeding programmes, are in danger of being lost or seriously diluted by admixture with new foreign types, without regard to the need for maintaining adequate flocks and herds in their original unimproved state. FAO convened a Study Group of nine specialists in Rome in November, 1966, to consider the evaluation, utilization and conservation of animal genetic resources. This involved discussion of methods of evaluation, and of some special techniques which might be adopted in programmes of conservation and utilization.

The Study Group recognized five different categories of animals, the first having the greatest significance to the most people: a. those farm animals which were the major sources of the world's animal protein and/or draft power on farms. This group included cattle, water buffaloes, sheep, goats, swine, poultry, horses and donkeys; b. animals adapted to special circumstances, such as the camel under desert conditions, the llama, alpaca, guanaco, vicuña under Andean conditions, and the yak which was adapted to the highlands of Central Asia; c. animals within the above categories which had special characteristics, such as resistance to a particular disease or parasite; the dwarf *brachyceros* cattle of West Africa were mentioned as a type in danger of extinction but which had a high degree of tolerance to trypanosomiasis; d. animals within the above categories which were threatened with extinction or with severe dilution through the introduction of outside blood, and for which adequate data were not available to evaluate their potential usefulness, but for which there was sufficient promise to justify their retention until evaluation procedures could be carried out; e. small domestic animals which had been proven useful in many circumstances, such as domesticated rabbits, cuy, cobayo and other wild species, which had special potentials either in their wild form or for crossing with domesticated animals or for possible domestication.

If the priorities are to be established between these on an objective basis, it becomes essential to evolve adequate measures of productivity as a basis for critical evaluation. It was the task of the geneticist to estimate how much of the variation in productivity is due to heredity and how much to environment.

The special techniques which may be suitable for application in programmes for the conservation of genetic resources include blood groupings (genetic polymorphisms in the blood and other tissues and fluids of farm animals, which might be useful in studies of population dynamics, but which could in no way replace direct evaluation procedures); storage of semen (use of deep-freezing technique, giving storage for at least 15 years in cattle, one month in sheep, and some days for the other animals); super-ovulation and storage of ova (techniques of interest only if a simple transplantation method is available; results already obtained with storage of ova justify further efforts to improve duration). It is concluded that semen preservation associated with storage of ova and a simple technique for transplantation could provide the cheapest way of maintaining animal populations over the years.

The Study Group recommended that FAO prepare a handbook to describe experimental designs and procedures, and to outline the yardsticks and measurements to be used. Documents presented included:
'Methods of evaluation that may be applied and research needed to aid in establishing priorities' — P. Mahadevan, Dean, Faculty of Agriculture, University of the West Indies, St. Augustine, Trinidad, West Indies.
'Blood grouping as a means of revealing similarities and differences between animal populations' — J. Rendel, Department of Animal Breeding, Agricultural College, Uppsala, Sweden.
'Genetic aspects of characteristics to be considered in relation to the utilization of indigenous livestock' — H. Newton Turner, (formerly) Senior Principal Research Scientist, Division of Animal Genetics, CSIRO, Epping, New South Wales, Australia
'Storage of semen' — J. Pagot, Director, Institut d'Elevage et de Médecine vétérinaire des Pays tropicaux, Maisons-Alfort, France
'Blood grouping, particularly as a means of revealing similarities and differences' — J. Pagot.
'Superovulation and storage of ova' — J. Pagot
There is also a specialized bibliography of some 150 titles.

FAO convened a conference on pig production and diseases in the Far East in Bangkok in February, 1968. The Study Group on Animal Genetic Resources at its third meeting in Denmark in April 1971, considered pig breeding (FAO, 1971). Technical aspects covered at the latter meeting included methods of population comparison and selection, interaction of genotype and environment, methods of measuring stress, productivity of indigenous and exotic pig breeds in developing countries, pig breeding with special reference to oestrus synchronization, ovulation control and importation of boar semen, and needs and possibilities for gene pool conservation.

208

The native pigs of the Philippines seem to have good prolificacy, but their growth rate performance compares unfavourably with pigs of exotic breeds. The crossbreds of exotic and indigenous breeds are usually intermediate to the parent breeds with regard to daily gain and feed conversion. It appears also that, under tropical conditions, provided good management and nutrition are being applied, the exotic breeds perform nearly as well as in their countries of origin. Testing facilities to assess the genetic merit of exotic stock should be made available and the spread of high production potential should be carried out through the application of artificial insemination, using semen of exotic boars. Although methods for the preservation of boar semen have yet to be developed further (e.g. freezing), artificial insemination could play a beneficial role in areas where large populations are concentrated near cities.

Some reference was made to the various types of swine native to different parts of China, and in particular to their apparently high level of fertility, early sexual maturity and superior mothering ability. Unfortunately few precise data are available. Some Chinese swine and some other native types do not necessarily maintain these traits when raised under more sophisticated conditions, but this observation is also based on limited data.

The indigenous stock of the Philippines have been improved considerably through upgrading with Berkshire, Duroc Jersey, Tamworth, Landrace, Large White and Hampshire. There is a clear trend towards larger production units in several countries of the Far East, particularly in the Philippines, Taiwan and Singapore.

In spite of continuing neglect, haphazard mating, excessive slaughter and high calf mortality, the genetic stocks of the water buffalo (*Bubalus bubalis*) represent a great potential reservoir for planned breeding and crossbreeding (W. Ross Cockrill, personal communication, 1974). The eighteen milking breeds, which include Jafarabadi, Kundi, Mehsana, Murrah, Nili, Ravi, Nagpuri and Pandharpuri all offer possibilities for greatly increased milk output. The water buffalo is also a potential meat producer; buffaloes bred and reared for slaughter at 14 to 16 months of age can reach a weight of 310 to 330 kg. and provide a dressed carcass of up to 58 per cent (see Ross Cockrill, 1974).

4.2.3 Assessment of climatic factors in relation to animal production

The following paragraphs represent extracts from the paper presented to the Conference on Agroclimatological Techniques at the University of Reading, July, 1968, by P. Mahadevan.

Traditionally, the main climatic factors on which attention has been focused are: temperature, humidity, solar radiation and length of day. The effects of these factors on animal production have been studied by measurement of a. the animal's ability to promote heat loss by such means as increased evaporative cooling from the body surface; b. its ability to reduce heat production by lowering its metabolic rate through more efficient utilization of energy; and c. its ability to withstand a rise in body temperature, or a sequence of compensatory reactions such as alkalosis.

The animal that is unable to promote heat loss shows its lack of tolerance of climatic stress by increased respiratory activity, and by rise in body temperature and pulse rate. The usual method of measuring tolerance, therefore, has been to study respiratory rates, rectal temperatures and pulse rates of animals under different levels of exposure to various climatic conditions. Studies of the relative precision of the different measures of heat tolerance indicate that rectal temperature is usually the more reliable criterion.

Attempts to assess the ability of animals to reduce heat production have not been as extensive as the importance of the problem warrants. The technical difficulties associated with the measurement of the various fractions of the animals' total heat production have not permitted critical studies to be made. Generally, however, any environmental stress will modify the requirements of the animal for energy, water and salt; the need for protein, iron and calcium will also be increased under conditions of excessive sweating. If these increased requirements could be measured precisely under different conditions of temperature and humidity, the data could then be applied to calculating total nutrient requirements under stress conditions.

Assessments of the ability of animals to withstand a rise of body temperature have been based on experiments, observations and trials in the field, and partly on experiments in climatic chambers. In recent years experiments in climatic chambers have been designed to study the effects of various environmental combinations such as (a) 24^{0} C. temperature and 90 per cent humidity, and (b) 35^{0} C. temperature and 25 per cent humidity, on feed and water consumption and on milk production.

Physiological characteristics which have been studied on the assumption that they were associated with tolerance of climatic stress in farm animals include coat and skin character, surface area/body weight ratios, blood composition, hormonal activity and metabolic rate, but the results have not been conclusive.

Mahadevan (op. cit.) concludes with statements of the kind of data that require to be collected, the possibility of selecting animals adapted to different climatic environments, and the importance of changing the environment to suit the selected animals.

An FAO Expert Panel on Animal Breeding and Climatology recommended, inter alia, that the investigation of the following aspects should be encouraged: 1. adaptation to heat and cold, resistance to disease and the ability to thrive on low quality feeds from the point of view of choosing or creating breeds suitable for particular environments; 2. much more precise information about the effect of adaptation to heat and cold, disease and low quality feed on the output of animal products from a given enterprise should be collected, and 3. methods of testing the adaptation of animals to heat and cold, disease and poor.

quality feed need to be evolved and the genetic relations of these characters should be determined. Simple tests and indices of adaptation which could be used by commercial breeders need to be worked out.

To support these studies, FAO was asked to publish handbooks of animal shelter designs for adverse climates (hot, humid tropics, hot or cold arid or semi-arid without irrigation, hot arid and semi-arid with limited irrigation), technique of study of heat tolerance in domestic animals and investigation of effects of breeding and climate on economic production.

4.2.4 Assessment of incidence of animal diseases

The FAO/WHO/OIE Animal Health Yearbook for 1962 contains a report (pp. 284—313) of a preliminary survey of the economic losses caused by animal diseases, which was supported by the FAO Freedom from Hunger Campaign. Even in countries with intensive veterinary activities (France, Ireland, Italy, Britain and U.S.A.), losses still range between 15 and 20 per cent of the total value of annual animal production. In less developed countries but with some veterinary services, losses of between 30 and 40 per cent are common. The tables in the Yearbook show that losses due to certain individual diseases are very high, with the major epizootic diseases, foot-and-mouth disease, swine fever and Newcastle disease taking first place. Losses from chronic and intercurrent diseases (tuberculosis, brucellosis, mastitis) or parasitic diseases appear less dramatic but are still enormous. Even so, there is a danger that they may be underestimated. Only when some of the developed countries began systematic surveys did the true importance of such conditions become apparent to them. It is only with such data that it is possible to show the enormous economic and other benefits which would accrue from greater investment in the prevention, control or eradication of animal diseases in general.

One of the main difficulties experienced in analysing losses from disease is that of classification. Reports considered in the FAO survey distinguish between direct and indirect losses, but there is no general agreement as to which types of loss fall into each category. Some authorities appear to have used *causality* as their criterion, while others have used the *visibility, calculability* or *assessability* of losses, their *significance for livestock owners* or a combination of several of these approaches. To add to the confusion, there exists on the part of the general public a tendency to regard *indirect losses* as more remote, in some way less real and consequently less important than *direct losses*.

Among the most popular methods of assessing disease losses is that of determining average losses per head through sampling surveys (Denmark, Netherlands, Britain). The incidence of disease and average loss per affected animal are determined either by rough estimate or by sampling survey. Determination of total losses may be based on both incidence and average losses in relation to the number of livestock in the affected area or country (Iran, Turkey), or on the incidence factor, the proportion of loss in the production of affected animals, and the total production of meat, milk, wool, etc. (Peru). Other methods of ascertaining total loss consist of determining the losses due to the various factors separately and adding all the results together (U.S.A., France, Japan, Israel, Federal Republic of Germany); or of making the assessment on the basis of the difference between the actual output and the production which could be theoretically expected from an existing livestock population (Ireland).

Nothing further has yet been done by the only competent international agency able to produce the information necessary for this assessment in terms of financial loss. Perhaps it is too big a task (W. Ross Cockrill, personal communication, 1974); perhaps the veterinarians 'are reluctant to believe that the collection and analysis of information on diseases repays the time and energy involved'; or they may agree with an article in the *Lancet* in 1937 by one of the pioneers of medical statistics, Dr. A. Bradford Hill, stating that 'it is exasperating, when we have studied a problem by methods that we have spent laborious years in mastering, to find our conclusions questioned, and perhaps rejected, by someone who could not have made the observations himself'.

Reports of actual incidence of animal diseases have been published since 1957 by FAO, in association with WHO and OIE, in an *Animal Health Yearbook*. This is comprised of tables compiled at FAO Headquarters from replies received to questionnaires distributed by the sponsoring organizations to the veterinary services of their member countries. There is a section on Asia, including the People's Republic of China. The diseases are arranged in tables designated by letters A – Q.

4.3 The human constituents

Planners in national and state capitals change existing or design new economic ecosystems for the production of more and better farm items of plant and animal origin. Only rarely do they assess the capacities of the controlling economic occupants of land, the rural men, women and children. Have they the physical strength to undertake the extra labour which is to be imposed upon them, and the mental capacity to learn and then correctly to apply new techniques in place of their reliable, indigenous methods? Will they respond through greater effort to whatever incentives or 'motivations' (the U.N. agency jargon) are placed before them, in order to induce them to produce more for the national good and for the better nutrition of urban communities, both rather remote, abstract concepts in much of rural Asia?

What further burden can one impose, for example, on the rural people of the 'inner core' of Indonesia, Java, Madura and Bali? These islands contain 65 per cent of the country's population and 7 per cent of its land area; a density of 930 people per square kilometre of cultivated land; the average per caput income in rural areas is little more than U.S.$50 per annum. Half of the population of Indonesia obtains an average of less than 1500 calories and 25

Table 4.8. Codes adopted in the *FAO Animal Health Yearbook*

Animal Group

01	av:	avian (poultry and other birds)
02	bov:	bovine (including buffalo)
03	cam:	camel
04	can:	canine
05	cap:	caprine
06	con:	rabbit
07	eq:	equine
08	fau:	wild fauna
09	ov:	ovine
10	sui:	swine
13	pel:	fur-bearing animals

Incidence of Disease

–	Not recorded; obviously not present
(–)	Not recorded; probably not present
	Year last occurrence: below symbol.
	0000: never.
?	Suspected but not confirmed
(+)	Exceptional occurrence
+	Low sporadic incidence
++	Moderate incidence
+++	High incidence
...	No information available
+\	Disease much reduced, but still exists
+ϕ	Confined to certain regions
+⇐	Mostly in imported animals
+!	Disease only recently recognized in country
+∿	Seasonal occurrence
+. .	Disease exists; distribution and incidence entirely unknown

Control of Disease

test	Systematic testing under official control scheme
Cn	Control of non-vertebrate vectors
Cr	Control of wildlife reservoirs
P	Prohibition of imports from infected countries
Q	Quarantine, movement control and other precautions at frontier and inside the country
Qf	Quarantine and other precautions at frontier
Qi	Quarantine measures and movement control inside the country
S	Slaughter policy
T	Treatment (therapeutic and preventive)
Tp	Preventive treatment
Tt	Therapeutic treatment
V	Vaccination
✶	Notifiable disease

213

Table 4.9. Assessment of rural economic ecosystems

Population	Density expressed as pressure on the land, nutrition density, persons per unit area of the staple food crop
Social and agrarian structure	Settlement pattern, land tenure, inheritance, leadership, lot viable, man hour per type of production; sociology of water use
Economic status	Owner/tenant/landless labour; subsistence status, average return from labour, indebtedness
Communications	Road or water systems to market towns or other communities
Social environment	Housing, water supplies, sanitation, health services
Seasonality	Climate, employment, diet, communication to market, rest period(s), disease(s)
Vulnerable groups	Proportion of population, nutrition, health, morbidity, mortality, age of participation in adult activities
Education	Availability and accessibility; to what level available; at what stage of education do children wish to leave land; parents' aspirations for their children
State assistance to cultivators	Credit, farm loans, improved seeds, improved livestock, artificial insemination centres
Extension	How many extension personnel per 10,000 cultivators? Number and location of demonstration centres, field plots. Evaluation of results
Progress	Acceptance of new farm techniques, and health and nutrition practices; who is more responsive (men or women), why, and in response to what incentives; resistance, taboos.

grammes of protein per head per day. The infant mortality rate is estimated at 125 per thousand live births, and may be assumed to be higher in rural areas. The inner core is characterized by severe overcrowding in rural areas, serious under-employment with consequent low wages, and tiny farm size (about half a hectare) with no room for expansion, since the present extent of cultivation is already endangering the total environment. It is to obtain the basic data necessary for a planned attempt to extricate these rural people from their straitjacket of poverty of land and other resources that the Rural Dynamics Study has been proposed by the Agro-Economic Survey (Chapter 7).

In Table 4.9 are given some of the socio-economic aspects which should be included in a survey of the human factor in rural economic ecosystems.

Size of land holding

A fundamental decision which has to be made at the start relates to the size of the land unit which will guarantee maximum production per hectare of land and per operating man unit. This is something which must be decided by objective experimentation in the major crop and livestock producing ecosystems of Asia. These experiments must be free of the liberal, supposedly socialistic emphasis on land reform, so popular in certain national and international circles, meaning every cultivator on his own plot of land, however small, inefficient and uneconomic. Because of the success achieved by the People's Republic of China — efficient communal farming on 95 per cent of the cultivated land, with private plots for the farmers' own use limited to 5 per cent — this system should be given a fair trial throughout the rest of Asia. Experiments should be designed to show whether such alienation of land from private ownership can be made acceptable to the cultivators in rural economic ecosystems with a less totalitarian political basis. It is necessary to consider the national welfare versus the perfectly justifiable feeling of security of the cultivator when he is growing food for the subsistence of his own family on his own land.

In the Philippines an innovative approach to rural development has begun on an integrated area basis. The area chosen is an elongated plain comprising 312,000 ha. of rich land cutting through the heart of Camarines Sur and part of Camarines Norte and Albay in southern Luzon, with the main emphasis on Camarines Sur. 90 per cent of its irrigated land is planted to high-yielding varieties of rice with a high level of technology, producing an annual surplus. A critical problem is the structure of agricultural tenancy, which is characterized by a few large estates at the one extreme and a large number of small land-owners, each having tenants, at the other. There are about 40,000 rice and maize share-cropping tenants within an average size of 1 ha. are Some 42,000 ha are affected annually by floods.

Long term success in the project hinges on the organization of these share-tenants into 2,500 compact farms. A compact farm is essentially a production-oriented organization of farmers operating as a single management unit, under the leadership of an elected farm co-ordinator. The unit serves as a channel for supervised credit and for dissemination of farm technology from a training centre. It also acts as a credit guarantee group and water-management unit. A federation of compact farms forms the nucleus of the Barrio Co-operative and ultimately for marketing and processing co-operatives.

The project involves the conversion of marshes into 760 ha. of fishponds, and fish pens will be constructed on the three large lakes in the basin for the economic advancement of the 1,500 small-scale fishermen in the vicinity. Cattle production will be stimulated by the construction of a new highway,

215

and swine production is planned in eight towns to supplement income from farming.

An example of the recurring controversy on land reform may be taken from the *Far Eastern Economic Review* for 8 July, 1974, where H. R. von Uexkull (Tokyo) puts the case of the much-maligned 'plantation ecosystem' with special reference to Negros Occidental, Philippines. Over 75 per cent of the sugar plantation workers' children are at school, against 50 per cent for the nation as a whole. Medical care is incomparably better than in other parts of the country, and labour conditions on many of the bigger, well-managed plantations are also above average. A correspondent (*Review*, 20 June) has suggested that the poverty of the 1,700,000 plantation workers would be alleviated if each were given one hectare of land.

An average plantation may produce at least 125 piculs per hectare. The national average is only 80 piculs, implying that many submarginal farms produce only 60 piculs or less, because of lack of capital, know-how and other matters at the farm scale. 'If a plantation that is currently producing 125 piculs was divided into 1 hectare units, within a year or two production would drop to 80 piculs or less, which means that even at the present record price for sugar, the 1 hectare farmer's income would not be Pesos 3,200 but Pesos 2,200, even if we assume that his production costs would be only Pesos 45 per picul instead of Pesos 60, and provided he can transport and market his crop with the same efficiency as a larger plantation.

'But, just a year and a half ago, the sugar price was only Pesos 90 a picul. At that price the 1 hectare farmer's net income would be only Pesos 165!. . . very soon there would be no cheap sugar available on the domestic market for the low-income consumer and no sugar available for export' (von Uexkull).

The situation on plantations described above may be compared with a case study of a pineapple plantation in South Johor, Malaysia (Neville, 1964), particularly the sections on labour organization and wages and the estate settlement (information is not available at time of writing on the social organization on the Del Monte estates in the Philippines). Whether or not the majority of the employees live in a nucleated settlement on the estate, the pineapple estates lack the additional focus that a processing unit provides on the rubber (Ooi, 1961), and larger oil-palm estates where processing plant, smokehouses or mills usually provide a major feature of the estate nucleus. There are, however, fundamental similarities which distinguish all Malaysian estates from the smallholdings (size, capital investment, overall organization, labour organization and wages, physical layout of plantation and estate settlement and the problems of research, production economics and marketing). The pineapple estates in Johor were latecomers in the field of agricultural enterprise, after the era of indentured labour, and most of the labour is from local sources (over 80 per cent Chinese, 15 to 20 per cent Malay).

But if land reform is to become general policy, what then is to be the optimal size of holding, the 'lot viable', on which a cultivator may produce enough of all basic nutrient requirements for his family plus some food or cash crop for sale? It is to be assumed that the ideal holding should be of a mixed nature; monoculture is an economically dangerous practice, whether of rice, rubber or any other crop, and a one-crop economy does not provide for balance in minimum nutritional requirements. It is impossible to generalize from one country to another, or from one economic ecosystem to another. Whyte (1964, pp. 78 et seq.) describes experiments laid down at the Indian Agricultural Research Institute, New Delhi, the Institute of Agriculture, Anand, Gujarat, and at Agricultural Schools in old Bombay State (now Maharashtra). These were designed to show the optimal size of holding on which a cultivator could grow foodgrains and legumes for his family and fodder crops for his cattle. A statistical analysis of results of comparisons between mixed farms and controls was made by Finney & Panikkar (1953).

Harrisson (1970) quotes a letter dated February, 1960, on this subject from the Director of Agriculture, Sarawak, in which he states that no great success had been achieved in experiments in a number of countries. Desirable or planned objectives in rubber-producing ecosystems are quoted:

Malaya	7 acres rubber, 2 acres padi, 2 acres fruit
Sri Lanka	3 acres rubber, 2 acres padi
Sabah	10 acres rubber, 3 acres padi
Sarawak	5 acres high-yielding rubber, 2 acres irrigated wet padi, 0.5 acre for house plus vegetable/fruit/coconut garden; as wet padi acreage becomes scarce, it will be replaced with 5 acres upland for cash crops other than rubber, e.g. coffee, fruits, coconuts, pineapples.

Barlow & Chan (1969) of the Rubber Research Institute of Malaysia have reviewed the pattern of size of holding and associated factors in both the small-holding and estate sectors of the West Malaysian rubber industry, using figures collected between 1960 and 1966. It is stressed that all that can be studied are certain desirable adjustments in size, and that the attainment of one single optimum is not possible. The main criteria in working towards such adjustments are the most profitable use of national resources, the provision of a reasonable income and long-term security for rubber workers, and the need to ensure the continued existence of groups of progressive producers.

On these criteria, it is concluded that the maintenance of the existing, large estate sector is desirable. Where subdivision occurs and the estate cannot be bought over by the workers, steps should be taken to ensure that the units formed are family-owned and operated farms of 6–7 acres (2.4 to 2.8 ha.). In the smallholding sector the prevalence of large numbers of persons earning very low incomes indicates that accelerated steps should be taken to open up new land for development, perhaps linking this with the consolidation of existing holdings. For rubber smallholdings a size goal of 6–8 acres (2.4 to 3.2 ha.) of high-yielding material is felt to be reasonable under anticipated circumstances and able to provide earnings per supported person exceeding those on large estates. An important feature is that under these stated 'size goal' conditions a considerably greater number of persons can also be supported per hectare than on estates.

In assessing the rural Asian as an economic resource in plans for greater production, it is necessary to recognize that man is a highly variable animal in respect of energy and response to education and incentive. On centralized land development schemes at Nakashibetsu, Hokkaido, it was found that one third of the farmers made full 100 per cent use of the land and facilities provided, one third only 50 per cent, while one third were intermediate in these respects. This is a factor of dilution which has to be considered in translating results from experimental plots and pilot farms to the field.

Incidence of disease

Rural health is a major factor governing the capacities of the crop and animal husbandmen and their families to undertake new and onerous field work in development areas. One may assess these capacities on the basis of defined ecological zones. In a mountainous country like Nepal, for example, one village in 1,500, working out at a ratio of one person per 1,500, was selected as the sampling unit for a national health survey (Worth & Shah, 1969) (see also Chapter 7 under Nepal).

It would theoretically be helpful to know what size of human sample and/ or type of formula are acceptable as statistical indications of the incidence of some specific disease sufficiently debilitating to affect the capacity to work. Models have been designed for this purpose for a number of tropical diseases, especially malaria (Macdonald, 1973). The Royal Tropical Institute, Amsterdam (J. W. L. Kleevens, personal communication, 1975) considers that it is always possible to make assessments of incidence of disease, but their validity depends on many factors. Paul Chen (personal communication, 1975) confirms this view, adding that pragmatic factors often reduce the accuracy of assessments. There is, with schistosomiasis, for example, no simple way of doing this (P. F. Basch, personal communication 1974). The presence of the disease can be surveyed by two principal methods, the skin test and stool examination; the former is simpler, faster and less reliable, the latter more quantitative in relation to actual egg output and public health significance. Samples are usually taken of schoolchildren, because they have high infection rates and are accessible at their school. But one cannot proceed from such surveys and estimate the percentage infection in an area unfamiliar to the investigator. Neither do the models published of the epidemiology of diseases such as malaria, stable or unstable, and schistosomiasis permit prediction of percentage infection (but see also WHO, 1965).

Standard textbooks on tropical diseases such as that of Adams & Maegraith (1966) describe the global distribution of diseases and the specific ecological circumstances in which the diseases and/or their carriers are to be found. It is perhaps because of the wide variation between individual human beings and communities in respect of nutritional status, natural or acquired immunity, susceptibility influenced by concurrent infection with other disease patterns of activity in relation to exposure, etc., that it has not been found possible to take small statistically significant samples, and to extrapolate accurately from these to a much larger population.

In these circumstances one has to turn for information on prevalence of disease to the studies of the medical geographers, such as Learmonth (1958) on the medical geography of inter-War British India, and Learmonth (1965) and Misra (1970) on health in post-Independence India (see Fig. 4.3 for areas in the subcontinent where malaria was prevalent before the introduction of preventive measures).

Descriptions of surveys of the occurrence of tropical diseases, indicating methodology of survey and the relations between the environment and disease, appear regularly in the specialist medical literature. Reference may be made, for example, to the ecology of the malaria vectors (Sandosham, 1970), of the vectors of filariasis (Macdonald, 1967) and its occurrence in Malaysia (Wilson, 1961; 1967) and south Sulawesi (Partono et al., 1972); dengue in Thailand (Pant et al., 1973); and schistosomiasis in Sulawesi (Carney, Masri, Salludin & Putralli, 1974). From the Amazon Basin in South America, we have the example of the increase of helminthiasis with increase in population, and localized improvement following improvement of environmental sanitation (Schwaner & Dixon, 1974).

Rural nutrition
The dietary habits, nutritional practices and standards and minimal nutritional requirements of the rural Asians have been discussed fully elsewhere (Whyte, 1974b). The information that is required for an assessment of nutritional status has been summarized (Table 4.10) by WHO (1963). In *FAO Nutrition News-letter,* Lörstad (1971) discusses the concept of recommended intake and its relation to nutrient deficiency, and presents a method to enable the calculation of the percentage of a population which is deficient in a given nutrient, assuming that the distributions of intake and requirement are both Gaussian. Later, the same author (1974) discusses statistical estimates of the incidence of undernutrition, based on the actual distribution of the intake and assumed Gaussian distribution of requirement. Survey methodology and the methods of processing and storing data collected in food consumption surveys have been applied in the National Food and Nutrition Programme, Zambia (Lörstad, 1969) – p. 27, *Nutrition Newsletter* 12/1 – See also publications on field studies and food habits, quick survey technique for children (Blankhart, 1971); guide for personal interviews and questionnaire (den Hartog, 1973); manual for study (Guthe & Mead, 1945); clinical and anthropometric methods (Jelliffe, 1966).

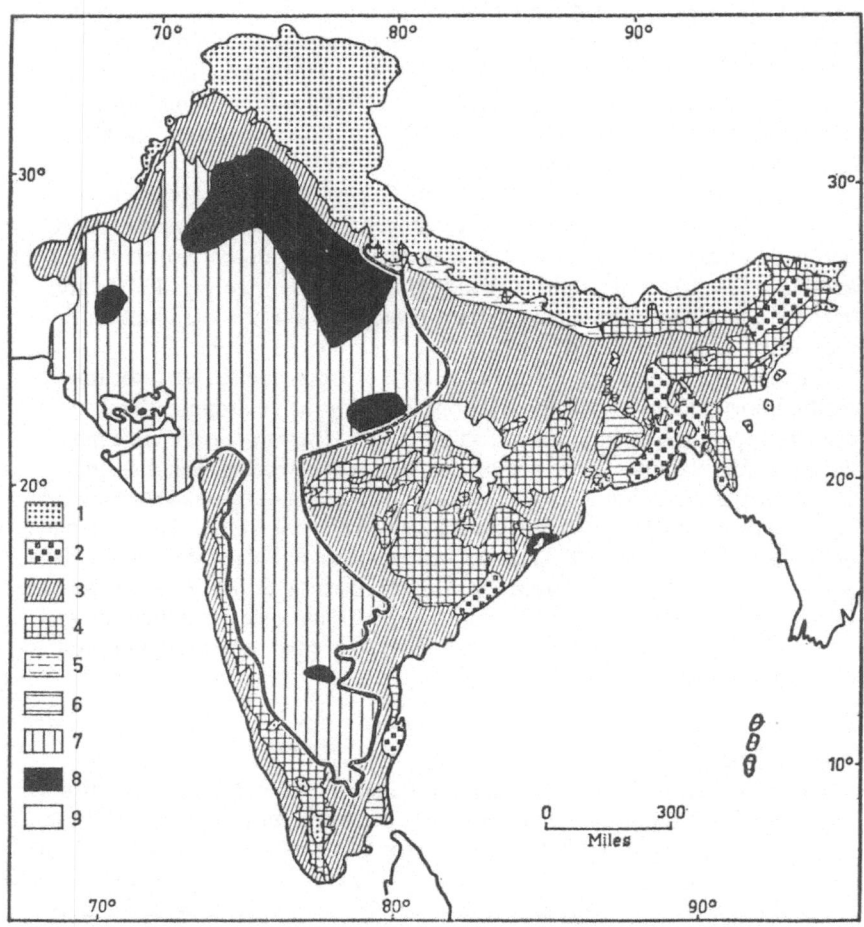

Fig. 4.3. South Asian Subcontinent, Incidence of Malaria as governed by environmental factors (from Spate & Learmonth, 1967, based on the work of S. R. Christophers & J. A. Sinton, in Annual Report of the Public Health Commissioner for 1940 (Delhi, 1941); note that this refers to a period before the introduction of preventive measures. Key: 1. Areas over 5,000 ft. (non-malarious); 2. Known healthy plains (spleen rate under 10%); 3. More or less static moderate to high endemicity, intensity depending on local factors – seasonal variations moderate, fulminant epidemics unknown; 4. Hyperendemic jungly hilly tracts and *terai*; 5. Probably hyperendemic hill areas; 6. Hyperendemic other than hills; 7. Variable endemicity of drier areas, usually with autumnal rise in fever incidence (potential epidemic areas), spleen rate low except in years following epidemics or in special local circumstances, much affected by irrigation conditions; 8. Known areas liable to fulminant epidemicity (diluvial) malaria, spleen rate high during and immediately after epidemics, slowly falling to low rates in c. 5 years; 9. unsurveyed. The heavy line marks the broad division between endemic and epidemic areas.

220

Table 4.10. Information needed for assessment of nutritional status (WHO, 1963).

Sources of information	Nature of information obtained	Nutritional implications
(1) Agricultural data Food balance sheets	Gross estimates of agricultural production Agricultural methods Soil fertility Predominance of cash crops Overproduction of staples Food imports and exports	Approximate availability of food supplies to a population
(2) Socio-economic data Information on marketing distribution and storage	Purchasing power Distribution and storage of foodstuffs	Unequal distribution of available foods between the socio-economic groups in the community and within the family
(3) Food consumption patterns Cultural-anthropological data	Lack of knowledge, erroneous beliefs and prejudices, indifference	
(4) Dietary surveys	Food consumption	Low, excessive or unbalanced nutrient intake
(5) Special studies on foods	Biological value of diets Presence of interfering factors (e.g. goitrogens) Effects of food processing	Special problems related to nutrient utilization
(6) Vital and health statistics	Morbidity and mortality data	Extent of risk to community Identification of high-risk groups
(7) Anthropometric studies	Physical development	Effect of nutrition on physical development
(8) Clinical nutritional surveys	Physical signs	Deviation from health due to malnutrition
(9) Biochemical studies	Levels of nutrients, metabolites and other components of body tissues and fluids	Nutrient supplies in the body Impairment of biochemical function
(10) Additional medical information	Prevalent disease patterns, including infections and infestations	Interrelation of state of nutrition and disease

BIBLIOGRAPHY

ABRAHAM, T. P., – 1966 – Investigations on field experimental techniques with rice crop. 2. Sampling in field experiments for estimation of plant characters and incidence of pests and disease. *Ind. J. Agric. Sci.* 36: 180–192.

ABRAHAM, T. P., P. ISRAEL & G. VEDAMURTHY, – 1963 – Sampling for estimation of stemborer infection in rice. *Ind. J. Agric. Sci.* 33: 174–185.

ADAMS, A. R. D. & B. G. MAEGRAITH, – 1966 – Clinical Tropical Diseases (4th ed.). Blackwell Scientific Publications, Oxford. 582 pp.

AHUJA, L. D., R. P. GOSWAMI & S. S. KUCHHAWAH, – 1965 – Estimation of body weight of Zebu cows from heart girth measurement. *Ann. Arid Zone* 4: 17–23.

ALLDEN, W. G., – 1970 – The effects of nutritional deprivation on the subsequent productivity of sheep and cattle. *Nutrition Abstr. Rev.* 40: 1167–1184.

AUSTIN, A., V. K. HANSLAS & H. D. SINGH, – 1969 – Improvement of cereal proteins by genetic and agronomic means. *Proc. Nutrition Soc. India* no. 7: 1–12.

AUSTIN, A., V. K. HANSLAS, H. D. SINGH & P. RAGHAVIAH, – 1970 – Protein survey of improved Indian wheat (*Triticum aestivum* L.) varieties. *Indian J. Agric. Sci.* 40: 302–308.

ALLAND, A., Jr., – 1969 – Ecology and adaptation to parasitic diseases. In 'Environment and Cultural Behaviour,' ed. A. P. Vayda, pp. 80–89. Natural History Press, New York.

BAKER, H. G., – 1974 – The evolution of weeds. *Ann. Rev. Ecol. Systematics* 5: 1–24.

BARLOW, C. & CHAN Chee-Kheong, – 1969 – Towards an optimum size of rubber holding. *J. Rubb. Res. Inst. Malaya* 21: 613–653.

BEHRMAN, J. R., – 1969 – Supply Response in Underdeveloped Agriculture: A Case Study of Four Major Annual Crops in Thailand, 1937–1963. University of Pennsylvania Press, Philadelphia. 466 pp.

BLANKHART, D. M., – 1971 – Outline for a survey of the feeding and the nutritional status of children under three years of age and their mothers. *J. Trop. Pediat. Environmental Child Health* 17: 175–186.

BOFINGER, V. J. & J. L. WHEELER (eds.), – 1975 – Developments in the Design and Analysis of Field Experiments. Proceedings of Symposium at University of New England. Bull. 50. Commonwealth Bureau of Pastures and Field Crops, Hurley. (about) 175 pp.

BOSE, A. B., S. P. MALHOTRA, L. P. BHARARA & C. S. K. JOHORY, – 1966 – Socioeconomic aspects of animal husbandry. *Ann. Arid Zone* 5: 72–80.

BRENCHLEY, G. H., – 1964 – Aerial photography for the study of potato blight epidemics. *World Rev. Pest Control* 3: 68–84.

BRENCHLEY, G. H., – 1968a – Aerial photography for the study of plant diseases. *Ann. Rev. Phytopathology* 6: 1–22.

BRENCHLEY, G. H., – 1968b – Aerial photography in agriculture. *Outlook on Agriculture* 5: 258–265.

BYRNE, G. F., C. W. ROSE, J. E. BEGG, B. W. R. TORSSELL & H. G. McPHERSON, – 1971 – Instrumentation for Crop-Environment Measurement in a Tropical Savannah Climate. Tech. Paper no. 32. Division of Land Research, CSIRO, Canberra, 19 pp.

CARNEY, W. P., S. MASRI, SALLUDIN & J. PUTRALLI, – 1974 – The Napu Valley, a new schistosomiasis area in Sulawesi, Indonesia. *Southeast Asian J. Trop. Med. Pub. Hlth* 5: 246–51.

CHAMPION, H. G., – 1969 – The effect of human population on the forests of the Indian Sub-Continent. In 'Problems of Land Use in South Asia,' ed. U. Schweinfurth & M. Domrös, pp. 19–28. Yearbook of the South Asia Institute, University of Heidelberg. Otto Harrassowitz, Wiesbaden.

CHANG, T. T., G. C. LORESTO & O. TAGUMPAY, – 1974 – Screening rice germ plasm for drought resistance. *SABRAO J.* 6: 9–16.
CHANG, T. T., S. D. SHARMA, C. ROY ADAIR & A. T. PEREZ, – 1972 – Manual for Field Collectors of Rice. IRRI, Los Baños. 32 pp.
CHIARAPPA, L., H. C. CHIANG & R. F. SMITH, – 1972 – Plant pests and diseases: assessment of crop losses. *Science* 176: 769–773.
CHIENGMAI UNIVERSITY, THAILAND, – 1970 – Seminar on Shifting Cultivation and Economic Development in Northern Thailand. Report.
CIMMYT (International Maize and Wheat Improvement Center), – 1969 – Prospects for better protein in wheat, rye and triticale. *CIMMYT News, Mexico:* 4 (7–8): 1–3.
COCKRILL, W. R., – 1968 – The buffaloes of Borneo. *The Veterinarian* 5: 93–103.
COCKRILL, W. R (ed.), – 1974 – The Husbandry and Health of the Domestic Buffalo. FAO, Rome, 993 pp.
COLE, R., – 1959 – Temiar Senoi agriculture: a note on aboriginal shifting cultivation in Ulu-Kelantan. *Malayan For.* 22: 260–271.
COLWELL, R. N., – 1956 – Determining the prevalence of certain cereal crop diseases by means of aerial photography. *Hilgardia* 26: 223–286.
DEOSTHALE, Y. V. & V. S. MOHAN, – 1970 – Locational differences in protein, lysine, and leucine content of sorghum varieties. *Indian J. Agric. Sci.* 40: 935–941.
DEOSTHALE, Y. V., V. NAGARAJAN & K. C. PANT, – 1970 – Nutrient composition of some varieties of ragi (*Eleusine coracana*). *Indian J. Nutrit. Dietet.* 7: 80–84.
DEOSTHALE, Y. V. & K. C. PANT, –1970 – Nutrient composition of some red rice varieties. *Indian J. Nutrit. Dietet.* 7: 283–287.
DEOSTHALE, Y. V. & K. C. PANT, – 1971 – Nutrient composition and amino acid pattern of some high yielding maize varieties. *Indian J. Nutrit. Dietet.* 8: 244–248.
DEVENDRA, C. & M. BURNS, – 1970 – Goat Production in the Tropics. Commonwealth Agricultural Bureaux, Farnham Royal. 190 pp.
DILLON, J. L., – 1968 – The Analysis of Response in Crop and Livestock Production. Pergamon Press, Oxford. 135 pp.'
DYKE, G. V., – 1974 – Comparative Experiments with Field Crops. Butterworth, London. 222 pp.
EMPSON, D. W. (ed.), – 1968 – Integrated pest control. (Papers presented at conference of advisory entomologists). *J. Appl. Ecol.* 5: 489–516.
ETIENNE, G., – 1973 – India's new agriculture: a survey of the evidence. *South Asian Rev.* 6: 197–213.
ETIENNE, G., – 1974 – The Global Process of Rural Development in China. Asian Documentation and Research Center: Graduate Institute of International Studies, Geneva. mimeo. 30 pp.
EVANS, G. C., – 1972 – The Quantitative Analysis of Plant Growth. Blackwell, Oxford, London, Edinburgh, Melbourne. 734 pp.
FERRARIS, R., – 1973 – Pearl millet (*Pennisetum typhoides*). Review Series no. 1. Commonwealth Bureau of Pastures and Field Crops, Hurley. (343 references).
FERWERDA, F. P. & F. WIT (eds.), – 1969 – Outline of Perennial Crop Breeding in the Tropics. Misc. Papers. no. 4. Land bHogesch., Wageningen. 511 pp.
FINNEY, D. J. & M. R. PANIKKAR, – 1953 – Experimental tests of mixed farming in India. *Indian J. Agric. Sci.* 23: 269–281.
FOOD AND AGRICULTURE ORGANIZATION OF THE UNITED NATIONS, – 1965 – Poultry Feeding in Tropical and Subtropical Countries. FAO Agricultural Development Paper no. 82. FAO, Rome, 96 pp.
FAO, – 1966 – Proceedings of FAO Symposium on Integrated Pest Control, 1965. FAO, Rome.
FAO, – 1967 – Report of the FAO Study Group on the Evaluation, Utilization and Con-

223

servation of Animal Genetic Resources, Rome, 21–25 November, 1966. FAO, Rome. mimeo. 31 pp.

FAO, – 1968 – Poultry Keeping in Tropical Areas. FAO, Rome. 57 pp.

FAO, – 1969 – Report of the Second Meeting of the FAO Expert Panel on Animal Breeding and Climatology, Rome, 25–29 November, 1968. FAO, Rome, mimeo. 90 pp.

FAO, – 1970a – Amino-Acid Content of Foods and Biological Data on Proteins. FAO, Rome. 285 pp.

FAO, – 1970b – Crop Loss Assessment Methods: Manual on the Evaluation and Prevention of Losses by Pests, Disease and Weeds. Commonwealth Agricultural Bureaux, Farnham Royal. 276 pp. Supplement, 94 pp.

FAO, – 1971 – Report of Third Ad Hoc Study Group on Animal Genetic Resources (Pig Breeding). FAO, Rome. mimeo. 21 pp.

FAO, – 1972 – Vegetable Research in South-east Asia. Third Meeting of the Technical Advisory Committee of the Consultative Group on International Agricultural Research. FAO, Rome. 66 pp.

FAO/IAEA, – 1970 – Mutation Breeding for Disease Resistance. Proceedings of a Panel, 12–16 October, 1970. FAO, Rome. 244 pp.

FAO/U.S. DEPARTMENT OF HEALTH, EDUCATION AND WELFARE, – 1972 – Food Composition Table for Use in East Asia. FAO, Rome/National Institutes of Health, Bethesda, Md. 334 pp.

FAO/WHO/OIE, – 1971 – Animal Health Yearbook, 1970. FAO, Rome. 334 pp.

FRENCH, M. H., – 1970 – Observations on the Goat. FAO Agricultural Studies no. 80. FAO, Rome. 204 pp.

FUKUI, H. & E. TAKAHASHI, – 1971 – Rice culture in the central plain of Thailand. *Tonan Ajia kenkyu* 8: 518–533.

GARLICK, J. P. & R. W. J. KEAY (eds.), – 1970 – Human Ecology in the Tropics. Symposia of the Society for the Study of Human Biology, Pergamon, Oxford. 112 pp.

GOLLEY, F. B. & H. K. BUECHNER, – 1968 – Practical Guide to the Study of the Productivity of the Large Herbivores. IBP Handbook no. 7. Blackwell, Oxford and Edinburgh. 308 pp.

GOMEZ, K. A., – 1972 – Techniques for Field Experiments with Rice: Layout, Sampling, Sources of Error. IRRI, Los Baños. 46 pp.

GREENE, H., – 1966 – Irrigation in arid lands. In 'Arid Lands,' ed. E. S. Hills, pp. 255–271. Methuen/UNESCO, London and Paris.

GUILBRIDE, P. D. L., & H. SILLAU, – 1970 – The adaptation of cattle to high altitude. *SPAN* 13: 177–179.

GUTHE, C. E. & M. MEAD, – 1945 – Manual for the Study of Food Habits. Report of the Committee on Food Habits. Bull. Nat. Res. Council. no. 111. Washington. 142 pp.

HALL, D. W., – 1970 – Handling and Storage of Foodgrains in Tropical and Subtropical Areas. FAO Agricultural Development Paper no. 90. FAO, Rome. 364 pp.

HARRISSON, T., – 1970 – The Malays of South-West Sarawak before Malaysia: A Socio-Ecological Survey. Macmillan, London. 671 pp.

den HARTOG, A. P., – 1973 – A Guide for Field Studies on Food Habits. Food Policy and Nutrition Division, FAO, Rome. mimeo. 35 pp.

HAUSHERR, K., – 1969 – Agricultural colonization in the Kapatagan Basin, Lanao del Norte, Mindanao, Philippines. In 'Problems of Land Use in South Asia,' eds. U. Schweinfurth & M. Domrös, pp. 100–116. Yearbook of the South Asia Institute, University of Heidelberg. Otto Harrassowitz, Wiesbaden.

HILL, T. M., – 1973 – A study of ruminant animal production potential to alleviate protein deficiencies of human diets in Java and Madura. Thesis, University of Illinois, Urbana. mimeo. 108 pp.

HSIEH, S. C. & V. W. RUTTAN, – 1966 – Technological, institutional and environ-
mental factors in the growth of rice production: Philippines, Thailand, Taiwan.
Department of Agricultural Economics, University of Minnesota. mimeo.
HUTCHINSON, K. J., – 1971 – Productivity and energy flow in grazing/fodder con-
servation systems. *Herbage Abstr.* 41: 1–10.
INGRAM, J. C., – 1971 – Economic Change in Thailand, 1850–1970. Stanford
University Press. 352 pp.
INTERNATIONAL RICE RESEARCH INSTITUTE. Annual Reports, IRRI, Los Baños.
IRRI, – 1972a – Rice, Science and Man: Papers presented at the 10th Anniversary
Celebration of the International Rice Research Institute. IRRI, Los Baños. 163 pp.
IRRI, – 1972b – Production of Seedlings: Tropical Rice Grower's Handbook. IRR1.
Los Baños. 24 pp.
IRRI, – 1974 – Planting Rice. IRRI, Los Baños. 10 pp.
JANICK, J., R. W. SCHERY, F. W. WOODS & V. W. RUTTAN,– 1970 – Plant Agricul-
ture. Freeman, San Fransisco. 246 pp.
JELLIFFE, D. B., – 1966 – The Assessment of the Nutritional Status of the Community.
WHO Monograph ser. no. 53. WHO, Geneva. 271 pp.
KACHROO, P. (ed.), – 1970 – Pulse Crops of India. ICAR, New Delhi. 334 pp.
KASASIAN, L., – 1971 – Weed Control in the Tropics. Leonard Hill, London. 307 pp.
KAUL, A. L., M. S. NAIK, R. D. DHAR, V. MEHDI & P. RAGHAVIAH, – 1969 –
Genetic and environmental variation in the quality of high yielding rice. *Proc. Nutrit.
Soc. India* no. 7: 36–44.
KRADER, L., – 1966 – Social life in the arid zones. In 'Arid Lands,' ed. E. S. Hills,
pp. 405–420. Methuen/UNESCO, London and Paris.
KROBER, O. A., – 1969 – Nutritional quality in pulses. *Proc. Nutrit. Soc. India* no.
7: 45–48.
LARSON, K. L. & J. D. EASTIN (eds.), – 1971 – Drought Injury and Resistance in
Crops. A Symposium. Crop Science Soc. America, spec. publ. no. 2. Madison.
88 pp.
LEARMONTH, A. T. A., – 1958 – Medical geography in Indo-Pakistan: a study of
twenty years' data for the former British India. *Indian Geogr. J.* 33: 1–59.
LEARMONTH, A. T. A. – 1965 – Health in the Indian Sub-Continent, 1955–1965.
Department of Geography, Australian National University, Canberra.
LEE, D. H. K., – 1966 – The individual in the desert. In 'Arid Lands,' ed. E. S. Hills,
pp. 387–404. Methuen/UNESCO, London and Paris.
LING, K. C., – 1972 – Rice Virus Diseases. IRRI, Los Baños. 134 pp.
LIONBERGER, H. F. & H. C. CHANG. – 1970 – Farm Information for Modernizing
Agriculture: the Taiwan System. Praeger, New York. 454 pp.
LITTLE, E. C. S., – 1964 – Agricultural aviation potentials for Asia (fertilizer applica-
tion). *J. Roy. Aeronautical Soc.* 68: 127–133.
LIU, Jung-Chao, – n.d. – China's Fertilizer Economy. Aldine, Chicago. 169 pp.
LODGE, G. A. & G. E. LAMMING (eds.), – 1968 – Growth and Development of Mam-
mals (Cows, Sheep and Pigs). Proc. 14th Easter School of Agricultural Science,
Nottingham. Butterworth, London, Plenum, New York.
LÖRSTAD, M. H., – 1969 – National Food and Nutrition Programme – Zambia. Food
Consumption Survey of Rural Zambia – Methodology. ESN:DP/ZAM/69/512. FAO,
Rome.
LÖRSTAD, M. H., – 1971 – Recommended intake and its relation to nutrient deficiency.
Nutrition Newsletter 9: 18–31.
LÖRSTAD, M. H., – 1974 – On estimating incidence of undernutrition. *Nutrition News-
letter* 12: 1–11.
McARDLE, A. A., – 1965 – A Hand Book for Poultry Officers in India. Ministry of
Food and Agriculture, New Delhi. 164 pp. + 21 pp. appendix.

225

McARDLE, A. A., – 1966 – Poultry Management and Production (2nd ed.). Angus and Robertson, Sydney. 742 pp.

McARDLE, A. A., – 1972 – Methods of poultry production in a developing area. *World Animal Review* 2: 28–32.

McARDLE, A. A., & J. N. PANDA, – 1966 – A Poultry Guide for the Villager (4th ed.). UNICEF, New Delhi. 43 pp + 49 pp. supplement.

MACDONALD, G., – 1973 – Dynamics of Tropical Disease. (A Selection of Papers of the late George Macdonald, with a Bibliographical Introduction and Bibliography, eds. L. J. Bruce-Chwatt and V. J. Glanville). Oxford University Press, London. 310 pp.

MACDONALD, W. W., – 1967 – Aspects of mosquito ecology in south-east Asia. Paper to British Ecological Society, autumn meeting, on Ecology of Human Diseases and their Vectors. University College, London.

McDOWELL, R. E., – 1966 – Problems of Cattle Production in Tropical Countries. Cornell International Agricultural Development Mimeograph 17. Ithaca, New York. 24 pp.

McDOWELL, R. E., – 1968 – Climate versus man and his animals. *Nature (Lond.)*:218: 641–645.

McDOWELL, R. E., – 1969 –Effective Planning for Expanding Livestock Production in Developing Countries. Cornell International Agricultural Development Mimeograph 32. Ithaca, New York. 12 pp.

MACFADYEN, A., – 1963 – Animal Ecology (2nd ed.). Pitman, London. 344 pp.

MAHADEVAN, P., – 1968 – The relations between climatic factors and animal production. In 'Agroclimatological Methods: Proceedings of the Reading Symposium.' pp. 115–122. UNESCO, Paris.

MAKALIWE, W. H. & A. PARTADIREJA, – 1974 – An economic survey of East Nusatenggara. *Bull. Indonesian Econ. Studies* 10: 33–54.

MASON, I. L., – 1972 – The history and biology of humped cattle. Paper to Symposium on Zebu Cattle Breeding and Management, 6–10 August, Venezuela. FAO, Rome. mimeo. 16 pp.

MASON, R. R., Y. T. MACK & K. C. TSUI, – 1968 – A review of pig farming in Hong Kong and of factors limiting expansion. *Agric. Sci. Hong Kong* 1: 29–48.

MATTERN, P. J., J. W. SCHMIDT & V. A. JOHNSON, – 1970 – Screening for high lysine content in wheat. *Cereal Science Today* 15: 409–411.

MISRA, R. P., – 1970 – Medical Geography of India. National Book Trust, New Delhi. 205 pp.

MOHAN, V. S. & Y. V. DEOSTHALE, – 1969 – Varietal difference in protein quality of cereals and millets. *Proc. Nutrition Soc. India* no. 7: 23–30.

MOSEMAN, A. H. (ed.) – n.d. – Agricultural Sciences for the Developing Nations. Radhakrishna Prakashan, Delhi. 221 pp.

MURDOCK, C. P., C. S. FORD, A. E. HUDSON, R. KENNEDY, L. W. SIMMONS & J. M. WHITING, – 1950 – Outline of Cultural Materials. Human Relations Area Files, New Haven. 3rd ed.

MYA MAUNG, – 1971 – Burma and Pakistan: A Comparative Study of Development. Praeger, New York. 164 pp.

NELSON, O. E., – 1969 – Genetic modification of protein quality in plants. *Advanc. Agron.* 21: 171–194.

NEVILLE R. J. W., – 1964 – The plantation in Malaya: Case Study of a pineapple plantation in south Johore. *Tijdschrift voor Econ. en Soc. Geografie* 55: 57–69.

ODEND'HAL, S., – 1972 – Energetics of Indian cattle in their environment. *Human Ecology* 1: 3–22.

OOI, Jin-Bee, – 1961 – The rubber industry of the Federation of Malaya. *J. Trop. Geogr.* 15: 59.

OU, S. H., – 1972 – Rice Diseases. Commonwealth Mycological Institute, Kew. 368 pp.

226

OU, S. H., – 1973 – A Handbook of Rice Diseases in the Tropics. IRRI, Los Baños, 58 pp.

PAKISTAN, POULTRY INSTITUTE, – 1972 – A Guide to Poultry Practice – Poultry Diseases. Poultry Research Institute, Karachi. 90 pp.

PANT, C. P., S. JATANASEN & M. YASUNO, – 1973 – Prevalence of *Aedes aegypti* and *Aedes albopictus* and observations on the ecology of dengue haemorrhagic fever in several areas of Thailand. *Southeast Asian J. Trop. Med. Pub. Hlth.* 4: 113–121.

PARTONO, F., HUDOJO, SRI OEMIJATI, N. NOOR, BORAHINA, J. H. CROSS, M. D. CLARKE, G. S. IRVING & C. F. DUNCAN, – 1972 – Malayan filariasis in Margolembo, south Sulawesi, Indonesia. *Southeast Asian J. Trop. Med. Pub. Hlth* 3: 537–47.

PELZER, K. J., – 1945 – Pioneer Settlement in the Asiatic Tropics. International Secretariat, Institute of Pacific Relations, New York/American Geographical Society, Special Publication no. 29. 290 pp.

PELZER. K. J., – 1963 – Mass migrations and resettlement projects in south-east Asia since 1945. In 'Proceedings Ninth Pacific Science Congress,' vol. 3, pp. 189–194. Secretariat of 9th Pac. Sci. Congr., Bangkok.

PENNY, D. H. & M. SINGARIMBUN, – 1972 – A case study of rural poverty. *Bull. Indonesian Econ. Studies* 8: 79–88.

POLUNIN, I., – 1960 – The effect of shifting agriculture on human health and disease. In 'Symposium on the Impact of Man on Humid Tropics Vegetation,' ed. Administration of Territory of Papua and New Guinea/UNESCO Science Co-operation Office for South-east Asia, pp. 388–93. UNESCO, Djakarta.

PROTEIN ADVISORY GROUP, – 1973 – Nutritional Improvement of Food Legumes by Breeding. Proceedings of a Symposium, Rome, 3–5 July, 1972. PAG, New York. 389 pp.

PURDUE AID SORGHUM PROJECT, – Protein and amino acid composition of World Collection of Grain Sorghums. Purdue University, Lafayette.

PURSEGLOVE, J. W., – 1968 – Tropical Crops: Dicotyledons, Wiley, London, Sydney, New York. 719 pp.

RAMASASTRI, B. V. & P. SRINIVASA RAO, – 1969 – Some studies on the nutritive value of rice varieties and pulses. *Proc. Nutrition Soc. India* no. 7: 13–22.

RAO, Y. S., P. ISRAEL, Y. R. V. JAGANNADHA RAO & H. BISWAS, – 1971 – Studies on nematodes of rice and rice soils. 3. Sampling and extraction methods for nematodes. *Oryza* 8: 65–74.

REH, E., – 1967 – Manual on Household Food Consumption Surveys. FAO Nutritional Studies no. 18. FAO, Rome. 96 pp.

van ROY, E., – 1971 – Economic Systems of northern Thailand. Cornell University Press, Ithaca. 312 pp.

RUTTAN, V. W., A. SOOTHIPAN & E. C. VENEGAS, – 1966 – Technological and environmental factors in the growth of rice production in the Philippines and Thailand. *Rural Econ. Problems* 3: 63–107.

SALEEM, M. T. & S. REHMAN, – 1970 – Screening wheat varieties for protein content. *J. Agric. Res. Pakistan* 8: 233–239.

SANDOSHAM, A. A., – 1970 – Malaria in rural Malaya. *Med. J. Malaya* 24: 221–226.

SARDANA, M. G., R. K. KHOSLA & U. M. B. RAO, – 1971 – Sampling technique for estimation of incidence of pests and diseases in paddy crop. *Oryza* 8: 1–14.

SCHOLZ, F., – 1969 – Zum Feldbau des Akha-Dorfes Alum, Thailand. In 'Problems of Land Use in South Asia,' ed. U. Schweinfurth & M. Domrös, pp. 88–99. Yearbook of South Asia Institute, University of Heidelberg. Otto Harrassowitz, Wiesbaden.

SCHWANER, T. D. & C. F. DIXON, – 1974 – Helminthiasis as a measure of cultural change in the Amazon Basin. *Biotropica* 6: 32–37.

SILCOCK, T. H., – 1970 – The Economic Development of Thai Agriculture. Cornell University Press, Ithaca. 250 pp.

SIVARAJASINGAM, S., T. K. MUKHERJEE & D. ISHAK, – 1974 – The genetic performance of local Indian dairy cattle and its crossbreds. *SABRAO J.* 6: 39–45.

SIVASUBRAMANIAN, S., R. SWAMINATHAN & V. SRINIVASAN, – 1968 – Study of variation in yield components of rice varieties. *Madras agric. J.* 55: 177–81.

SORENSON, J. L., – 1969 – The social bases of instability in rural southeast Asia. *Asian Survey* 9: 540–545.

SOUTHEAST ASIAN JOURNAL OF TROPICAL MEDICINE AND PUBLIC HEALTH, – 1971 – 9th SEAMEO TROPMED Seminar on Epidemiology, Prevention and Control of the Endemic Diseases in Southeast Asia and the Far East. Tokyo & Osaka, 6–14 July, 1971. *Southeast Asian J. Trop. Med. Pub. Hlth* 2: 426–429.

SPATE, O. H. K. & A. T. A. LEARMONTH, – 1967 – India and Pakistan: Land, People and Economy. (3rd ed.). Methuen, London. 439 pp.

SPENCER, J. E. & W. L. THOMAS, – 1971 – The geography of health and disease. In 'Asia, East by South: A Cultural Geography,' J. E. Spencer & W. L. Thomas, pp. 108–119. Wiley, New York, London, Sydney, Toronto.

SRINIVASA RAO, P., – 1970 – Studies on the nature of carbohydrate moiety in high yielding varieties of rice. *J. Nutrition* 101: 879–884.

STANTON, W. R. & A. WALLBRIDGE, – 1969 – Fermented food processes. *Process Biochem.* April (unpaginated reprint).

SUKHATME, P. V., – 1972 – Protein strategy and agricultural development. *Indian J. Agric. Econ.* 27: 1–24.

SWAMINATHAN, M. S., – 1973 – The plant population explosion: an attempt by scientists to offset the effects of the human population explosion. *Ceres* no. 32: 11–15.

TRIBE, D., – 1970 – Animal ecology, animal husbandry and effective wildlife management. In 'Use and Conservation of the Biosphere,' pp. 123–141. UNESCO, Paris

UCKO, P. J. & G. W. DIMBLEBY (eds.), – 1969 – The Domestication and Exploitation of Plants and Animals. Duckworth, London. 581 pp.

UNIVERSITY OF CAMBRIDGE, DEPARTMENT OF APPLIED BIOLOGY, – 1971 – Animal Production – a Study of Animal Systems. Memoir no. 43. Cambridge. 20 pp.

VINCENT, J. M., – 1970 – A Manual for the Practical Study of Root-Nodule Bacteria. IBP Handbook no. 15. Blackwell, Oxford and Edinburgh. 176 pp.

WEBER, K. E., – 1969 – Shifting cultivation among Thai peasants: some working hypotheses. In 'Problems of Land Use in South Asia,' eds. U. Schweinfurth & M. Domrös, pp. 67–87. Yearbook of South Asia Institute, University of Heidelberg. Otto Harrassowitz, Wiesbaden.

WHYTE, R. O., – 1964 – The Grassland and Fodder Resources of India (2nd ed.). ICAR Scientific Monograph no. 22. ICAR, New Delhi, 553 pp.

WHYTE, R. O., – 1968 – Land, Livestock and Human Nutrition in India. Praeger, New York. 309 pp.

WHYTE, R. O., – 1970 – Livestock planning for Monsoon Asia. *SPAN* 13: 37–41.

WHYTE, R. O., – 1971 – Grazing in the land ecosystems of India. *Ann. Arid Zone* 10: 110–119.

WHYTE, R. O., – 1973 – An environmental interpretation of the origin of Asian cereals. *Ind. J. Genetics and Plant Breeding* 33A.

WHYTE, R. O., – 1974a – Tropical Grazing Lands: Communities and Constituent Species. Junk, The Hague. 222 pp.

WHYTE, R. O., – 1974b – Rural Nutrition in Monsoon Asia. Oxford University Press, Kuala Lumpur, London, New York, Melbourne. 296 pp.

WHYTE, R. O., – 1974c – The evolution and status of the Gramineae in Asia. *Plant Genetic Resources Newsletter* no. 31.

WHYTE, R. O., – 1975a – The Gramineae of western monsoon Asia: their antiquity and evolution. *Biotropica* 7:

WHYTE, R. O., – 1975b – An environmental interpretation of the origin of Asian food legumes. *Indian J. Genetics and Plant Breeding* 35:

WHYTE, R. O. & M. L. MATHUR, – 1966 – The buffalo in India. *Indian Dairyman* 18: 161–180; also as chapter in *The Husbandry and Health of the Domestic Buffalo* (ed. W. Ross Cockrill), FAO, Rome.

WHYTE, R. O. & M. L. MATHUR, – 1968 – The Planning of Milk Production in India. Orient Longmans, New Delhi. 221 pp.

WIJERATNE, W. V. S., – 1965 – Milk, meat and egg production in Ceylon. *Trop. Agriculturist* 121: 63–68.

WILSON, T., – 1961 – Filariasis in Malaya – a general view. *Trans. Roy. Soc. Trop. Med. Hyg.* 55: 107.

WILSON, T., – 1967 – The epidemiology of filariasis in Malaya. Paper to winter meeting of British Ecological Society, 4–6 January.

WORLD HEALTH ORGANIZATION, – 1963 – Expert committee on Medical Assessment of Nutritional Status. WHO Technical Report Series no. 258. WHO, Geneva. 67 pp.

WHO – 1965 – Expert Committee on Bilharziasis, 3rd Report. WHO Techn. Rep. Ser. no. 299

WORTH, R. M. & N. K. SHAH, – 1969 – Nepal Health Survey, 1965–1966. University of Hawaii Press, Honolulu. 158 pp.

WRIGLEY, G., – 1969 – Tropical Agriculture: the Development of Production (2nd ed.). Faber and Faber, London. 376 pp.

YANG, H. P., – 1971 – Fact Finding with Rural People: A Guide to Effective Social Survey. FAO Agricultural Development Paper no. 52. FAO, Rome. 138 pp.

YANG, M. M. C., – 1970 – Socio-Economic Results of Land Reform in Taiwan. East-West Center Press, Honolulu. 555 pp.

YOSHIDA, S., D. A. FORNO, J. H. COCK & K. A. GOMEZ, – 1972 – Laboratory Manual for Physiological Studies of Rice (2nd ed.). IRRI, Los Baños. 70 pp.

YOUNG, V. R., – 1970 – Protein Nutritional Aspects of Plant Breeding. Document to 17th PAG Meeting. 1.16/4. PAG, New York. 11 pp.

ZEUNER, F. E., – 1963 – A History of Domesticated Animals. Hutchinson, London. 560 pp.

5 THE USE OF LAND IN RURAL ECOSYSTEMS

5.1 Distinction between types of land use.

For the purposes of assessment of potential, it becomes essential to know the relative intensity of use of land by man. In the more densely populated parts of rural Asia, we are concerned in many cases with land in which any recognizable degree of equilibrium between man and the total land or environmental resources has long since disappeared, if expressed in terms of loss of soil or of soil fertility, desiccation, waterlogging or flooding, or other evidences of imbalance. The objective of research and demonstration on this land is to evolve and introduce methods whereby some degree of return back to the lost equilibrium may be achieved, within the limits imposed by pressure of human populations, by low crop yields and livestock productivity, and by chronic undernutrition and malnutrition in man.

Although much of the cultivated land of Asia may not be actively deteriorating in this way, it has nevertheless reached a stage of stagnation in crop yields and livestock productivity. With the inexorable increase of human population living on a fixed resource, with a progressive increase in nutrition density per unit area of land, the cultivators adopt more fertility-removing practices in an attempt to produce more food, or they change from a superior to an inferior foodgrain or other source of carbohydrate more adapted to soils of lower fertility.

On uncultivated Asia also, the land has lost its equilibrium, or is hovering in an uneasy state of balance between correct use and over-use. Excessive deforestation of climax vegetation has led not only to the permanent destruction of the timber resource as such, but also to the loss of its capacity to conserve water from torrential tropical rainfalls for the benefit and protection of cultivated and other rural and urban resources lower down in the same catchment (see also 2.5 for the possible effect of misuse of land systems on a microscale on the dynamics of general circulation in the atmosphere — macroclimate). In the more arid areas, badly managed and excessive grazing and browsing by domestic livestock, cutting for fuel and clearance for cultivation have all been contributory factors.

The older surveys of these types of land, which might be called overdeveloped, tend to be statements and mappings of a static situation, without any serious attempt to suggest how the rural peoples involved might escape or evolve from their social and economic impasse. Contrasting with these types of surveys are those of a dynamic, integrated and forward planning which have been adopted by the CSIRO Division of Land Research on land which

230

might be called underdeveloped, or not yet developed, leading towards a stable man/land equilibrium, in Northern Australia and Papua New Guinea; also the surveys conducted in parts of Africa by the Land Resources Division of the British Overseas Development Administration; or the pre-investment surveys conducted by the Food and Agriculture Organization, generally in association with the International Bank for Reconstruction and Development (Chapter 7).

The Australian Division of Land Research states that the nature of present land use does not provide satisfactory criteria for distinguishing land types. Present land use may not be a true indication of future land-use possibilities. It is only when land use has been very intensive for a long time and has become itself an inherent feature of land, closely correlated with other inherent features, that it is a useful distinguishing feature for basic surveys. Nor do the Australians consider Man in his primitive subsistence or economic ecosystems to the extent that may be thought essential in Asia. 'One natural resource factor, the most important of all, the human occupants of a country to whose benefit this work is directed, will be considered only indirectly' – a decision which would no doubt lead to heated debate were an anthropologist to be included in an integrated survey team (but see 6.2).

Reference should be made in passing to what Griffin (1973) has called 'the first and basic model from which the majority of land use theories have been derived', the so-called von Thünen theory (1826). Von Thünen's primary objective was to determine the relation between the intensities and types of agricultural production and the available markets (see also Hall, 1966). The core of the theory is represented by some basic assumptions: the ideal site consists of completely rational (optimizing) economic behaviour, an isolated state, a single control city, settlement in villages away from the control city, and a racially homogeneous population, uniform topography, uniform climate and soil fertility, and a relatively uniform and primitive transportation system. 'Von Thünen's hypothesis that intensity of land use decreases with increasing distance from a central place is almost a truism; few would dispute the validity or logic of a statement so easily observed. Nevertheless, at a certain level of generalization the theory breaks down' because of the great complexity introduced by factors such as differences in soil fertility, climatic conditions, road patterns, etc. 'Perhaps von Thünen presented a valid theory that is extremely difficult to test empirically – or perhaps he presented not a theory but a method for analysing land-use intensities over space' (Griffin, 1973).

Although there are few parts of the Asian landscape which have physical and cultural characteristics similar to those proposed by von Thünen, his model of approximately concentric rings around major centres of population, the demand for specialized products, and the higher urban purchasing power applies to a greater or lesser extent to parts of China and the countries of the South Asian (Indian) subcontinent. A concentric pattern of decreasing inten-

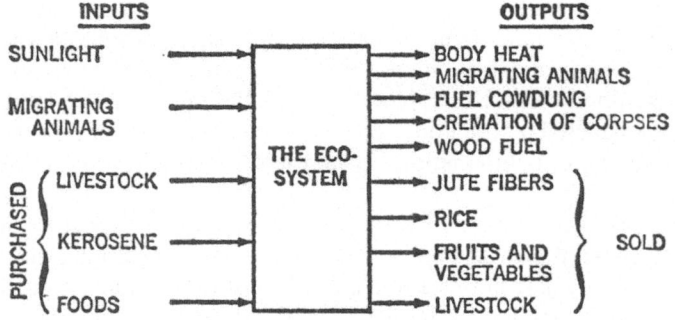

Fig. 5.1. India, West Bengal: Model for rural ecosystem in this densely populated area (Parrack, 1969).

sity of land use has been proposed for the milk procurement areas in India. The radius of the inner zone of intensive fodder cultivation in systems of alternate husbandry is governed by the distance over which it is safe in the tropics to transport milk to a rural chilling station or to an urban milk processing plant. The outer zones would grow fodder and harvest natural grass hay for transporting to the inner zone, would grow fodder seeds, and would maintain rest camps for dry cows and heifers until they came into milk again and were returned to the inner zone. A similar pattern could evolve around the production of fresh vegetables and of poultry products. In fact, a survey of potential in areas of long-established agriculture could well start with a study of the practical and economic possibilities of introducing these catalytic forms of land use and production into an otherwise stagnant economy.

5.2 Intensive land-use patterns

Using the People's Republic of China as an example, reference may be made to Chapter 5 of Buchanan (1970) in which he reviews the attempts which have been made to map and categorize the complex mosaic of types of land found in this country. The survey of land use in China began with the pioneer work of J. Lossing Buck (1937) and his colleagues. Their work was based on a study of 16,786 farms in 68 localities in the eastern agricultural region of China, and covered not only land use, but also size of holding, farm labour, marketing and prices and rural standards of living, nutrition and health.

According to Buchanan (1970, pp. 177–85) the latest and most detailed classification of China into its basic agricultural regions is that carried out by the Chinese Academy of Science in 1962. This was based on a graded classification from four major regions (or first category regions) to a total of 129 regions of the fourth category (Teng, 1963, in Buchanan, p. 177). It is considered that a scientific demarcation of the basic regions, by emphasizing the main issues and showing the direction of development, provides a better idea of what future development is actually possible: to the extent that this classification is based in part on regional trends and an evaluation of future potentialities it is a dynamic classification as opposed to the essentially static classification of earlier workers in this field' (Buchanan, 1970).

Some explanations of the first to fourth category regions of Teng (1963) is necessary (see Buchanan, 1970, pp. 186–90 for provisional list of these regions). The first category regions are based on the characteristics of the crop land/forest/pastoral land complex, on the environmental factors which influence this, and on the historical development of agriculture. In defining regions of the second category, emphasis is on the *potential* number of harvests which can be obtained in a given period, and this indicates to some extent the possibilities or limitations of future development. The third and fourth category regions are based on the combination of crops and livestock, on the methods of farming and the level of production.

Chen Cheng-siang (1970) has described the agricultural regions of China on a broader scale, and M. Y. Nuttonson has provided an appendix of general observations on the physiography and climate of China, with a map of North American climatic analogues (see 3.4.8). Chen Cheng-siang (1970) recognizes three major regions: monsoon region in eastern China, north-western arid region, and south-western cold-high region, and divides each into subregions (Fig. 5.2). According to Chen, the monsoon region (zone of seasonal winds from opposite directions) has the following characteristics (in percentage of total for China): total land area, 50.6; cultivated land 96.1; agricultural population 98.1; food production 97.0; cotton production 96.5

Huang Ping-wei (1961 – from Buchanan, 1970, p. 313) recognizes three major or higher stage units based on natural conditions which it is impossible or very difficult to alter:

	percentage of total land area
eastern monsoon sector	46
Mongolian/Sinkiang highlands	27.3
Chinghai/Tibetan highlands	26.7

and also a series of secondary or lower stage units based on conditions such as soil type or microtopography which are more readily transformed by man (see Buchanan, 1970, pp. 317–20). Huang's description of the tectonic, climatic and historical characteristics of the eastern monsoon sector are given in Table 5.1 quoted from Huang Ping-wei (1961) via Buchanan's translation (1970, p. 314).

In this part of China, the influence of man is very great. Almost the whole area fit for ploughing is in use, natural forests have all been felled, soil erosion

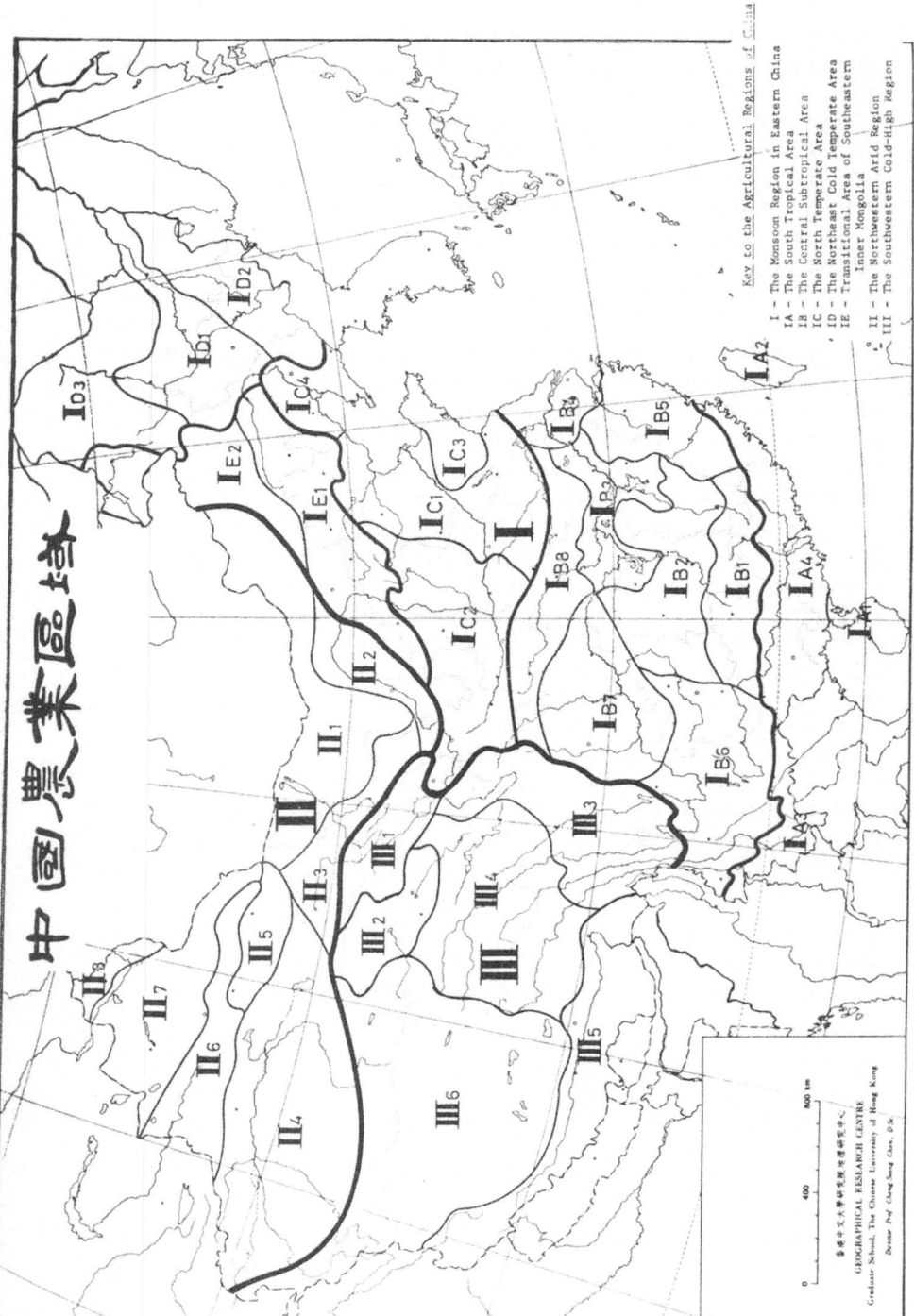

中國農業區域

Key to the Agricultural Regions of China

I – The Monsoon Region in Eastern China
IA – The South Tropical Area
IB – The Central Subtropical Area
IC – The North Temperate Area
ID – The Northeast Cold Temperate Area
IE – Transitional Area of Southeastern
Inner Mongolia
II – The Northwestern Arid Region
III – The Southwestern Cold-High Region

GEOGRAPHICAL RESEARCH CENTRE
香港中文大學研究院地理研究中心
Graduate School, The Chinese University of Hong Kong
Director Prof Chang Sung Chen, D.Sc.

0   400   800 km

Fig. 5.2. People's Republic of China: Agricultural regions recognized by Chen Cheng-siang (1970).

Table 5.1. People's Republic of China: Tectonic, climatic and historical characteristics of the eastern monsoon sector (from Huang Ping-wei, 1961, reproduced by Buchanan, 1970).

Percentage of area of country	46
Major factors defining regional natural differences	Variation in amount of warmth and of temperatures, depending on latitude (northwards from the line of the Tsingling range–river Hwai), also variation in conditions of humidity with varying distance from the sea.
Neo-tectonic movements and topography	Small uplifts; eastwards of the line Chinchow-Cheng-chow-Peking-Hopei subsidence predominates. Absolute heights over the greater part of the territory are not more than 1,000 m. In the regions of subsidence a large percentage of the area is below 500 m. and there is a spacious plain of accumulation.
Climate	Great influence of summer monsoon, comparatively high humidity.
Hydrography	Surface water predominantly from rain supply, sufficiently abundant ground water.
Exogenous processes	Continual weathering, river erosion and accumulation; abrasion and accumulation along the sea coast; weathering by frost in high-latitude mountain regions, aeolian erosion and accumulation in some regions.
Soils	In comparison with the other two sectors, soil profiles comparatively well developed; mechanical composition relatively fine; high humus content; small quantity of dissolved salts; territorial differences often significant.
Vegetation	Predominantly forests, partially steppes.
Phytocoenoses	Ancient coenoses were destroyed little by the Quaternary glaciation. Flora very varied. Geography of plants very complex.
Natural historical factors	Slight evolution from the Quaternary glaciation conditioned the multiformity of biological species; several plants from the end of the Mesozoic and Tertiary periods have been preserved. Ancient red crusts of weathering distributed very widely south of the river Yangtze.
Influence of human activity	Man's influence very great. Almost all the area fit for ploughing is in use, natural forests have been mostly felled; soil profiles destroyed by erosion. In a varying degree waters, micro-climate and micro-topography have been modified.
Basic problem of land utilisation and transformation of nature	This is the most important agricultural region in China, here agrotechnics are gradually being improved; chemical techniques are being applied; irrigation, mechanisation and electrification of agriculture is increasing. Hill and mountain ranges occupy more than half the territory. A wide development of forestry and animal husbandry is necessary.

235

widespread, hydrology, microclimate and microtopography modified. It is in this region that the major efforts are now being made, through soil conservation measures, improvement of agrotechniques, application of fertilizer, irrigation, mechanization and electrification. Uncultivable hill and mountain ranges occupy more than half the land, and there is a need for extensive reforestation and the development of hill-slope animal husbandry on the Japanese pattern. Within this region have been recognized 'areas of high and stable yield' (Donnithorne, 1970) which receive priority for investment grants and agricultural credit, for rural electrification, and for the allocation of fertilizers and agricultural machinery. It is not known what survey techniques and criteria were adopted in the delimitation of these special areas, which resemble the areas selected for Intensive Agricultural Development Projects by Government of India and Ford Foundation in India.

The most densely populated area on the Chinese mainland is the British Crown Colony of Hong Kong. As in Singapore and only to a slightly lesser extent, urbanization is progressively eliminating the residual area of land still devoted to agriculture, horticulture and animal husbandry (Table 5.2). Wong Ming-hung (1974) has proposed five land classes (see also Tregear, 1958). Conditions in Hong Kong will apply increasingly to the many examples, actual and developing, of the effect of massive urbanization on the use of land around such conurbations throughout Asia. Wong (op. cit.) refers to the study of Edwards & Wibberley (1971). These workers have assessed the loss of agricultural area through urbanization and industrialization in terms of agricultural production in Britain. Because the better land is generally taken for non-agricultural purposes, in terms of area alone the loss is of 1 per cent per decade, but in terms of production potential, this loss will amount to almost 2.2 per cent per decade.

In Hong Kong and the New Territories, over 70 per cent of the total land area is classified as woodland, grassland and scrublands and badlands, and only 11 per cent as arable, including orchards and market gardens. Under these conditions, and even accepting the greater initial costs involved, urban

Table 5.2. Hong Kong: Agricultural land use (in sq. km.). (Agriculture and Fisheries Department, 1972).

Year	Fresh Water Paddy	Brackish Water Paddy	Market Garden Crops	Field Crops	Orchard	Fish Pond	Abandoned/ Fallow	Total
1968	50.56	8.00	35.85	11.73	6.42	8.00	15.13	135.74
1969	51.02	7.69	38.07	10.15	6.32	8.37	13.84	135.48
1970	49.13	6.89	39.50	8.62	6.24	9.12	15.38	134.89
1971	45.69	–	40.53	8.00	6.42	10.00	22.90	133.51
1972	37.71	–	41.80	7.12	6.35	10.88	28.08	131.93

and industrial development should, wherever possible, be located on higher elevations, leaving the lower lands for cultivation. This principle has been widely followed in rural Japan until quite recently.

The Soil Conservation, Research, Demonstration and Training Centre, Dehra Dun, U. P., has divided India into 20 land resource regions, by synthesizing information regarding soils, water, climate, topography, vegetation, land use, etc. (Gupta, Tejwani, Mathur & Srivastava, 1970). The names of the regions indicate their location and important characteristics, with respect to climate, or vegetation, or soil (Table 5.3). In the further subdivision of these land resource regions into 186 land resource areas, the most important criterion has been land use (forest land, irrigated land, dryland, etc.).

In presenting these conclusions at the Seminar on Soil Survey and Soil Fertility Research in Asia and the Far East, New Delhi, 1971, Dewan (FAO, 1971) supported the idea of a land resources map on a suitable scale (1:250,000 to 1:500,000) for the use of planners and executors of development programmes in India. The procedure and methods to be adopted in the preparation of land resource inventories would have to be carefully worked out

Table 5.3. India: Land resource regions (Gupta, Tejwani, Mathur & Srivastava, 1970)

Name of land resource region	Area Sq. km.	Percentage of the total land resource area
Northern Himalaya snow-clad region	116,000	3.56
Northern Himalaya alpine grass and meadow region	98,250	3.02
Northern Himalaya forest region	131,750	4.08
Punjab-Haryana alluvial plains region	101,250	3.10
Upper Gangetic alluvial plain region	200,000	6.15
Lower Gangetic alluvial plain region	145,500	4.41
Northeastern Himalaya alpine grass and meadow region	16,000	0.50
Northeastern forest region	161,000	4.97
Assam valley region	88,500	2.73
Rajasthan desert region	191,000	5.87
Rann of Kutch region	46,500	1.44
Gujarat alluvial plain region	62,750	1.94
Mixed yellow, red and black soil region	115,750	3.58
Black soil region	673,500	20.73
Eastern red soil region	573,500	17.65
Gangetic delta region	25,250	0.77
Western coastal region	61,000	1.88
Southern red soil region	347,750	10.72
Eastern coastal region	93,500	2.90
Andaman, Nicobar and other islands	8,000	0.25
Total area	3,076,750 sq. km. or 307.7 million ha.	

(6.6). Field work would be done by the States, with correlation by a central organization to achieve uniformity and consistency.

5.3 Surveys according to type of land use and economy

5.3.1 Shifting cultivation

Reference has already been made to this type of land use, introducing man as a component in tropical ecosystems (2.7); for a comprehensive list of local and tribal names for the system in India/Pakistan/Nepal, Sri Lanka, Andaman/ Nicobar Islands, Burma, Thailand, Malaysia, Indonesia, Philippines, China, Korea and Japan (and the African and American tropics), see Appendix B to Spencer (1966). Shifting cultivation has been variously called bush fallowing, slash and burn agriculture, field/forest rotation, brûlis, swithen or swivven (old northern English dialects), kaingin (Philippines), ladang (Indonesia and Malaysia), chena (Sri Lanka), taungya (Burma, where Leach, 1964 distinguishes between monsoon taungya and grassland taungya), djum (Assam), chitemene (parts of Africa), milpa (Central America). The term taungya is also used by forest officers in India to describe the controlled system of alternation of agriculture with silviculture. This depends upon the annual coupe being placed under agricultural crops and forest species almost simultaneously; crop cultivation ceases when the trees have become established (Chaturvedi & Uppal, 1953), and the agriculturists may move elsewhere, or take employment with the Forest Department and its ancillary industries.

Those who have made a special study of the subject (Pelzer, 1945; Conklin, 1961; Spencer, 1966) conclude that the practice of swidden seeks to preserve the natural equilibrium of the environment; maintains the general structure of the pre-existing natural ecosystem, yielding a larger economic return per unit area than any form of settled agriculture in the hill and mountain country in which it is adopted; this is especially so where tribesmen try to integrate their swidden technology into the jungle ecology on a continuing basis (van Roy, 1971, p. 48).

It would be important to know the relative rates of erosion on land under shifting cultivation and other systems of land use. Writing on the Iban methods, Freeman (1970) states his belief that when only one crop is taken from virgin land, there is little erosion. Reference is made to a Dutch experiment in West Java (Gonggryp, 1941); two adjacent parcels of forest land on rather steep slopes at an altitude of 1,800 metres were selected for study. One piece of land was terraced, the other left in its natural state. Freeman quotes Pelzer (1945): 'During the first year, both parcels lost very little soil because the humus layer protected it; but in the second year the unterraced parcel lost 5,065 kg. of silt per square metre, whereas the terraced parcel lost only one half as much, or 2.524 kg.' Under Sarawak conditions, if virgin rain forest is felled, fired and farmed for one season only and then allowed to recuperate, adequate regeneration takes place (Freeman, 1970). If thereafter the resulting secondary growth (*daman*) is brought into cultivation for one year only at intervals of 12 to 15 years, the land may be so utilized virtually indefinitely. But 'the Iban are prodigal of natural resources'.

On the basis of his work with the Yagaw Hanunoo of southeastern Mindoro Island in the Philippines, Conklin (1957) has evolved an ethnoecological approach to shifting agriculture. Some of the most frequent and problematic statements and assumptions made by various authors are placed below, beside Conklin's (1969) proposed rephrasings, based on the conclusion that the swidden farmer sometimes knows more about the interrelations of local culture and natural phenomena than ethnocentric writers from the temperate zone realize.

1a. Swidden farming is a haphazard procedure involving an almost negligible minimum of labour output. It is basically simple and uncomplicated (Cook, 1921).
1b. Swidden farming follows a locally-determined, well-defined pattern and requires constant attention throughout most of the year. Hard physical labour is involved, but a large labour force is not required (Conklin, 1969).
2a. Usually, and preferably, swiddens are cleared in virgin forest (rather than in areas of secondary growth). Tremendous loss of valuable timber results (Cook, 1921).
2b. Where possible, swidden making in second-growth forest areas (rather than in primary forests) is usually preferred (Conklin, 1969).
3a. Swidden fires escape beyond cut-over plots and destroy vast forest areas. One author states that from 20 to more than 100 times the swidden area itself are often gutted by such fires (Cook, 1921).
3b. Swidden fires are often controlled by firebreaks surrounding the plot to be burned. Accidents happen, but greater damage may result from hunting methods employing fire in an area having a long dry season than from swidden clearing *per se* (Conklin, 1969).
4a. Swidden techniques are everywhere the same. Such features as lack of weeding and the use of a single inventory of tools are practically universal (Hutton, 1949).
4b. Many details of swidden technique differ from area to area, and with changing conditions. Weeding is assiduously accomplished in some regions. Fencing is considered requisite if domestic cattle are kept, less so where such animals are rare. Wooden hand implements are very simple and are used only once. Metal cutting implements and harvesting equipment, however, vary greatly from region to region (Conklin, 1969).
5a. Stoloniferous grasses such as 'notorious *Imperata*' are abhorred as totally useless pests by all groups whose basic economy is swidden agriculture (Hutton, 1949).
5b. Even the most noxious weeds, in one context, may serve the local economy admirably in another. *Imperata*, if dominant, restricts swidden opportunities, but its total loss causes similar hardships for those depending on it for pasture and thatch (Conklin, 1969).
6a. Swiddens are planted with a single (predominant) crop. Any given swidden can thus be said to be a rice or a maize or a millet field or the like. Hence, it is possible to gauge the productivity of a swidden by ascertaining the harvest yield of a single crop (Hutton, 1949).
6b. Swiddens are rarely planted with single or even with only a few crops. Hence the productivity of a swidden can be determined only partially by an estimate of the harvest yield of any one crop (Conklin, 1969).
7a. Furthermore, it is possible to gauge the efficiency (i.e. relative to some other method

239

of agriculture) of a given swidden economy in terms of its one-crop yield per unit of area cultivated (Hutton, 1949).

7b. It appears that the efficiency of swidden farming can be ascertained — relative to some other type of economy — only by taking into account the total yield per unit of labour, not per unit of area (Conklin, 1969).

8a. Swiddens are abandoned when the main crop is in. 'The harvest ends the series of agricultural operations' (Gourou, 1953).

8b. Because of intercropping, the harvest of one main swidden crop may serve only to allow one or more other crops to mature in turn. Plantings and harvests overlap usually for more than a full year, and frequently continue for several years (Conklin, 1969).

9a. There is no crop rotation in swidden agriculture. Instead, soil fertility is maintained only by the rotational use of the plots themselves. The duration of the rotational cycles can be determined by the time interval between successive clearings of the same plot.

9b. Swidden intercropping, especially if wet season cereals are alternated with dry season leguminous crops, amounts to a type of crop rotation, even if on a limited scale. Cycles of field 'rotation' cannot be meaningfully assessed by merely determining the number of years which lapse between dates of successive clearings. The agricultural use of the swidden plot following initial clearing may have continued for one, several, or many years.

10a. Not only is fertility lost, but destructive erosion and permanent loss of forest cover result from reclearing a once-used swidden after less than a universally specifiable minimum number of years' fallowing (set by some authors at 25 years — Gourou, 1953). It is claimed that 'dangerous' consequences of more rapid rotation often result from native ignorance.

10b. It is difficult to set a minimum period of fallowing as necessary for the continued, productive use of swidden land by reclearing. Many variables are at work. A reasonable limit seems to be somewhere between eight and fifteen years, depending on the total ecology of the local situation. Swidden farmers are usually well aware of these limitations (Conklin, 1969).

Following the approach of Conklin (1957), Gourou (1956) — swidden is practised on very poor tropical soils, has a limiting technique using only the axe, is marked by low density of population and low level of consumption — and Pelzer (1945) — marked by absence of tillage, low labour input, no utilization of draft animals or manuring, absence of private land-ownership — Geertz (1968) compares two types of ecosystems, in inner and outer Indonesia, the sawah (irrigated padis) and the swidden. When genuinely adaptive, swidden maintains the general structure of the pre-existing natural ecosystem into which it is projected, rather than creating and sustaining one organized along novel lines and displacing novel dynamics. Swidden agriculture operates in the same way as the tropical forest ecosystem, in the same supernatant, plant-to-plant direct cycling manner. 'In their contrasting responses to forces making for an increase in population — the dispersive, inelastic quality of the one and the concentrative, inflatable quality of the other — lies much of the explanation for the uneven distribution of population in Indonesia and the ineluctable social and cultural quandaries which followed from it.' Kellman (1969) has examined some environmental components of shifting cultivation in upland Mindanao.

The gross characteristics of a swidden system, and therefore the aspects to be incorporated in a specialized survey, have been listed by Spencer (1966 and

240

personal communication, 1973). The diagnostic criteria which are significant in an analysis of the system are given in greater detail by Spencer (1966, pp. 181–6).

1. Practised chiefly by simpler cultures of small total population, but occasionally used by almost anyone to whom the cropping system appears expedient.
2. Human labour chiefly operative, using a few hand tools primarily, but power tools occasionally.
3. Labour patterns frequently cooperative, but involving many variations in working-group structure.
4. Clearing of fields primarily by felling, cutting, slashing and burning, and using fire to dispose of the vegetative debris after drying; in special situations, such as perennially very humid localities, debris may not be fired but may be allowed to rot as a wet mulch.
5. Frequent shifting of cropped fields, normally in some kind of sequence, with land control resting in specified social groupings under customary law, but sometimes occurring under other legal institutions of land control.
6. Many different systems of crop planting in given fields, but both multiple cropping and specialized monocropping present.
7. Use of annual and short-term food crops predominant, but important use of long-term shrub and tree crops common, and use of textile crops happens occasionally.
8. Use of crops primarily for subsistence, but admitting that exchange patterns may reach total sale of whole product.
9. Use of permanent dooryard, village or near-homestead gardens frequent among groups using permanent or near-permanent settlement sites.
10. Yields per acre and per man-hour normally compare with those of permanent-field agriculture within regions in which comparison is correctly made, but yields are often below those of mechanically powered permanent-field agriculture.
11. Small annual cropped area per caput, but comparable to that of other non-powered sedentary cropping systems.
12. Use of vegetative cover as soil conditioner and source of plant nutrients for cropping cycle.
13. When system is efficiently operated, soil erosion no greater than soil erosion under other systems that are being efficiently operated.
14. Soil depletion no more serious than that under other systems of agriculture when operated efficiently.
15. Details of practice vary greatly, depending upon the physical environment and the cultural milieu.
16. Transiency of residence common but not universal, with many patterns of residence according to evolutionary level of detailed system employed and preference of culture group.
17. Operative chiefly in regions where more technologically advanced systems of agriculture have not yet become economically or culturally possible, or in regions where the land has not yet been appropriated by people with greater political or cultural power.
18. Destructive of natural resources only when operated inefficiently, and not more inherently destructive than other systems of agriculture when these are operated inefficiently.
19. A residual system of agriculture largely replaced by more complex systems, except where retention or practice is expedient.

In north Thailand, from 3 to 4 million hectares of forest in catchment areas are subject to shifting cultivation; the cycle has become too short here, and also in the northeast and south, due to increase in human population caused by high birth rates and immigration (doubling every twenty years). Of

Indonesia's 120 million hectares of forest land, about 16 million have degenerated to *Imperata* grassland, and on a similar area the periods of rest between cultivations have become too short. A comparable situation exists in the Philippines, especially on Mindanao. In east Malaysia, the problem has not yet become acute (Conway & Romm, 1973).

The intensification and stabilization of cropping on land used for shifting cultivation require a thorough knowledge of such land. It is theoretically possible to change from an extensive to an intensive form of land use with a higher output per unit area by the use of such modern techniques as mechanization, higher-yielding varieties, and fertilization. Apart from the fact that communities of shifting cultivators do not usually have the necessary financial resources nor the physical strength to undertake such intensification, many examples of failure may be cited as due to the absence of correct classification and evaluation of land (Moormann, 1973). Classification of land for its potential suitability for a change to an intensive form of agriculture is indicated (6.3). For this purpose, a thorough knowledge of crop performance is necessary. These data may be available for some crops, but not for others. Too often, results from research stations and experimental plots are given without a quantitative description of the physical environment, thus making extrapolation to field conditions difficult. In addition, negative results of great potential value to the land planner for bench mark sites are frequently not even published.

5.3.2 Subsistence economies

The forms of agriculture that come under this general head are not individually definable types of land use; rather are they types of rural economy that have become associated with several different types of land use; the basic characteristic is the production of food by the cultivator for his own household, to provide a seed stock for the coming year, and to meet the needs of less fortunate or otherwise employed neighbours, so that a village, and particularly a tribal village 'can maintain itself with minimal resort to external economic relations' i.e. as a rural ecosystem.

Klatt (1970) has stated that problems arise in Asia in particular where only 0.2 ha. of farmland is available per head of population, where less than 250 kg. of grain is produced per caput, where the diet provides no more than 2,200 calories and 50 gm. protein per day, where the population increases by more than 2 per cent per year, and where well over half the population is engaged in agriculture, but contributes less than half the nation's total product. These conditions apply over wide areas of subsistence farming in monsoonal and equatorial Asia.

The production of grain is the chief concern of a subsistence farmer. Other food crops usually play only a marginal role. The cultivator's freedom of choice is greatly restricted. ' In these respects many millions of Asia's one thousand million subsistence farmers do not qualify as innovators and entrepreneurs. Yet without their participation, economic, social and political development is bound to remain stunted' (Klatt, op. cit.).

This assessment is confirmed by the ILO (1973) report on employment and economic status in Asia (Pakistan, India, Bangladesh, Indonesia, Thailand, Philippines), in relation to small-scale subsistence agriculture, but excluding workers on large-scale plantations growing tea, coffee, rubber, sugar, or pineapple. In spite of steadily increasing populations, no substantial new avenues of employment other than small-scale agriculture carried out on traditional lines are forthcoming. Authorities concerned with the region are becoming increasingly aware that any meaningful development has to look at the rural areas first, or, in the words of President V. V. Giri of India in his final address on Independence Day, 1974: ' It is my firm belief that the only durable solution to the challenge of poverty and unemployment lies in the purposeful development of rural India . . . measures to translate this idea into action have been halting and tardy. There has to be a massive programme of settling village communities on land; every inch of land that is available should be utilized.'

But the ILO report notes that not only the landless agricultural labourers who work exclusively for wages (10 per cent of the agricultural labour force in Thailand, Pakistan and Japan, 20 per cent in the Philippines, 25 per cent in India), but a considerable proportion of the smaller cultivating peasants (owner-cultivators, leasehold tenants, sharecroppers with no occupancy rights), lived in the 1960's either very precariously on the margin of subsistence or well below it.

A land use survey may accurately define the location, type and area of farming land from primitive swidden to various forms of settled agriculture. The planners may be well aware of what technical actions could be applied to improve and diversify production of crop and livestock products. But in subsistence economies, it is the human factor which is the most intractable. Man is trapped in a type of rural limbo, between the primitive hunter, collector and swidden operator, who are still component parts of the biological land ecosystem, and the successful economic farmer who produces commodities for export to urban markets. Thus, from Table 5.4 (Wharton, 1970) it can be seen that the physical input factors that would be the subject of a land use survey and of a subsequent planning exercise are a relatively insignificant part of all the factors affecting agricultural development in a subsistence economy.

A conference held in February/March 1965 at the East-West Centre, University of Hawaii (Wharton, 1970) considered the major changes which must take place in the developing world as part of the process of agricultural development, especially the steps which must be taken to promote a change from subsistence to commercial agriculture. And in Asia there is increasing evidence that this must be done on a basis of a declining rural economy and resource, caused by excessive and ever-increasing pressures of human popula-

243

tion on the land. For example, Rahim (in Wharton, 1970), writing of the Comilla Project in East Pakistan (now Bangladesh) states: 'In Bengali folklore we find farming described as a source of wealth and prosperity followed over generations and earning plenty of food and a good wife. A good farmer possessed several pairs of bullocks. . . . The situation today is totally different. The Bengali farmer has lost everything. He has not followed the path of progress. He is bound by the old traditions. To earn a minimal amount of food for subsistence is his biggest problem in life. . . . How to modernize traditional farming? How to do it quickly?'

The Comilla experiment was started by the East Pakistan Academy for Rural Development in 1960, with Michigan State University playing a key role in training and research. A thana with an area of 107 square miles was selected, with a population of 217,297 persons, a density of 2,031 persons per square mile (Census 1961). The greater proportion of the thana is a plain growing rice (80 per cent of the total cropped area). Most of the countryside is inundated during the rainy season to a depth of a few feet. Over 80 per cent of the land lies fallow in the dry winter season, although water from rivers and underground resources would be available if credit and the necessary expertise were provided.

A group of authors (Raper et al., 1970) has attempted to condense a massive set of documents dealing with the complex of institutions at the Comilla project. The reviewer of this volume (Papanek, *Journal of Asian Studies* 31, pp. 217–20), states that its importance 'does not lie in its evaluation or analysis of the material, but in the fact that a wealth of material about the Comilla experiment may now be brought to a new audience in a convenient form, and since the development of viable alternatives for rural economic growth is one of the world's major problems, this contribution to the field is to be welcomed.'

Thus, in attempting in his conclusion (Wharton, 1970, pp. 455–67) to define the items on a research agenda, the main emphasis is on sociology (the nature of subsistence farmers, their rational values, motives, attitudes and aspirations), economics (the dual peasant, resource efficiency, use of labour, response to incentives), and institutional forces. It was generally agreed at the Conference that the technology of subsistence agriculture was low, and that it was fairly constant or changed rather slowly through time. But 1. more evidence is needed on the variation in technological levels and economic performance among subsistence farmers and the factors influencing these variations, 2. most of the work on the new technology has been undertaken by sociologists rather than economists, in spite of the emphasis given by the latter, 3. little is known about the role of research in development, especially in the physical and biological sciences, 4. to what extent are the new technologies geared too much to the requirements of commercial rather than subsistence farmers?

Are the models of change too general to be useful in providing guides to

244

Table 5.4. Subsistence economy: classification of factors affecting agricultural development (Wharton, 1970).

Physical Input Factors

1. Nonhuman physical inputs
   - a. Land
   - b. Climate
   - c. Seeds
   - d. Water
   - e. Fertilizer
   - f. Pesticides
   - g. Structures
   - h. Work animals
   - i. Other animals
   - j. Tools and machinery
   - k. Fuel and power other than animal power
2. Labor.

Economic Factors

1. Transport, storage, processing, and marketing facilities for products.
2. Facilities for the supply and distribution of inputs including credit.
3. Input prices, including interest rates.
4. Product prices, including prices of consumer goods.
5. Taxes, subsidies, quotas.

Organizational Factors

1. Tenure, land.
2. Farm size and legal form
3. General government services and policies
4. Voluntary and statutory farmers' organizations for:
   - a. Coordinating physical input use, e.g. irrigation associations, tractor stations.
   - b. Economic services, e.g. purchase, sale, credit associations and cooperatives.
   - c. Social services, e.g. health centers, schools, family planning centers.
   - d. Local government.
   - e. Diffusion of knowledge, e.g. adult-education classes, youth clubs.

Socio-Psycho-Cultural Factors

1. Integration of agricultural institution, practices, and values within the technosocial matrix of the nation.
2. Public administration factors, structure, values, mode of operation of the innovating bureaucracy.
3. Social structure, cultural values, and dynamics of peasant communities.
4. Process of sociocultural change, barriers, and motivations in the innovative sequence, functional harmony or disharmony in society as its constituent parts change.

Knowledge Factors

1. Organization of basic and applied research.
2. Diffusion of knowledge relating to:
   - a. Technical knowledge, e.g. agronomy, plant genetics, soil science, water management, agricultural engineering, pest control, home technology.
   - b. Economic knowledge, e.g. land economics, general economics, farm management.
   - c. Policy, e.g. politics, public administration, planning.
   - d. General education, e.g. literacy, adult education, mass communication.

action, and is separate diagnosis of each problem area required? Is the main problem a lack of economic opportunities, or rather a lack of aspiration?

5.3.3 Arid and semi-arid lands

The economy of Rajasthan for which the research programme of the Central Arid Zone Research Institute, Jodhpur (Table 5.5) is primarily intended is based mainly upon livestock and livestock products, the rainfall being too precarious and erratic for settled agriculture (Bose et al., 1966). In years of favourable rainfall, a bumper crop of bajra (finger millet or *Pennisetum typhoides*) is obtained, but such years are few and far between. It is therefore from necessity that the people in these arid areas have clung to the livestock industry as their essential means of livelihood. According to the 1956 livestock census, Rajasthan has about twelve million cattle, 8.7 million goats and 7.3 million sheep; the cattle population represents about eight per cent, and sheep and goats twenty per cent of the national total. The five dry districts of Jodhpur, Barmer, Jaisalmer, Bikaner and Churu, which constitute the desert proper, account for thirty-one per cent of the total sheep population of the State. Even under present conditions, the State contributes about 40 per cent of the total wool produced in India.

The entire livestock population in western Rajasthan subsists mainly on natural grasslands, the crop residues here playing little or no part. Lucerne cultivation is restricted to the vicinity of the towns, and is sold in markets, especially for feeding horses. Taking into consideration the livestock population and the availability of grazing land, the grazing incidence per cow unit (weigh-

Table 5.5. India: Central Arid Zone Research Institute, Jodhpur, Rajasthan.

Division	Responsibilities
Basic Resources Studies	Ecology, soil science, geomorphology, systematic botany, geology, hydrology, microbiology, cartography
Plant Studies	Silviculture, agrostology, range management, soil conservation, organic chemistry, horticulture, toxicology, plant protection
Wind Power and Solar Energy Utilization	Climatology, solar energy, wind power
Economics and Extension	Sociology, economics, extension
Animal Studies	Animal ecology (rodent control), animal physiology, animal nutrition
Soil−Water−Plant−Relationship	Plant physiology, agronomy, soil physics, soil fertility

ing about 225 kilograms) in desert districts is two hectares in Jodhpur district, 2.25 in Barmer, one in Jaisalmer, three in Bikaner and one in Churu. These figures represent two to six times what the particular type of grass cover should be expected to carry.

Matters relating to integrated survey in general, and to the techniques to be adopted in the individual disciplines of climatology, hydrology, geomorphology, soil and vegetation studies, animal studies, cartography, and socio-economic surveys of rural communities, settled and nomadic, are dealt with in the annual reports and occasional publications of the Institute, in *Annals of Arid Zone,* and other scientific journals.

In order to assess the resource potential of the arid lands, integrated surveys have been completed in some 45,000 sq. km. in the States of Rajasthan, Haryana, Gujarat and Karnataka. In addition, semi-detailed integrated surveys have been conducted in a Development Block, comprising 34,000 sq. km., mostly in western Rajasthan. Detailed surveys at the village level have also been made on 10,500 ha. (Central Arid Zone Research Institute, 1974). These surveys have provided information with regard to a. land use capability classes based on land use data, erosion hazards, type and intensity of dunes, soils; b. pasture types; c. tree vegetation; d. surface and groundwater resources, water quality, water potential zones and their capabilities to supply groundwater for village tanks and other purposes; e. socio-economic conditions of the people, including characteristics of population, social and economic correlations of caste groups, household structure, rural working force, class of farmers and size of holding, animal/vegetation/human relations, form of settlement, etc. (CAZRI, op. cit.).

5.3.4 Dryland agriculture

Rain-fed agriculture is practised on some 80 to 90 per cent of the cultivated area of monsoonal and equatorial Asia. On an unknown but considerable percentage of the rain-fed area, low and irregular rainfall and wide variation in the duration and intensity of the rainy season(s) are factors seriously limiting crop production. The classification of land based on primary criterion of rainfall reliability and distribution is an essential starting point in all projects involving land planning and adjustment of land use in the arid and semi-arid zones. Under non-irrigated conditions where seasonal drought and not cold is the limiting factor, the 'growing season' is the period over which 'effective rainfall' may be expected (see East Africa Royal Commission, 1955).

FAO has identified low-rainfall countries as those having measurable land areas that could be termed either arid or semi-arid according to Meigs' definition, or where seasonally low rainfall in a humid or subhumid environment presents critical problems. The developing country subgroup is divided into four categories, of which category II, covering 56 per cent of all low rainfall areas in developing countries, is regarded as the most useful source of information on which to base analysis.

A sequel to the high-yielding varieties programme in India was the increasing awareness of a rapidly growing gap in income potential between farms with and without irrigation (especially those receiving less than 1125 mm. rainfall per annum). From 1970/71, Government of India initiated research and development on dry farming, as follows: intensive research for evolving techniques which will help to give maximum returns from the available soil and water resources; practical application of results of available knowledge of soil and moisture conservation, cultivation of drought-tolerant and short-duration crop varieties, new techniques of fertilizer application, etc. (state of knowledge summarized in *A New Technology for Dry Land Farming,* IARI, 1970).

Under an ICAR Co-ordinated Research Project on Dryland Agriculture, initiated in June 1970 with collaboration of the Government of Canada, there were established 15 main research centres, 8 subcentres and one special centre, and a co-ordinating cell located at Hyderabad. These research centres are situated in typical agro-climatic zones with moisture index varying between -80 to -20 per cent, and covering different soil groups: alluvial, red, black and lateritic.

Parallel with the research programme, there is a Development Programme on 24 pilot projects, each of about 3,200 ha., located in close proximity to the ICAR research centres. Master plans were drawn up, based on soil fertility and land use capability survey, hydrological survey and contour survey. Bench mark surveys were made to determine conditions at the start of the project; these are to be repeated at yearly intervals to assess farmers' progress in farming methods and socio-economic conditions. Farm plans are prepared for each farm family, based on soil and water resources, management skills in crop and animal husbandry, and financial resources.

The widening gap between irrigated and rainfed agriculture, an increasing population pressure on the land, a recurring cycle of frequent droughts, and the lack of suitable technology to ensure dependable harvests add new urgency to efforts to meet long range needs for increased levels and dependability of food production in the low rainfall, unirrigated, semi-arid tropics of the world. These factors provided compelling reasons for the creation in July 1972 of the International Crops Research Institute for the Semi-Arid Tropics (ICRISAT) with headquarters at Hyderabad, India.

ICRISAT is the newest of the six international agricultural research institutes, each of which is concerned with the improvement and increased production of one or more crops on a worldwide basis. ICRISAT is supported by the Consultative Group on International Agricultural Research and it is the first of its kind which has been set up for evolving a technology suited for the rainfed areas of the semi-arid tropics. The Institute has an international Governing Board. It has been set up at Hyderabad in India on 1,360 ha. land which has been donated by Government of India.

The primary research objectives of ICRISAT are: 1. serving as a world centre to improve the genetic potential for grain yield and the nutritional quality of sorghum, pearl millet, pigeonpea and chickpea; 2. developing farming systems which will help to increase and stabilize agricultural production through better use of natural and human resources in the seasonally dry, semi-arid tropics; 3. assisting national and regional research programmes through co-operation and support and contributing further by sponsoring conferences, operating training programmes on an international basis and assisting extension activities (J. S. Kanwar, personal communication, 1974).

The classification of Troll & Paffen (1965) has been accepted as the general climatic description of the world region of concern in the farming systems research programme of ICRISAT: V3: Wet and dry tropical climates with 4.5 to 7 humid months ($P > PE$) and 5 to 7.5 arid months ($P < PE$). V4: Tropical dry climates with 2 to 4.5 humid months in the warm season and 7.5 to 10 arid months. V4a: Tropical dry climates with the humid months in the cooler season. To a tropical ecologist, it appears that within the scope of the 'semi-arid tropics' is included a whole range of ecoclimates which may not be compatible within a single research scheme. The inclusion of much of Thailand within its geographical scope is particularly open to question.

The following regions of the world can be identified (Krantz & Kampen, 1973); a 400 to 900 km. wide strip across Thailand, Burma, India and Pakistan; a 900 km. belt across Africa from Senegal to Kenya; a 1,300 km. belt from Angola to southern Tanzania; scattered areas from northeast Brazil to central Mexico; northern Argentina; northern Australia.

The original report of R. Cummings (October, 1971) on the basis of which ICRISAT was established to work on sorghum, *Pennisetum typhoides*, pigeon pea and chickpea, and to study their related farming systems, also recommended that ICRISAT should be closely linked with research programmes for these same crops and farming systems in similar ecological and cropping areas in other parts of the world, principally Africa, south Asia and Latin America. At the fourth meeting of the Technical Advisory Committee of FAO Consultative Group on International Agricultural Research, Washington D.C., August, 1972, it was proposed that ICRISAT should be linked to four co-operating centres in Africa (Alemaya in Ethiopia, Serere in east Africa, Samaru in Nigeria and Bambey in Senegal). Each of these represents markedly different ecological zones.

5.3.5 The special characters of deltaic land use

Deltas have played a most important role in the land use history and anthropology of Asia (UNESCO, 1966) (see 3.1). People settled first in the hinterlands and moved down into the delta lands when they had acquired more knowledge of the control of floods and other environmental factors (J. Büdel in UNESCO, 1966). It is important that the land surveyor and manager should view deltas as ecosystems and emphasize the factors that interact within their boundaries (D. Mueller-Dombois in UNESCO, 1966). Landscape surgery will then be restricted to mere modifications, to improve upon nature within its own framework. However, delta ecosystems cannot be understood as isolated units, since they are constantly influenced by forces operating in the catchment areas. The management of high forests in particular will have significant effects on the natural or unnatural processes within the delta itself. Ohya (in UNESCO, 1966) reported to the same symposium on a comparative study of the geomorphology and flooding in deltaic areas in Taiwan, Thailand, Burma and India (Gangetic plain) (3.1).

Although the cultivated lands in the deltas of monsoonal and equatorial Asia are now used primarily for padi cultivation, it is probable that this is a comparatively recent development (G. A. W. van de Goor in UNESCO, 1966), and that rice was preceded by crops reproduced by vegetative methods (Sauer, 1952). Fukui (1971) describes a situation in the Chao Phraya river basin in Thailand (3.5). Although the deltaic region represents the rice bowl of Thailand, there is little possibility of introducing intensive methods up to the level required by high-yielding varieties of rice, a situation in marked contrast to the non-deltaic areas.

5.3.6 Irrigated agriculture

Kung (1971), writing from Plant Production and Protection Division, FAO, has presented a comprehensive picture of irrigation agronomy, including irrigation and drainage for the individual field as well as water management for the community as a whole. There are many works dealing with the planning, design, construction and operation of an irrigation system, but none so far on the effective use of water at the farm level in Monsoon Asia. The present study does this from the point of view of plant physiology, crop requirements, irrigation practices, co-operative management, cropping patterns, agronomic research and demonstration and extension.

We are reminded by Kung that 'rice culture at present dominates agriculture in monsoon Asia. About 95 per cent of all the production is consumed in the region. The seasonal rhythm of rainfall, with precipitation concentrated in the growing season and a relatively dry period at the time the crop matures, is best adapted to the rice plant and its cultivation. A consequence of high water requirements is that rice culture tends to be concentrated on level or gently sloping lands having soil characteristically heavy in texture. Provision, control and retention of adequate water supplies are easiest in such situations. It is also easy to collect and distribute water for lowland rice in moist regions without building vast and expensive irrigation projects. In the flat plains and the deltaic areas, the construction of level and flood-

250

able padi fields is simplest and least costly for reasons both of topography and soils. A hot moist climate and wide stretches of land seasonally flooded or easy to irrigate by simple, inexpensive means are all natural factors favourable to rice culture. Other crops such as maize, sorghum, sugarcane, soybean, groundnut, jute, kenaf and cotton are grown upland in the predominantly lowland rice region, where farming lands are not levelled, diked and flooded.

'Padi fields in monsoon Asia are mainly rainfed, with some floodfed, while fields for upland crops are mainly rainfed. The rainfed padis simply catch and hold water as it falls. The floodfed padis in or near river basins receive and retain water brought to them from the rise in level of the stream. Near the sea, high ocean tides regularly force fresh river water into the padis. In certain areas, farming lands are supplied with water in addition to rainfall or flood by various artificial means. Water is often pumped or lifted from streams, lakes, open tanks and wells. Diversion ditches are built from streams and farming lands fed by gravity. Large storage dams permit water to be fed to fields by gravity through a network of canals and ditches. In general, relatively small areas are provided with such facilities for artificial irrigation. The greater part of farming land is still entirely dependent upon rainfall for natural irrigation, with small areas having a casual type of water control. The possibilities for extending irrigation are therefore very promising. Conservation of irrigation water through proper handling is becoming more and more important as demands on water supplies increase with the increase in population.

'Water management is important in both naturally and artifically irrigated fields. Irrigation systems must conform with certain basic agronomic principles if the objectives are to be accomplished economically and efficiently. Irrigation involves much more than the building of structures and the application of water to the soils. In a sense Asian farmers who till the lands generally know through experience how to manipulate water for crop cultivation; but by no means all of them understand crop requirements and the soil−water−plant relationship with the result that crop yields are extremely low and land and water are wasted' (Kung, 1971).

In recent years, agriculturists in Asian countries have assembled some information on the water requirements of rice through field measurements, but more is needed regarding total consumptive use and daily rates for major upland crops of the region. Kung quotes for general reference in Tables 5.6 and 5.7 figures estimated empirically by specialists who have worked in the region for years.

'Supplementary irrigation introduced little change in cropping patterns, as it is designed to supply water only to the main crop grown in the monsoon season with no provision for supplying water in the dry period. This type of irrigation mainly insures crop production against losses when summer rainfall is uncertain; it does not affect the growing season of the lands.

251

Revolutionary changes in cropping patterns can be expected only after the completion of a multipurpose project which supplies water the year round, extending the growing season of the lands from five to six months to the entire year and making it possible for farmers to grow one or two additional crops after the harvest of the main crop. In other words, under irrigation conditions, the same lands can be utilized two or three times in one year (Fig. 5.3). In places where rainfall is evenly distributed and there are ample facilities for irrigation, many types of multiple cropping are practised' (Fig. 5.4) (Kung, op. cit.).

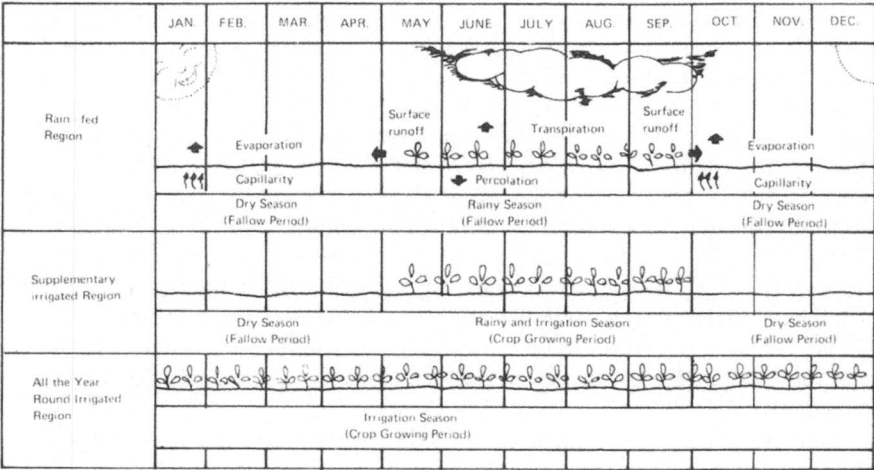

Fig. 5.3. Diagram showing how rainfall and irrigation affect crop growing seasons and how water may be lost from soil (Kung, 1971).

In principle, with sufficient irrigation in the wet season, a gradual change of cropping pattern is possible on either padi fields or uplands, first to double cropping if certain essential prerequisites are met, and only later to triple cropping when cultivators have come to accept double cropping. Two factors determine the order of crops in the cropping pattern, reaction to daylength and response to standing water. Farmers are being asked to break with a centuries-old tradition of single cropping. A new system can be introduced only if due regard is given to selection of crops, cultivars, cropping patterns and suitable land for their adoption. Kung's advice is as follows:

Criteria for Selecting Crops
The selection of crops should first of all conform to the land classification and the existing crop distribution in the areas. For optimal yields, due considera-

252

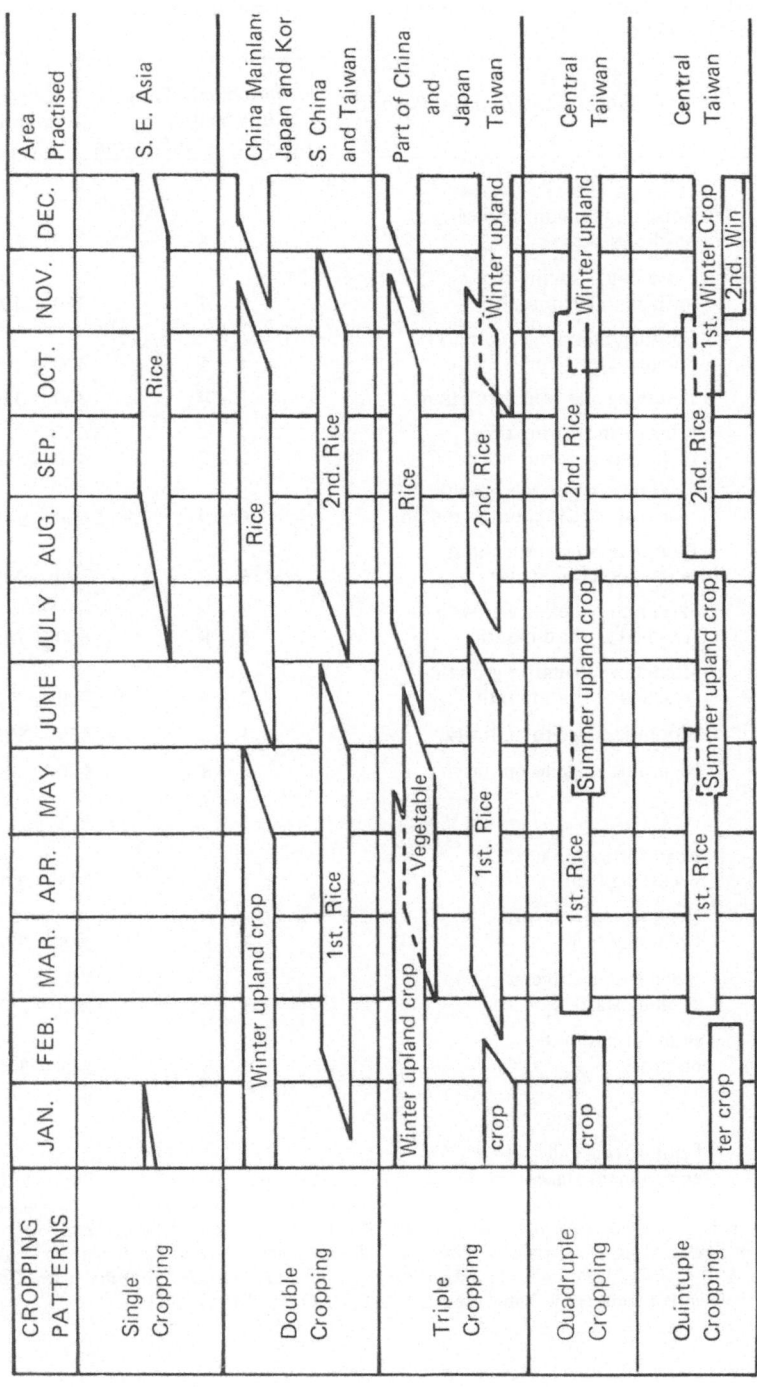

Fig. 5.4. Common cropping patterns practised in Asian rice-producing regions (Kung, 1971).

Table 5.6. Critical water demand periods, approximate daily consumption and total water consumption for major upland crops (Kung, 1971).

Crop	Critical Water Demand Periods	Approximate Daily Consumption	Total Water Consumption
		(.........millimetres.........)	
I. *Field Crops*			
Wheat	Booting, blooming and early heading stages	5 – 6	370 – 500
Maize	Tasselling – silking stages until grain becomes firm	5 – 7	350 – 400
Sorghum	Booting, blooming and milky-dough stages	4 – 5	300 – 400
Soybean	Blooming and seed formation	2 – 4	300 – 350
Groundnuts	Prior to and during time pods start to form	2 – 5	400 – 500
Sugarcane	Maximum vegetative growth (reached at fifth month) and on	6 – 9	1 600 – 1 870
Cotton	Blooming stage until about ½ to ¾ bolls are mature	6 – 9	500 – 900
Jute	Maximum vegetative growth (reached at third month)	6 – 8	600 – 700
Kenaf	Maximum vegetative growth (reached at third month)	2 – 4	300 – 450
Sesame	Blooming stage to maturity	4 – 5	450 – 525
Castor bean	Blooming stage to maturity	6 – 8	600 – 740
II. *Vegetables*			
Lettuce	Head formation until becoming firm	4 – 5	245 – 370
Cabbage	Head formation until becoming firm	4 – 5	245 – 370
Carrots	When root enlargement swelling starts	3 – 5	200 – 245
Onions	Bulb formation to maturity	4 – 5	245 – 370
III. *Fruit*			
Orange	Fruit setting and enlargement stages	3 – 4	740 – 970

Sources: 1. Morris E. Bloodworth, *Some Principles and Practices in the Irrigation of Texas Soil,* Texas Agricultural Exp. Sta. 2. M. R. Balakrishnan, *Diversification of Agriculture under Irrigation,* Thailand, ETAP, FAO No. 1318, 1961. 3. Peter Kung, *Irrigation Farming and Multiple Cropping in the Ganges-Kobadak Project Area,* Pakistan (now Bangladesh), ETAP, FAO No. 1456, 1962.

Table 5.7. Estimated water requirements for upland crops grown in southeast Asia (Kung, 1971).

Crops	Vietnam	Thailand	Burma	Bangladesh	India	Pakistan
	(........................ *millimetres* ....)					
I. Cereals						
Wheat	--	--	710	380	380	330 – 510
Barley	--	--	--	--	360	380
Maize	500	350 – 400	530	380	460	460
Sorghum	--	--	--	380	660	410
Millet	--	--	--	--	760	330
II. Legumes						
Soybeans	500	300 – 350	640 – 660	380	--	--
Groundnuts	400	400 – 500	530 – 640	380	660	410
Pea	--	--	--	--	310	--
Gram	--	370 – 400	430 – 480	150	--	180 – 250
Lentil	--	--	--	150	--	--
Dhaincha	--	--	--	150	--	--
Khesari	--	--	--	150	--	--
III. Sugar and Starch Crops						
Sugarcane	1 500	1 600 – 1 870	--	1 780	2 410	1 070 – 2 030
Sweet potato	550 – 650	--	--	--	690	--
Irish potato	--	--	--	510	--	560
IV. Fibre Crops						
Cotton	1 000	500 – 900	690	--	1 070	460 – 970
Jute	550	600 – 700	--	760	--	--
Kenaf	--	300 – 450	--	--	--	--
V. Oil Seeds						
Sesame	750	525	480 – 610	--	--	--
Rape Seeds	--	--	--	150	280	250 – 430
Linseeds	--	--	--	150	330	--
VI. Others						
Tobacco	600	--	--	760	990	790
Green Manures	200	300 – 600	--	150	--	330
Egypt Clover	--	--	--	--	--	1 320 – 1 880
Chillies	--	--	--	--	990	--
Vegetables	--	400 – 500	--	--	--	560
Tomato	500	--	--	--	--	--

Suggested by: *Y. H. Djang*, 1961; *M. R. Balakrishnan*, 1961; *W. R. Nelson*, 1958; *Peter Kung*, 1962; Handbook of Agriculture, 1961; *Otto Schiller*, 1956.

Applied to: West Bink Thuan Project; Chao Phya Project; Mu River Project; Ganges-Kobadak Project; Hyderabad; Panjab.

tion should be given to the plant and soil relations. In general, tobacco, watermelon and groundnuts are recommended for light soils, and rice, jute and wheat for heavy ones. Sugarcane is good on medium soils with good drainage. Suggested criteria for selecting main and second crops are given. *Main crop* (to be grown in the wet season or the monsoon season): a. It gives good irrigation response and consumes large amounts of water. b. It is a traditional crop preferred by the growers. c. It takes no more than six months for full growth. d. It has a stable market demand.

255

Second crop (to be grown in the dry season): a. It has a shorter growing period than four months. b. It is easy to grow and consumes less labour and water. c. It has a potential market demand. d. It is nonsensitive to photoperiodism.

Criteria for Selecting Cropping Patterns

A sound cropping pattern should be aimed at producing the amount of food for humans and livestock and also cash crops for economic improvement in rural areas. The following criteria are suggested. a. It should take into account the need to maintain and improve soil fertility and to limit insect and disease damage. b. It should provide for an even spread of labour throughout the year. c. The monsoon season should be used for crops which consume more soil moisture. d. The main crop should be planted so that its harvest period will come in the dry period (especially important for crops such as rice and cotton).

Recommended cropping patterns are: food crop − cash crop; food crop − fodder crop; cereals − legumes; long duration crop − short duration crop; deep rooted crop − shallow rooted crop.

Criteria for Selecting Suitable Areas for New Cropping Patterns

The land classification maps are the guides for allotting lands for different cropping patterns. Topography generally gives a rough guide to soil textural classes, as the soils in highlands are mostly light in texture and heavier in lower lands.

1. Existing or potential rice area:
a. Lowest depression with very heavy soils where drainage is a problem. Single cropping of rice may remain the major crop for a rather long period.
b. Lower land with heavy soil and sufficient water resources. Rice − rice (double rice is also recommended on saline or acidic soils where quantity of water is sufficient).
c. Medium to lower lands with medium to heavy soils and water resources are moderate. Upland crop − rice.
2. Higher land, with light to medium soils and moderate water resources. Dry crop − upland crop (Fig. 5.5).

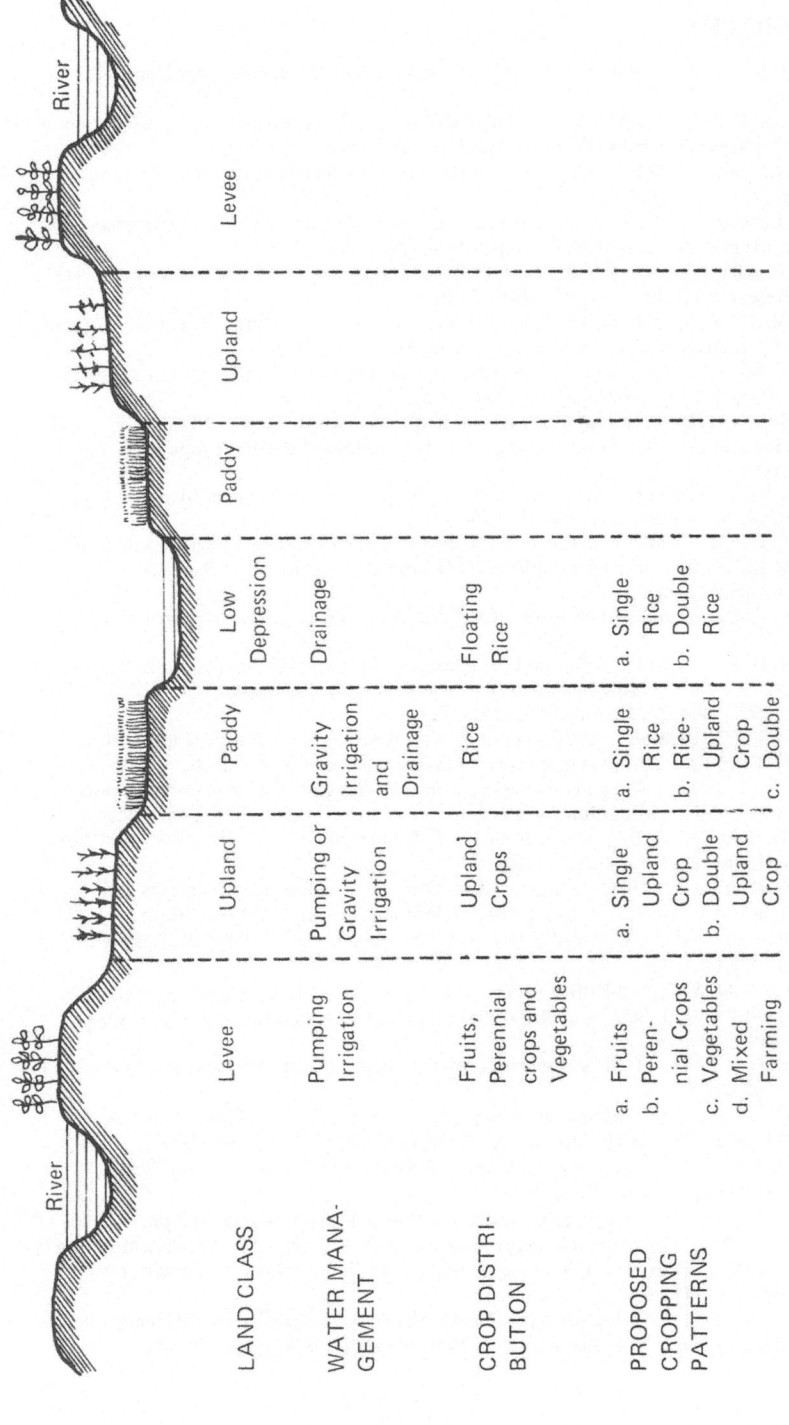

LAND CLASS	Levee	Upland	Paddy	Low Depression	Paddy	Upland	Levee
WATER MANAGEMENT	Pumping Irrigation	Pumping or Gravity Irrigation	Gravity Irrigation and Drainage	Drainage			
CROP DISTRIBUTION	Fruits, Perennial crops and Vegetables	Upland Crops	Rice	Floating Rice			
PROPOSED CROPPING PATTERNS	a. Fruits b. Perennial Crops c. Vegetables d. Mixed Farming	a. Single Upland Crop b. Double Upland Crop	a. Single Rice b. Rice-Upland Crop c. Double Rice	a. Single Rice b. Double Rice			

Fig. 5.5. Crop distribution and proposed cropping patterns under different topographical conditions (Kung, 1971).

257

BIBLIOGRAPHY

ABROL. I. P., – 1971 – Salinity control in irrigated agriculture. *Agri. Digest* no. 22: 35–40.

BOSE, A. B., S. P. MALHOTRA, L. P. BHARARA & C. S. K. JOHORY, – 1966 – Socio-economic aspects of animal husbandry. *Ann. Arid Zone* 5: 72–80.

BUCHANAN, K., – 1970 – The Transformation of the Chinese Earth. Bell, London. 336 pp.

BUCK, J. Lossing, – 1937 – Land Utilization in China. Commercial Press, Shanghai. 494 pp. Reprinted Paragon Book Reprint Co., New York, 1964.

CENTRAL ARID ZONE RESEARCH INSTITUTE, – 1974 – Fourteen Years of Arid Zone Research (1959–1973). CAZRI, Jodhpur. 56 pp.

CHATURVEDI, M. D. & B. N. UPPAL, – 1953 – A study in shifting cultivation of Assam. Indian Council of Agricultural Research, Res. Ser. no. 2. 20 pp.

CHEN, CHENG–SIANG, – 1970 – The Agricultural Regions of China. American Institute of Crop Ecology, Silver Spring, Md. 47 pp.

COMMONWEALTH ASSOCIATION FOR SURVEYING AND LAND ECONOMY, – 1972 – Surveying and Land Economy in Asia. Report of Seminar, Kuala Lumpur, 21–23 February.

CONKLIN, H. C., – 1954 – An ethno-ecological approach to shifting cultivation. *New York Acad. Sci. Trans. 2nd Ser.* 17: 133–142.

CONKLIN, H. C., – 1957 – Hanunoo Agriculture: A Report on an Integral System of Shifting Cultivation in the Philippines. FAO Forestry Development Paper no. 12. FAO, Rome. 217 pp.

CONKLIN, H. C., – 1961 – The study of shifting cultivation. *Current Anthropology* 2: 27–61.

CONKLIN, H. C., – 1969 – An ethnoecological approach to shifting agriculture (reproduction of Conklin 1954). In 'Environment and Cultural Behaviour,' ed. A. P. Vayda, pp. 221–233. Natural History Press, New York.

CONWAY, G. & J. ROMM, – 1973 – Ecology and Resource Development in Southeast Asia. The Ford Foundation: Office for Southeast Asia, Bangkok. 82 pp.

COOK, O. F., – 1921 – Milpa agriculture: a primitive tropical system. Annual Report of the Smithsonian Institution for 1919. Government Printing Office, Washington.

DALRYMPLE, D. G., – 1971 – Survey of Multiple Cropping in Less Developed Nations. U. S. Department of Agriculture. 108 pp.

DONNITHORNE, A., – 1970 – China's grain: Output, Procurement, Transfers and Trade. Chinese University of Hong Kong, Economic Research Center. 36 pp.

EAST AFRICAN ROYAL COMMISSION, – 1955 – 1953 – 1955 Report. H.M.S.O., London. 482 pp.

EDWARDS, A. M. & G. P. WIBBERLEY, – 1971 – An agricultural land budget for Britain, 1965–2000. School of Rural Economics and Related Studies, Wye College, University of London.

ETIENNE, G., – 1973 – India's new agriculture: a survey of the evidence. *South Asian Rev.* 6: 197–213.

FISHER, C. A., – 1972 – Indonesia – a giant astir. *Geographical J.* 138: 156–165.

FOOD AND AGRICULTURE ORGANIZATION OF THE UNITED NATIONS, – 1971 – Soil Survey and Soil Fertility Research in Asia and the Far East. World Soil Resources Reports no. 41. FAO, Rome. 229 pp.

FREEMAN, D., – 1970 – Report on the Iban. Athlone Press, London. 317 pp.

FUKUI, H., – 1971 – Environmental determinants affecting the potential dissemination of high-yielding varieties of rice: a case study of the Chao Phraya river basin. *Tonan Ajia kenkyu* 9: 348–374.

GEERTZ, C., – 1968 – Agricultural Involution: the Processes of Ecological Change in Indonesia. University of California Press, Berkeley and Los Angeles. 176 pp.

258

GONGGRIJP, L., – 1941 – Het erosie-onderzoek. *Tectona (Buitenzorg)* 34: 200–220.
GOUROU, P., – 1945 – Land Utilization in French Indo-China. Institute of Pacific Relations. 3 vols. mimeo.
GOUROU, P., – 1953 – The Tropical World: Its Social and Economic Conditions and its Future Status. Longmans Green, New York.
GOUROU, P. – 1956 – The quality of land use of tropical cultivators. In 'International Symposium on Man's Role in Changing the Face of the Earth,' ed. W. L. Thomas, Jr., pp. 336–349. University of Chicago Press, Chicago.
GREENE, H., – 1966 – Irrigation in arid lands. In 'Arid Lands,' ed. E. S. Hills, pp. 255–271. Methuen/UNESCO, London and Paris.
GRIFFIN, E., – 1973 – Testing the von Thünen theory in Uruguay, *Geogr. Rev.* 63: 500–516.
GROOME, J. St. J., – 1970 – Land use surveys as an aid to development. *SPAN* 13: 185..
GUPTA, S. K., K. G. TEJWANI, H. N. MATHUR & M. M. SRIVASTAVA, – 1970 – Land resource regions and areas of India. *J. Indian Soc. Soil Sci.* 18: 187–198.
HALL, P. (ed.), – 1966 – Von Thünen's Isolated State, Oxford.
HILLS, E. S., – 1966 – Research and the future of arid lands. In 'Arid Lands,' ed. E. S. Hills, pp. 439–461. Methuen/UNESCO, London and Paris.
HOUSTON, C. E., – 1972 – Farm water management in the Far East. Technical Paper to Seminar on Water Management and Control for Agriculture, Tokyo, October, 1972. FAO, Rome. mimeo. 8 pp.
HUANG, Ping-wei, – 1961 – The Complex Natural Zonation of China. U.S.S.R. Academy of Sciences, Geographical Series 1, 25–39. Moscow.
HUTTON, J. H., – 1949 – A brief comparison between the economics of dry and irrigated cultivation in the Naga Hills and some effects of a change from the former to the latter. *Advancement of Science* 6: 26.
INDIAN AGRICULTURAL RESEARCH INSTITUTE, – 1970 – A New Technology for Dry Land Farming. IARI, New Delhi. 189 pp.
INTERNATIONAL LABOUR OFFICE, – 1973 – Mechanization and Employment in Agriculture: Case Studies from Four Continents. International Labour Office, Geneva. 192 pp.
JANZEN, D. H., – 1973 – Tropical agroecosystems. *Science* 182: 1212–1219.
KELLMAN, M. C., – 1969 – Some environmental components of shifting cultivation in upland Mindanao. *J. Trop. Geogr.* 28: 40–56.
KLATT, W., – 1970 – Problems of food and farming in Asia. *J. Roy. Central Asian Soc.* 57: 4–18.
KRANTZ, B. A. & J. KAMPEN, – 1973 – Water management for increased production in the semi-arid tropics. Paper to National Seminar on Water Resources in India and their Optimum Utilization in Agriculture. Water Technology Center, IARI, New Delhi. 20 pp.
KUNG, P., – 1971 – Irrigation Agronomy in Monsoon Asia. FAO, Rome. 106 pp.
LAFONT, P. B., – 1967 – L'agriculture sur brûlis chez les Proto-Indochinois des hauts plateaux du Centre Viet-Nam. *Cah. d'Outre-Mer* no. 77: 37–48.
LEACH, E. R., – 1964 – Political Systems of Highland Burma. University of London. 324 pp.
LIN, S., – 1972 – Asian agrarian reform – a perspective study tour report. *Land Reform, Land Settlement and Cooperatives* 2: 33–51.
McGINNIES, W. G., B. J. GOLDMAN & P. PAYLORE (eds.), – 1968 – Deserts of the World. University of Arizona, Tucson. 188 pp.
MOORMANN, F. R., – 1973 – Classification of land for its use capability and conservation requirements. Paper to FAO/SIDA Regional Seminar on Shifting Cultivation and Soil Conservation in Africa, Ibadan, Nigeria, 2–21 July, 1973. FAO, Rome. mimeo. 11 pp.

PARRACK, D. W., – 1969 – An approach to the bioenergetics of rural West Bengal. In 'Environment and Cultural Behaviour,' ed. A. P. Vayda, pp. 29–46. National History Press, New York.

PELZER, K. J., – 1945 – Pioneer Settlement in the Asiatic Tropics: Studies in Land Utilization and Agricultural Colonization in Southeastern Asia. International Secretariat, Institute of Pacific Relations, New York/American Geographical Society, Special Publication no. 29. 290 pp.

PELZER, K. J., – 1958 – Land utilization in the humid tropics: agriculture. In 'Proceedings Ninth Pacific Science Congress,' vol. 20, pp. 124–143. Reproduced 1964 as Reprint Series no. 7, Yale University Southeast Asia Studies.

PELZER, K. J., – 1968 – Man's role in changing the landscape of southeast Asia. J. Asian Studies 27: 269–279.

RAPER, A. F. et al., – 1970 – Rural Development in Action: The Comprehensive Experiment at Comilla, East Pakistan. Cornell University Press, Ithaca. 374 pp.

ROY, E. van, – 1971 – Economic Systems of Northern Thailand: Structure and Change. Cornell University Press, Ithaca. 289 pp.

SAUER, C. O., – 1952 – Agricultural Origins and Dispersals. Amer. Geogr. Soc., New York. 110 pp.

SCHWEINFURTH, U. (with M. DOMRÖS) (eds.), – 1969 – Problems of Land Use in South Asia. Yearbook of the South Asian Institute, University of Heidelberg, 1968/9. Otto Harrassowitz, Wiesbaden. 138 pp.

SHAND, R. T. (ed.), – 1969 – Agricultural Development in Asia. Australian National University, Canberra. 360 pp.

SILCOCK, T. H., – 1967 – Thailand: Social and Economic Studies in Development. Australian National Uni. Press. Canberra. 334 pp.

SPENCER, J. E., – 1966 – Shifting Cultivation in Southeastern Asia. University of California Publications in Geography vol. 19. Berkeley and Los Angeles. 247 pp.

STEWART, G. A. (ed.), – 1968 – Land Utilization. Macmillan, London.

SWAMINATHAN, M. S., – 1970 – The next step is dryland farming. Ceres 3(6): 59–60.

TENG, C. C., – 1963 – The demarcation of the agricultural regions of China. Acta Geographica Sinica, December.

von THÜNEN, J. H., – 1826 – Der Isolierte Staat in Beziehung auf Landwirtschaft und Nationalökonomie. Hamburg.

TREGEAR, T. R., – 1958 – Land Use in Hong Kong and the New Territories. University of Hong Kong Press.

TROLL, C. & K. H. PAFFEN, – 1965 – Karte der Jahreszeitenklimate der Erde. In English in 'World Maps of Climatology', eds. E. Rodenwaldt & H. J. Jusatz. Heidelberger Akademie der Wissenschaften, Berlin, Heidelberg, New York. pp. 19–28.

UNITED NATIONS EDUCATIONAL, SCIENTIFIC AND CULTURAL ORGANIZATION – 1966 – Scientific Problems of the Humid Tropical Zone Deltas and their Implications. Proceedings of the Dacca Symposium. UNESCO, Paris. 422 pp.

WALLACE. B. J., – 1970 – Shifting Cultivation and Plow Agriculture in Two Pagan Gaddang Settlements. Bureau of Printing, Manila. 117 pp.

WANG Gung-Wu (ed.), – 1964 – Malaysia: A Survey. Pall Mall Press, London. 466 pp.

WHARTON, C. R., – 1970 – Subsistence Agriculture and Economic Development. Frank Cass, London. 481 pp.

WHYTE, R. O., – 1966 – The use of arid and semi-arid land. In 'Arid Lands,' ed. E. S. Hills, pp. 301–361. Methuen/UNESCO, London and Paris.

WEITZ, R. & A. ROKACH, – 1968 – Agricultural Development: Planning and Implementation: Israel Case Study. Reidel, Dordrecht. 404 pp.

WIKKRAMATILEKE, R., – 1957 – Whither chena? The problem of an alternative to shifting cultivation in the Dry Zone of Ceylon. Geograph. Studies 4(2): 81–89.

WONG, Ming-hung, – 1974 – Land evaluation and soil fertility. J. Chinese University of Hong Kong 2: 449–464.

6 ASSESSMENT OF LAND CAPABILITY

6.1 Assessment of potential as basis for planning and investment

So far we have been concerned primarily with the problems of survey and planning on the intensively utilized cultivated land of Asia. But there still remain some limited areas of primary vegetation, much larger areas of secondary and terminal vegetation, and large areas of land with low population density and light use. It is to these lands that the techniques of the CSIRO Division of Land Research may be said to apply more particularly.

The so-called integrated approach (6.2) considers the whole process from initial survey and stocktaking of resources through assessment to initial planning of projects as one co-ordinated programme, a job for a team of specialists welded into a whole by a man experienced in the appraisal of projects (Christian & Stewart, 1968). These surveys are conducted on lands with relatively low population, on which the potential in terms of total productivity and diversity of products for export to other regions or countries has by no means been achieved through the unplanned development of the past. Where only limited information has been collected previously and the relative importance of different kinds of resources has not been established, concurrent examination of several resources can avoid costly duplication of field effort. Integrated surveys also provide opportunity for an on-the-spot interchange between specialists in different disciplines.

FAO has been concerned with pre-investment surveys in many parts of Asia, conducted by teams of specialists, covering the various developmental activities and providing a basis for coherent development programmes for specific geographical regions (see Chapter 7). These may be grouped as: 1. reconnaissance surveys of land, soil, water, forest and grazing resources and present patterns of land use, 2. detailed technical surveys, and 3. studies based on the technical findings and the socio-economic status of the region concerned.

FAO has introduced the parametric method of land evaluation (Riquier, 1974) which consists of: a. evaluating separately the different qualities and limitations of land and giving each a numerical value in relation to its importance, b. combining these different factors following a mathematical law designed to take account of the relative influences of interactions between the factors in order to obtain a final index, and c. using this index to place lands in a scale of values. Riquier considers that this method may be applied to classification in relation to fertilizer needs, suitability for irrigation, the supporting of heavy loads (roads, airfields), the production of wood from forestry sites, or simply to the assessment of agricultural potential in general. The method (which has certain advantages and disadvantages) is an attempt to make land evaluation quantitative and compatible with modern methods of calculation. The introduction of the concepts of yield and of productivity in a quantitative manner provides the basis for discussion between soil scientists and economists. The parametric method may be readily introduced into all other *holistic* methods of land classification to assess the agricultural value of soil in the true sense (Riquier, op. cit.).

King (1970) of the (British) Land Resources Division believes that some of the confusion which exists on the subject of land systems, due partly to different disciplinary approaches, may be overcome by establishing rigorous land system *descriptions*. These should be mutually exclusive and quantitative or semi-quantitative. The easiest way of satisfying these conditions is to describe land systems by means of parameters, of which the following geomorphic parameters are considered to be the most useful: 1. process; 2. altitude; 3. relief; 4. dominant geology (comprising lithology and strata altitude); 5. drainage pattern; 6. stream frequency; 7. characteristic plan-profile; 8. geomorphic position; 9. dominant facets; 10. characteristic facets; 11. characteristic variant; 12. land zone.

It is now accepted that it is the combination of the resource factors in an area rather than individual factors which determines the usefulness of areas, and also that finite areas differing in this way can be recognized and made the basis of the planning of development authorities (Christian & Stewart, 1968). These authors regard the following successive steps as a logical ideal sequence for the inventory and assessment of land resources (see also Figs. 6.4. and 6.5. from Brinkman & Smyth, 1973).

Regional reconnaissance survey: a. Segmentation of a region into areas with distinctive characteristics and definition of the important resource factors of these areas. b. Preliminary technical interpretation of these resource factors, their interrelationships, and implications for development, including possibilities, problems and limitations. c. Selection of those areas with the most promising combination of resource factors, and others which, for some reason deserve priority attention, with a broad indication of the forms of development likely to be feasible in each, and the particular resource factors which require further study for these objectives. Other areas are left for later consideration.

Post-reconnaissance activities. These will take a variety of forms according to circumstances, but will usually be one more of the following: a. Immediate development either because expediency demands it in spite of incomplete information or because experience in the area or similar areas indicates its feasibility. In these cases a preliminary resource inventory and study will at least help in guiding such developments in a way that will fit them into the broad framework of an acceptable over-all development plan. b. More detailed studies in the promising areas selected by the reconnaissance surveys, i.e., pre-project surveys, to provide: quantitative information on the nature and distri-

262

bution of resource factors pertinent to the project; information related to methods of use of those resource factors. c. Assessments of the social and economic feasibility and acceptability of exploiting the technical possibilites which have been demonstrated. d. More detailed examinations for precise planning of the implementation of projects, i.e., pre-development surveys.

Where large regions are involved even initial development is a long-term activity. In making an over-all development plan there is danger of being too rigid, for before long-term plans are implemented, new technical information and new economic or social situations will make changes in such plans essential. For example, in areas of rapidly increasing population, it is impossible to predict very far ahead the changes and increases in demand by the expanded populations particularly if there is also a progressive change from a rural to a secondary industry economy. To provide the necessary flexibility, it is essential first to have reliable information on the basic facts concerning natural resources, and second to maintain continuing investigations and periodic reassessments of the possibilities of resource use.

6.2 The integrated approach*

6.2.1 UNESCO and ITC, Enschede

The term 'integrated survey for development planning' is regarded as covering any survey that includes two or more related disciplines requiring co-ordination, taking into account the reasons for which the survey is being performed. The term has become something of a magic concept which evokes widely different ideas, according to the ITC/UNESCO Institute at Enschede, Netherlands. For a training institute, it is important to investigate what those who apply it consider integration to be. For many it is rather a vague term, but of high political value in connection with investigations for development projects; for others with a more definite concept of the meaning of the term, it is difficult to formulate their requirements for training purposes.

Surveys which are accepted as multidisciplinary or interdisciplinary are sometimes considered as integrated. The word integration indicates a unit which is more than the sum of its inherent parts. Opinions differ as to whether integration is to be established by specialists in different disciplines, or by generalists such as geographers, whose synthesizing of the whole field of a problem may be considered as integration. The higher the degree of specialization the more difficult it is to establish multidisciplinary activities (ITC/UNESCO Seminar, 1971).

Basing its concept not on functional premises and starting with the mutual

* For integrated surveys conducted by the Central Arid Zone Research Institute, Jodhpur, see 2.6, 2.7, 5.3.3.

relation and interdependence of the participating disciplines, the ITC/UNESCO Institute considers important characteristics for an integrated survey that 1. in the multidisciplinary unit, all specialized activities are directed to and focused on the purpose of the survey as a whole; 2. the activities in the participating disciplines and of the specialists are regularly adjusted to the findings and interim results of the other disciplines; 3. the activities of survey and investigation are formulated afresh and shifted towards a changing purpose of the survey, following consideration of results from earlier stages. 4. in an operational sense, adjustment underlines the need for full co-operation between specialists, execution of a project in the most economical way and avoidance of duplication.

Although the seminar held at the Central Arid Zone Research Insitute, Jodhpur, Rajasthan, in November, 1970, dealt in an introductory manner with integrated surveys in general, the main emphasis was on 'integrated surveys, range ecology and management' (UNESCO, 1971). Apart from the misuse of the term 'range' to apply to a livestock-cum-vegetation ecosystem completely distinct from that area in the western United States where the term is used and had its origin, a purist may question whether it is correct to consider integrated survey of only one type of land and land use in a vast area of western India where numerous other types of land use occur as a mosaic in what is admittedly still a predominantly uncultivated, i.e. range environment. The seminar was concerned with rangeland ecosystems, not with integrated surveys.

R. L. Wright discusses the role of integrated surveys in developing countries and of geomorphology and pedology in integrated surveys of rangelands; also methodology of integrated surveys; R. E. Winkworth deals with questions relating to the study of range ecosystems, primary production and measurement of biomass, water relations and autecology, the structure and function of rangeland ecosystems and management within the ecosystem framework; I. S. Zonneveld deals with pragmatic land classification, aerial photography in integrated surveys, field description of vegetation, and the role of men and animals in grazing land ecosystems.

But the human and livestock populations of western India will continue to increase, and competition for land resources between man and livestock will intensify. The construction of the Rajasthan Canal and the smaller irrigation schemes will reduce the range acreage, and there will be increasing demand for new land for dryland crops. Finally, range livestock should not be expected to live, produce and reproduce on free range grazing alone throughout the year. Correct animal nutrition calls for a balance between feed from grazing, conserved cut range hay, some cultivated fodders from dry and irrigated lands, and some grain supplement, all to provide a reasonably uniform plane of nutrition throughout the year, with a little extra for periods of pregnancy and the rearing of the young. There is no indication of these kinds of integrated long-term objectives in the report of this seminar.

Examples of surveys conducted throughout monsoonal and equatorial Asia,

264

with differing objectives and different degrees of 'integration' are given in Chapter 7.

6.2.2 Australia, CSIRO

Christian & Stewart (1953, 1968) state that integrated surveys as developed in Australia use the word 'land' to denote the complex of factors of individual significance (Chapter 3) which should be considered together (1.2). The Australian land research studies have shown that it is important to consider the whole process from initial survey and stocktaking of resources through assessment to initial planning of projects as one co-ordinated programme. Although integrated resource surveys are particularly relevant to the need to obtain balanced information about as many resources as possible for the purposes of planning and assessment of priorities, there are also practical and economic reasons for a joint effort (Christian & Stewart, op. cit.). Although this is especially true in areas where only limited information has been collected previously, it also applies where much information or experience is already available. Further a general appreciation of a region as a whole can be gained with greater ease (Stewart, 1962). A comprehensive reconnaissance survey will delineate areas having different potentials to better advantage than a survey giving more detailed information about one particular resource. The greatest aid to the survey and assessment of resources has been the development of aerial photography and the specialized techniques of photointerpretation (6.4).

The research emphasis of the Division of Land Research which was defined by Christian & Stewart (1968) was changed after the Land Evaluation Symposium held in Canberra (1968) (Stewart, 1968). More weight has been placed on objective and sustained methods of assessment of potential, and the provision of computer facilities has played an important part (J. McAlpine, personal communication, 1974).

There are three main components in this more recent work, the first being the incorporation of socio-economic data to the assessment of potential, the second considerably greater range in those biophysical factors of which account is taken, and the third stressing computer applications.

This work has not yet reached the stage of definitive publications. For example, it has not yet been decided whether to place considerable emphasis on detailed parametric studies, or whether these should be abandoned in favour of more rapid and somewhat less objective assessment of potential based on land units.

In extensive soil surveys and land resource surveys, information on soils obtained from auger holes or pits is extrapolated to aerial units delineated by landform or vegetation boundaries determined from air photos or by field observation. Scott & Speight (1971) have made quantitative comparisons between rather rigorously specified attributes and classifications of soil, landform and vegetation, measured on a fine grid-pattern in a particular landscape (in the

265

Australian Capital Territory), with the aim of discovering what kind of statements can reliably be made about this landscape and how the process of extrapolation can be improved. The results in themselves suggest more explicit criteria for the selection of sample sites for extensive soil surveys, with associated new requirements to observe or measure properties of dispersal areas. There is a need to establish the feasibility of using coarser landform data which are more likely to be available for large tracts of country.

This fundamental question of the extrapolation of site data to mapped regions has received little attention, because the methodology of the mapping and description of land systems has evolved in a haphazard manner; different organizations and survey teams have perceived the logical steps involved in different ways (Speight, Heyligers & Scott, 1973). A team within the CSIRO Division of Land Use Research has been concerned with demonstrating, for an area in Papua New Guinea, a land resource survey that makes use of automated techniques of data storage, retrieval and manipulation. This approach was prompted by recognition of the substantial amount of data that is collected and then wasted in the process of producing a printed report. All steps in the logical chain from observational data through to advice concerning proposed land use become subject to critical scrutiny. The flow of information involved in setting up a data bank is shown in Fig. 6.1a and b.

6.2.3 Great Britain: Land Resources Division, ODA

The history, organization, functions, work systems and projects of the Land Resources Division of the (British) Overseas Development Administration since its beginnings in 1946 are described in a number of publications from the Division. For example, Baulkwill (1972) discusses the scientific role of the Division in the following terms: 'The modern theory of land resource assessment and development is that the previously evolutionary process can be enacted and brought to a successful conclusion within a few decades by the use of a succession of increasingly detailed scientific surveys, beginning with a broad reconnaissance of land resources and ending with studies of the relatively small areas chosen for intensive development (Table 6.1). Essentially this scientific process is a progressively intensive classification of the suitability of the land for forestry and agriculture, the classification being based on such determining or indicative factors as climate, soil type, slope, drainage, existing vegetation and land use, accessibility and the proximity of markets. But this does not mean that all development must be intensive or must be preceded by intensive surveys. A broad reconnaissance may easily indicate certain types of improved land use (e.g. systems of extensive ranching or conservation of mountain forests to prevent erosion) which can be introduced without more detailed study. The classification is of course not conducted *in vacuo*, but is at all times set in the context of human needs and human resources which can be estimated by economic and social enquiry. It is thus implied that the

266

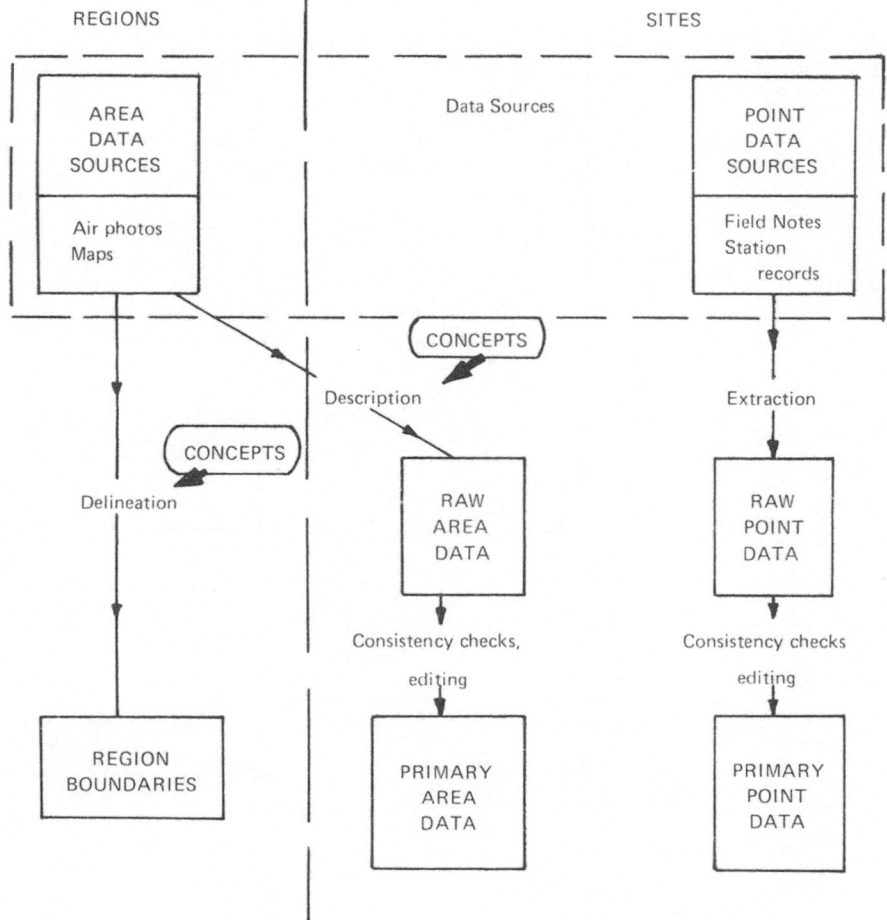

Fig. 6.1a. Flow of information involved in setting up a data bank; Part 1 (Speight, Heyligers & Scott, 1973).

classification will be put to practical use in the form of land development and that, whenever feasible, advanced or intermediate techniques will be employed to improve the return from human labour. From what has been said, it will be clear that assessment has to precede development and that the mere introduction of modern techniques such as the use of fertilizers or pesticides cannot be taken as evidence that the land is being used to the best advantage.

'The process of scientific resource appraisal described above can, of course, only be applied in its entirety to a little-populated and little-developed region. In practice most countries have been partially surveyed at several

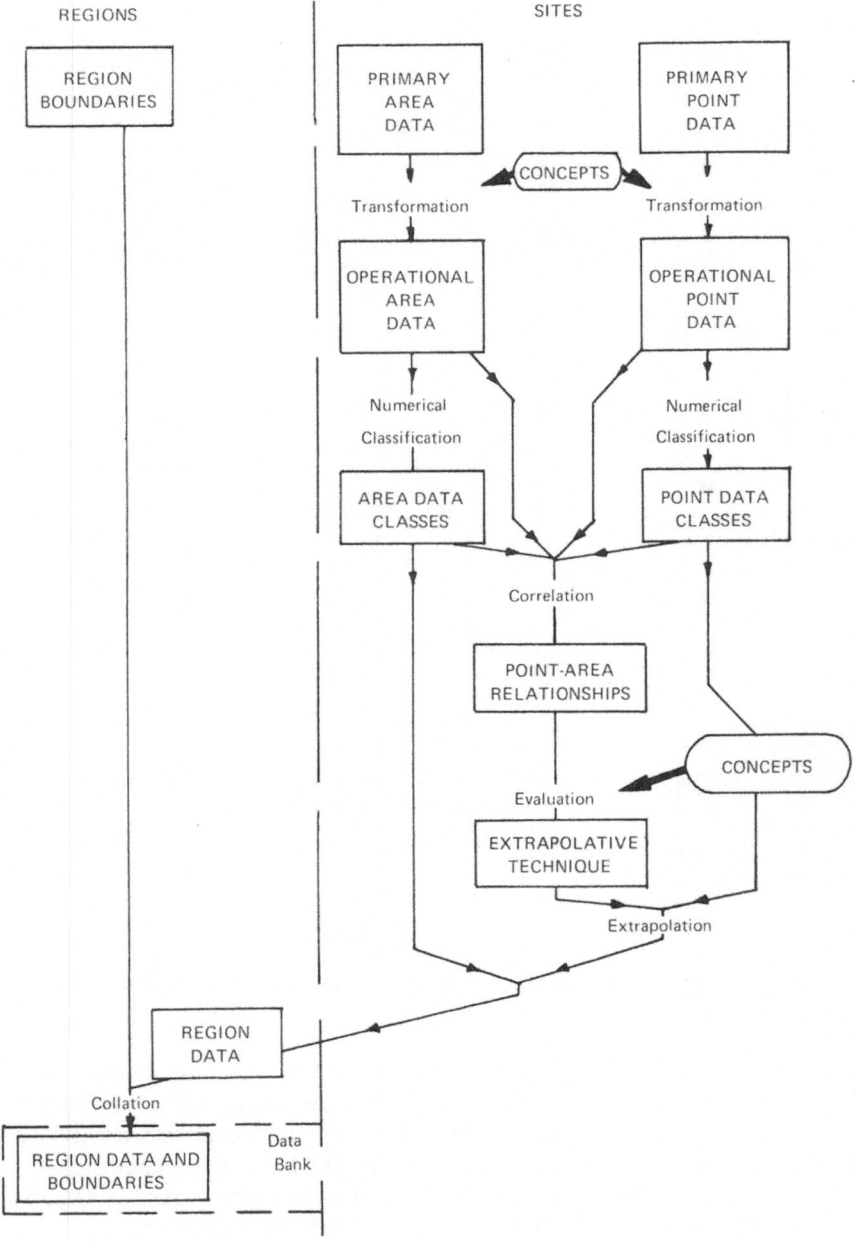

REGIONS   SITES

REGION BOUNDARIES

PRIMARY AREA DATA

PRIMARY POINT DATA

CONCEPTS

Transformation   Transformation

OPERATIONAL AREA DATA

OPERATIONAL POINT DATA

Numerical Classification   Numerical Classification

AREA DATA CLASSES

POINT DATA CLASSES

Correlation

POINT-AREA RELATIONSHIPS

CONCEPTS

Evaluation

EXTRAPOLATIVE TECHNIQUE

Extrapolation

REGION DATA

Collation

Data Bank

REGION DATA AND BOUNDARIES

Fig. 6.1b. Flow of information involved in setting up a data bank; Part 2 (Speight, Heyligers & Scott, 1973).

Table 6.1. Land Resources Division, Overseas Development Administration, Great Britain: The Principal types of survey (Baulkwill, 1972).

Survey type and typical area covered (mi²)	Objective	Typical map scale	Landscape analysis	Landscape components*				Economic analysis
				Geomorphology	Soil	Vegetation	Land use	
1. Reconnaissance. More than 2,000. May be as large as 100,000	National or regional inventory	1:250,000 and 1:500,000	Land systems or higher categories	Major relief units	Order, sub-order, great soil group or associations	Climatic and edaphic formation types	Agroecological groups	National or regional economy
2. Extensive. 1,000–10,000	Detailed inventory; broad assessment of agricultural potential	1:100,000 to 1:250,000	Land systems	Relief units or major landforms	Great soil groups, or associations of series	Climatic and edaphic formation types and plant associations	Land use systems and cultivation density	Regional economy and/or sector analysis
3. Intensive 500–5,000	Location and definition of development projects	1:25,000 to 1:100,000	Land systems and facets	Detailed landforms	Series or associations of series	Plant associations	Land use and farming systems, plus specific parameters	Sector analysis including market prospects and cost-benefit analysis of development projects
4. Development study. Usually more than 50	Resource analysis and development planning	1:10,000 to 1:25,000	Land facets and elements	Landform elements or slope units	Phases of series and/or selected parameters	Plant associations and species distribution	Specific parameters, eg crop distribution, field patterns	Detailed examination of development projects, cost-benefit analysis and commodity studies

* These components may be mapped individually in some surveys.

different stages or partially developed by previous trial and error. It follows therefore that Land Resource Division's role in resource assessment is almost invariably one of "completing the picture" both geographically and in terms of intensity of assessment.' Of the scientific disciplines used to hasten the process of resource assessment and development, the following are of special significance at one stage or another of the whole process: airphoto interpretation; integrated survey; landscape analysis/land systems/land facets (the last significant in terms of existing and potential land use and measurement of areas); land classification in terms of suitability for various types of agriculture and forestry.

The guidelines within which the soil scientists of the Land Resources Division generally operate have been described by Stobbs (1970). This statement might be brought up to date 'by saying that over the last three or four years the trend of our work has brought us increasingly into the field of specific development studies and away from the general geographical accounts based on small scale reconnaissance work. This has been accompanied by the expansion of the interdisciplinary teams with proportionately greater emphasis on surface and groundwater hydrology, soil moisture studies and socioeconomic studies of marketing problems, family organization, farming systems and the like. In turn, though the conventional soil, landform and vegetation surveys are still done, they occupy relatively less of the work force and time. The pace of a project may now well be tied to the collection of data from drilling programmes, flow and sedimentation records and meteorological data, rather than the progress of the soil surveyor. More modern equipment such as the Wallingford neutron probe and automatic weather station are in regular use, and the end product of the surveys adds to the conventional maps and descriptions a land capability assessment and, according to scale, more or less detailed statements on crop management, introduction and research problems; economic viability; and sociological constraints. Inevitably all this has repercussions on our past statements of methodology. Programmes of air photography, specific as to type and scale for the particular project, are becoming the norm. The water budget may be more significant than the soil map in determining land use and management practices and hence the parametric approach to soil mapping more important than the morphological, but as of now, no general statement governing the application of such techniques has been made. Each project is governed by its own terms-of-reference, and for this, recourse must be had to the individual Land Resource Studies, Supplementary and Technical Bulletins' (A. R. Stobbs, personal communication, 1974). For particulars of these publications, reference should be made to the Land Resources Division, Tolworth Tower, Surbiton, Surrey KT6 7DY, England.

Great Britain: University of Sheffield. Integrated surveys are an important requirement in the assessment of potential, especially where a fundamental reappraisal of land use may be rendered difficult through the existence of traditional systems (R. L. Wright, personal communication, 1974). Although geomorphology may here provide a valuable classificatory framework, it is in that context that most surveys have basic inadequacies (Wright, 1972a). Moreover, integrated surveys can provide only a preliminary stage in the procedure of evaluation and need to be followed by diverse co-ordinated research (Wright, 1972b).

Classification by *association* (desirable for most purposes), and by *subdivision* may be distinguished. The method of association calls for clearly defined primary units or taxonomic *individuals* (Wright, 1972a). Class delimitation and the selection of *differentiating characteristics* are also important. Examination of these aspects reveals fundamental deficiencies in many terrain surveys, especially with regard to classification, taxonomy and detailed geomorphological mapping. Quantitative differentiation in terms of specific terrain properties is suited to intensive surveys of small areas. For larger areas, such an approach needs to be reinforced by some system of land units based on the aggregation of suitable areal *individuals*. Geomorphic differentiation of biophysical conditions provides the framework for planning the intensive research effort into biophysical interrelations. These studies should be directed towards systems analysis of biophysical processes governing productivity, and should incorporate farming practices and socio-economic factors (Wright, 1972b).

Great Britain: University of Reading. See also Chapter 8.

A symposium was held in the Department of Agriculture and Horticulture, 15 to 18 September, 1974, a. to explore possible ways of studying whole agricultural systems, with special reference to systems methodology but not excluding any other approaches, and b. to define the role of bio-economic models in both agricultural research and practice and to chart areas in which biologists and economists can most usefully collaborate.
Contributors: Australia – G. W. Arnold, D. Bennet; Canada – S. C. Thompson; Great Britain – C. R. W. Spedding, J. N. R. Jeffers, G. R. Conway, G. A. Norton, P. J. Charlton, P. R. Street, J. Eadie, T. J. Maxwell, G. R. Allen; New Zealand – J. B. Dent; Portugal – F. Estácio; Trinidad – A. King; United Nations (IBRD) – G. F. Donaldson, A. C. Egbert; U.S.A. – G. M. van Dyne, G. S. Innes
The proceedings are published by Applied Science Publishers, Barking, Essex, England.

6.2.4 Malaysia

In 1964, a programme of evaluation of land use and natural resources was undertaken during the period of the First Malaysia Plan (1966–70), to provide data with an economic orientation for more realistic planning during the Second Malaysian Plan in West Malaysia. A section of the Economic Planning Unit in the Prime Minister's Department co-ordinated the work on behalf of

271

the Technical Sub-Committee of the National Development Planning Committee.

A complete aerial photographic reconnaissance of West Malaysia (129,500 sq. km.) was made for the year 1966 on a scale of 1:25,000, together with a supplementary reconnaissance over about 20,000 sq. km. on scales of 1:30,000 or 1:40,000 (Panton, 1966, 1970; see also Wong, 1971). The higher level photographic cover of the smaller area helped to speed up the rate of production of the few unpublished sheets in the standard 50-foot contoured, 1:63,360 topographic maps series which, together with 1:25,000 scale compilation sheets, comprise the base maps for most land resource surveys.

With this material, a survey of present land use was made by a joint Canada/Malaysia team in a period of three years. The coverage of reconnaissance, geological, soil and forest surveys was rapidly extended with improved accuracy, and maps for mineral potentiality, soil suitability and forest productivity based on economically acceptable criteria were drawn up by the responsible agencies on standard one-inch maps. A land capability classification map series was prepared as a guide for resource use planning. Statistical summaries of areas for each category were prepared from a sample based on the intersection points of the 1,000 yard (900 metre) grid squares which are superimposed on all standard one-inch maps (Alexander, 1964). A preliminary vegetation map is available (Wyatt-Smith, 1964), with descriptions of the types recognized.

Regional reports were prepared for each of the constituent States of West Malaysia, almost all of which, by a fortunate historical circumstance, comprise realistic physical planning units with boundaries marked by watersheds or main rivers (Panton, 1970). The programme is to be repeated in West Malaysia in the Second Plan, with improvements based on experience, and to obtain data which had been less reliable in the first surveys. Hydrological data are to be collected to facilitate planning of water use, present land use cover will be updated, more details obtained of land alienation patterns.

Out of a total area of 13 million hectares, some 6.35 million hectares are regarded as suitable for agriculture; 3.6 million hectares had been alienated for agricultural purposes, and 2.8 million hectares were actually cultivated in 1966. Some 3.2 million hectares are gazetted as forest reserve, but cross-tabulation of soil suitability data show that about one-third of this area is suitable for agricultural crops. To that extent the forest reserve category land is vulnerable, because of pressure of rural population and the higher economic return obtainable from land under agriculture. Thus it would appear that the long-term outlook for forestry is somewhat bleak, a statement with which Malaysian foresters would fully agree. An editorial in *Malayan Forester* vol. 34, p. 162, 1971, states that the forestry sector is speedily marching on a disastrous course that will reduce the nation to the category of the 'have-beens' in terms of forest wealth. Since forest land is State-owned, exploitation is a 'free-for-all scramble.' Abdul Ghaffar bin Hamid (1972) also talks of insecurity of forest land tenure

272

and the neglect of fundamental principles of conservation and silviculture, in the clearance of forest land for agriculture. But, states Panton (1970), the cross-tabulated data also reveal that within West Malaysia there are about 6 million hectares of productive forest of which only 2 million hectares appears suitable for alternative agricultural use. A further 2 million hectares consists of non-productive forest, for which a good case may be made in favour of gazettement as protective reserves.

These comparisons are regarded not only as indicating the inappropriateness of many of the existing forest reserve boundaries, but also that there is an opportunity of establishing a permanent productive-cum-protective forest estate of some 6.3 million hectares, in which the productive segment alone would be substantially larger than the present acreage of gazetted forest reserve.

Special emphasis has been given to a number (fifteen to date of Panton article) of 'natural resource development opportunity regions', in which the agricultural potentials are especially promising. The Jengka Triangle was the first of these to be developed, and three other regions with a total of 0.6 million hectares are to follow.

SUMMARY OF SPECIFICATIONS USED IN THE MALAYSIAN LAND EVALUATION PROGRAMME (Panton, 1970).

(Full details of each classification are included in the technical reports published independently by the relevant agencies.)

Base maps
Copies of 50 foot contoured, 1:63,360 scale (one inch to a mile) topographical maps with super-imposed 1,000 yard grids covering each of the 70 administrative districts in West Malaysia are supplied to all agencies co-operating in the land evaluation programme.

Present land use
This cover is compiled by stereoscopic interpretation on 1:25,000 scale rectified aerial photographs, according to a legend which is substantially the same as that recommended by the Commission on World Land Use Survey of the International Geographical Union, with slight modifications to suit Malaysian conditions. The land use boundary details are keyed to transparent compilation bases of the 1:25,000 national mapping series and traced onto dimensionally stable film, to which a certain amount of planometric and other detail is added to facilitate future field orientation of the final maps, and to permit acreage counts for land administrative units. A 3 acre minimum definition for the undermentioned 32 land use categories is obtainable using these techniques.

1 Urban and associated areas	7 Mixed horticulture
2 Estate buildings and associated areas	8 Agricultural stations
3 Tin mining	9 Rubber
4 Other mining	10 Oil palm
5 Power line right of ways	11 Coconut
6 Market gardening	12 Pineapple

13	Coffee	23	Padi
14	Tea	24	Shifting cultivation
15	Cocoa	25	Improved permanent pasture
16	Pepper	26	Lalang, unimproved coarse pasture and
17	Sago		scrub-grassland
18	Banana	27	Forest
19	Fibre crops	28	Scrub
20	Orchards (rambutans, durians, citrus,	29	Newly cleared land
	cloves, nutmegs, etc.)	30	Swamp, marshland and wetland forests
21	Fish and hyacinth ponds	31	Unproductive land
22	Annual or diversified crops	32	Unclassified

Land alienation and gazettement

Maps showing alienated (i.e. passed out on title by agreement to the private sector) and gazetted land. Details for all single or contiguous plots of any category exceeding ten acres are prepared by reduction from cadastral survey records, according to the undermentioned classification.

1 Land alienated for all country (agricultural land) purposes, including approved applications, but excluding land held on Temporary Only Lease and also land allocated for agricultural settlement schemes in course of development.
2 Land alienated on mining lease and mining certificate.
3 Land gazetted as Malay Reserve.
4 Land gazetted as Grazing Reserve.
5 Land gazetted as Aborigine Reserve.
6 Land gazetted as Forest Reserve.
7 Land gazetted as Game Reserve.
8 Land alienated as town or village land, which occurs within local authority areas (Municipality, Town Council, and Local Council areas).
9 Land reserved for government purposes other than that falling into the above categories.
10 Land covered by current mineral prospection permit.

Mineral potentiality

These maps are compiled from prospection records supplemented by interpretation made from geological maps in the non-prospected areas.

1 Probable mining land as deduced from prospecting results and geological evidence.
2 Area under mining lease or certificate, or area in which active mining is taking place.
3 Possible mining land as deduced from geological evidence.
4 Area which on geological evidence might contain mineral deposits.
5 Area for which no geological or other information is available.
6 Non-mining land.

Soil suitability

The criteria for this classification are the limitations to agricultural development, as determined from soil survey records and pedological and agronomic experience.

1 Soils with no limitations to agricultural development.
2 Soils with few minor limitations to agricultural development.
3 Soils with at least one serious limitation to agricultural development.
4 Soils with more than one serious limitation to agricultural development.
5 Soils with at least one very serious limitation to agricultural development.

274

Forest productivity

The boundaries for these categories are derived from forest type maps, prepared by photo interpretation methods with minimal ground checking.

1   Treated or regenerated forest or forest plantation.

1M Productive Mangrove Forest (under management).

2A Forest of high potential productivity with a total volume of 64 tons round timber for all species, including at least 40 tons round timber of commercial species per acre.

2B Forest of high potential productivity with a total volume of 64 tons round timber for all species, but including less than 40 tons round timber of commercial species per acre.

3A Forest of average potential productivity with a total volume of 48–64 tons round timber for all species, including at least 28 tons round timber of commercial species per acre.

3B Forest of average potential productivity with a total volume of 48–64 tons round timber for all species, but including less than 28 tons round timber of commercial species per acre.

4A Forest of marginal productivity with a total volume of 32–48 tons round timber for all species, including at least 16 tons round timber of commercial species per acre.

4B Forest of marginal productivity with a total volume of 32–48 tons round timber for all species, but including less than 16 tons round timber of commercial species per acre.

5   Forest of limited potential productivity with a total volume of 32 tons per acre for all species.

5M Unproductive Mangrove Forest (not under management).

Water resource

Lack of quantitative runoff data for most catchments has restricted the evaluation of this resource to a simple cartographic summary of the main utilized catchment boundaries.

1 Currently utilized catchment, providing water for existing schemes, including hydro-electric and potable and irrigation supplies.

2 Catchment proposed for future utilization, which will provide water for proposed schemes, including hydro-electric and potable and irrigation supplies.

3 Existing irrigation scheme area, being an area presently supplied with irrigation water for agricultural purposes.

4 Proposed irrigation schemes area, being an area which it is proposed will be supplied with irrigation water for agricultural purposes.

5 Isohyets, showing rainfall depths at 10 inch intervals.

Land capability classification

These maps are assembled by transfer of appropriate boundaries from the mineral potentiality, soil suitability and forest productivity maps. They are intended as a broad guide for resource use allocation purposes, and should be considered simply as aids to planning, because they indicate the most appropriate areas where particular development activities might be centred.

Class I   Land possessing a high potential for possible mineral development.

Class II   Land possessing a high potential for possible agricultural development with a wide range of crops.

Class III Land possessing a moderate potential for possible agricultural development because of a limitation in the range of crops.

Class IV   Land possessing a high potential for possible productive forest development.
Class V   Land posessing little or no mineral, agricultural or productive forest develop-
           ment potential, but suitable for possible alternative development purposes,
           such as protective forest reserves, water catchment areas, game reserves,
           recreation areas, etc.

Lee Peng Choong of the Economic Planning Unit (personal communication, 1973) states that these approaches and techniques continue to be followed in so far as planning for natural resource development and the formulation of land use policies are concerned. However, the development of natural resources forms only one part of the national development effort. 'It is Malaysia's objective progressively to restructure her economic base towards one where the secondary and tertiary sectors will play an increasingly dominant role in contributing to national income and employment; but within a perspective of the next two decades natural resource development will still be important, both to meet the short-term goals of providing employment quickly and to generate capital for growth in other sectors. Information generated by the natural resource and land use surveys are still being utilized for planning purposes and are currently being updated and refined. While the approaches and techniques remain the same, the interpretation of the potentials for natural resource development will change in the light of changing social and economic factors both nationally and world-wide, in accordance with changing demand for Malaysia's traditional export commodities in world markets and to changing demand for use of her land resources.'

'There has been a number of projects (Lee Peng Choong, personal communication, 1974) to develop major tracts of land in Malaysia, beginning with the Jengka Triangle, which was essentially a scheme to convert a large area of forests, about 100,000 acres, to agriculture. This was followed by the Johor Project, and one in southeast Pahang, in which the scope of planning was enlarged to ensure continuing economic expansion in the regions within a stable ecosystem. The areas involved were also larger, 2.5 million acres in southeast Pahang and 670,000 acres in Johor. The similar study for Sarawak covers 3.5 million acres.

'On the socio-economic front, the main emphasis, besides agriculture and other natural resources-based activities such as in timber and mining, was to plan for the development of settlements in such a way that they would be conducive to the development of urban-centred activities, activities which would provide the basis for a more diversified employment mix in the future, and which would, according to a universal trend towards urbanization, be the thing in demand. Practically, this means concentration of populations in larger centres, rather than widespread dispersal in small villages, as had been planned for Jengka, so as to provide a larger market for urban-type services, at the same time enabling the provision of a wider range of social services due to economies of scale. Special attention was also given to the establishment of Government institutions, for instance, branches of research institutions, so as

276

to lend greater sophistication to these embryo urban centres, thereby increasing their attractiveness for other development. The settlement pattern for Jengka is also being replanned according to the same concept.

'On planning for development in relation to the physical environment the techniques used are essentially similar. First, selection of the areas was based on existing information of the areas available from reconnaissance soil surveys, forest surveys, geological and topographical maps. All the areas selected had sizeable tracts of land believed to be suitable for agricultural development and are relatively accessible.

'Consultants engaged to plan these projects were required to carry out surveys of the resource potential in sufficient detail to locate areas suitable for the development of particular crops, to assess the economic viability of forestry projects, and to locate population centres, transport systems, water supplies, etc. They were also required to foresee changes in the ecosystem, particularly of the hydrological regime following development. The consultants were also required to make recommendations on the development of specific areas for recreation and tourism and for the conservation of unique natural biological communities or landscape features within each region, to form part of a national system of national parks.

'These were translated into a stage-by-stage programme for development corresponding to each five-year plan with projections of cash flows, population, labour, management requirements, etc. Recommendations were also required for the institutional organisation to oversee development of the region.'

Malaysia/Sabah: The State is engaged in the preparation of inventories of the land resources as a basis for assessing land capability for planning. A Technical Sub-Committee of the State Development Planning Committee is responsible. The various elements of the inventory which it was hoped to complete before the end of 1974 for incorporation in the Third Malaysia Plan are: geology and mineral resources; soils and suitability for agriculture; natural vegetation and timber resources; hydrology and water resources; present land use; topography and planimetry; land tenure and ownership; game and recreational resources.

The general approach to land capability classification adopted by the Economic Planning Unit of the Prime Minister's Department for West Malaysia has also been adopted in Sabah. The following resource groups have been recognized, in order of decreasing importance: mining, agriculture, forestry and hydrological and amenity areas (Malaysia, State of Sabah, 1973).

Land Capability Classes: the various natural resource groups are again interpreted into five land capability classes; these are set up so that land having the greatest theoretical alternative uses but always giving the highest return on development is in Class I and the least uses in Class V, with these uses becoming progressively less between these two classes, as illustrated in Figure 6.2. This indicates, for example, that although hydrological and wildlife areas can

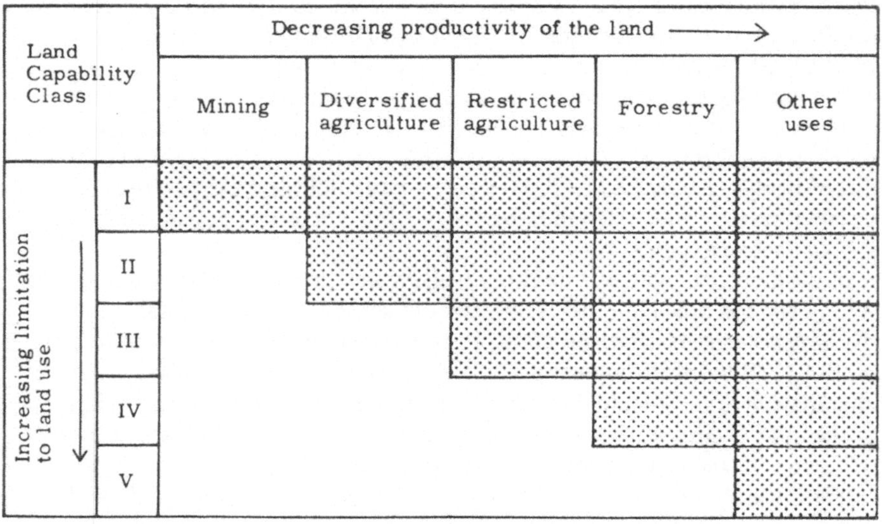

Land Capability Class		Decreasing productivity of the land ———→				
		Mining	Diversified agriculture	Restricted agriculture	Forestry	Other uses
Increasing limitation to land use ↓	I	░	░	░	░	░
	II		░	░	░	░
	III			░	░	░
	IV				░	░
	V					░

Fig. 6.2. The relation between land resources and land capability (Malaysia, State of Sabah, 1973).

be established theoretically in all five classes, the optimal use of the land will depend on adequate levels of minerals, or its agricultural crop potential or timber exploitation capacity, in that order of importance.

The system of classification employed at this level is identical to that employed in West Malaysia (Panton, 1966, 1970). Therefore Land Capability Class I., although not likely to occur extensively, has been retained in order to facilitate direct correlation on a national basis.

Land Exploitation Units. Any one area of land may have one or more resource which may be economically exploitable. It follows, therefore, that on a broader scale natural groupings of land occur, each having similar qualities and uses in having the same kinds of natural resource potentials. These are defined as land exploitation units and are essentially complementary to and fall within the five land capability classes recognized. The overall recommendation as to the future use of the land is defined at the class level. The relation between the resource groups and the other elements employed in the classification is set out in Table 6.2. Thus each unit has a class connotation followed by a suffix indicating the assigned unit.

The following land exploitation units are recognized:

1A   a high potential for mineral development and therefore best suited for mining.
IIA   a high potential for agriculture only.
IIB   a high potential for both agriculture and timber exploitation.
IIC   a high potential for agriculture and a marginal potential for timber exploitation.
IID   a high potential for agriculture and also a possible mining potential.

278

IIE	a high potential for both agriculture and timber exploitation and also a possible mining potential.
IIF	a high potential for agriculture, a marginal potential for timber exploitation and also a possible mining potential.
IIIA	a moderate potential for agriculture only.
IIIB	a moderate potential for agriculture and also a high potential for timber exploitation.
IIIC	a moderate potential for agriculture and also a marginal potential for timber exploitation.
IIID	a moderate potential for agriculture and also a possible mining potential.
IIIE	a moderate potential for agriculture, a high potential for timber exploitation, and also a possible mining potential.
IIIF	a moderate potential for agriculture, a marginal potential for timber exploitation and also a possible mining potential.
IVA	a high potential for timber exploitation only.
IVB	a marginal potential for timber exploitation only.
IVC	a high potential for timber exploitation and a possible mining potential.
IVD	a marginal potential for timber exploitation and a possible mining potential.
IVE	productive mangrove resources only.
VA	little or no potential for agriculture or forest resource exploitation, and best suited for protective or recreational purpose.
VB	little or no potential for agriculture or forest resource exploitation, but with a possible mining potential.

Malaysia/Sarawak: Panton (1970) stated that a programme of land evaluation was planned to begin in 1971 and to continue until the end of the Second Malaysian Plan in 1975.

The objectives and field survey techniques adopted in the soil survey of the central Sarawak lowlands (Scott, 1974) have already been discussed (3.3.2).

The Miri-Bintulu regional planning study is described briefly in Chapter 7.

6.2.5 Philippines

The eight maps in the North Central Ifugao Land Use Series, published by the American Geographical Society on behalf of Professor H. C. Conklin of Yale University (Conklin, 1972) represent the application of integrated survey and airphoto-interpretation in an anthropological study of the peoples of a mountainous area of the Philippines, who 'have long been known for their astonishing feats of engineering in the construction and maintenance of extensive rice terraces' (Conklin, 1967).

Using high resolution aerial photography and machine mapping, it is now possible with relatively few ground control points to determine contour differences of less than two metres from 1:12,000 to 1:20,000-scale stereopairs, thus overcoming the formidable obstacles faced by the unassisted ground surveyor and cartographer. The combination of government maps on 1:50,000 sheets for northern Luzon with these new aerial photographs brought to more than 1,000 the number of vertical aerial prints available for this study.

Table 6.2. The relation between resource suitability groups, land capability classes and land exploitation units (Malaysia, State of Sabah, 1973).

Land Capability Class	Land Exploitation Unit	Resource suitability groups		
		Mining	Soils	Forestry
I	IA	1-2	4-5	4-6, 8
II	IIA	4	1-2	4-6, 8
	IIB	4	1-2	1-2
	IIC	4	1-2	3
	IID	3	1-2	4-6, 8
	IIE	3	1-2	1-2
	IIF	3	1-2	3
III	IIIA	4	3	4-6, 8
	IIIB	4	3	1-2
	IIIC	4	3	3
	IIID	3	3	4-6, 8
	IIIE	3	3	1-2
	IIIF	3	3	3
IV	IVA	4	4-5	1-2
	IVB	4	4-5	3
	IVC	3	4-5	1-2
	IVD	3	4-5	3
	IVE	4	4-5	7
V	VA	4	4-5	4-6, 8
	VB	3	4-5	4-6, 8

In spite of many difficulties caused by logistics, weather and terrain, all photography and ground surveying necessary for the machine mapping phase of this project were completed in 1963. Four film-manuscript separations (topography, vegetation, drainage, and cultural detail) for map sections covering the survey area (at 1:5,000 scale, with 4-metre contours), and the focal area (at 1:2,500 scale, with 2-metre contours), were photogrammetrically plotted during 1963 and 1964. The required ground checking continued for a total of nine months at various periods between 1964 and 1966. Table 3 in Conklin (1967), is of interest, in that it shows the advantages and limitations of aerial photography in constructing maps of value in ethnographic studies.

The set of eight maps covers 96 sq. km. of agriculturally dominated landscape. The mapped area, which falls within the Province of Ifugao, divides roughly into two catchments, the Alimit on the north and east, and the Ibulao on the south and west. Elevations above sea level range from 600 to 2,000 m., extending well above the highest zone of wet-field terracing. The largest territorial units of traditional as well as contemporary social and cultural importance are the agricultural districts, each of which has a ritual parcel as its symbolic centre. These districts are most pertinent to local land utilization and are given major emphasis on the maps.

The eight maps show drainage courses (both natural and artificial), main course of irrigation channels, pond field terraces (where mainly rice, but also some taro and other secondary crops are grown), drained field terraces for dry crop cultivation, house terraces with dwellings, levelled settlement sites, swidden areas either cleared or under cultivation (where root crops, legumes and maize are grown), areas of light to intermediate second growth (mainly caneland), woodland areas (including woodland plots near settlement), exposed slopes or terrace embankment.

Many of the smaller irrigation ditches and most intermittent and submerged water areas are not indicated. Footpaths along pond-field bunds cannot be delineated; except where they pass between terrace areas, trail networks remain incomplete. Also, paths following water channels are not shown.

The shape and extent of the artificially *levelled* landforms (e.g. pond-fields) show little change from year to year, but shifts from one usage status to another among the *unlevelled* landforms (especially among swiddens and caneland or woodland) are frequent. Thus maps of the same region two years apart would show much greater land use change in slopeland areas than in terraced zones.

A projected *Ethnographic Atlas of Ifugao* is to include, in addition to the maps, photographs, diagrams and text describing Ifugao terrace construction, irrigation methods, rice cultivation and related economic activities, as well as methodology adopted in the project.

6.3 Categories and qualities of land

6.3.1 Site, land unit and land system

The first part of this section, apart from Tables 6/3 and 6/4, is quoted from Christian & Stewart (1968). The land surface at one location is regarded as the end product of evolution involving the action of a combination of physical and biological processes, acting on a particular type of geological parent material at certain rates for given periods in certain geological climates (Christian, Jennings & Twidale, 1957). In this process of evolution, particularly in the more recent phases, the shaping of the land form, the development of the soil profile, and of surface and sub-surface drainage features, and the occupation, utilization and modification of habitats by the biological components, have gone on simultaneously. Land is thus a dynamic feature, and when we describe the land complex at a site we are merely describing the state of the land surface at one moment in a slow evolutionary sequence. The period over which the individual processes have operated is as important as the processes themselves, or the material on which they have operated. In many places, man himself has also been a factor in this sequence, sometimes causing abrupt changes, sometimes through long periods of intensive land use, having an impact on the nature of land which must rank as an inherent feature of the land surface as important as any of the 'natural' genetic factors.

The Australian Land Research approach recognizes these units in the subdivision of landscape: the site, land unit and land system. The land system is the unit of mapping. At the reconnaissance-survey level, land systems are described in terms of their component land units which recur in association to form the land-system pattern. The site is the smaller identifiable unit. Land units may consist of a single site or a group of geographically associated sites.

Site

A site is a part of the land surface which is, for all practical purposes, uniform throughout its extent in land form, soil and vegetation. A small amount of variability occurs within a site, but it is of such low order that, at the level of the survey being made, the variation falls within the units of classification used in each discipline. It can reasonably be inferred that the various occurrences of a site are also likely to be similar in a number of less easily observed or measured characteristics which in practice could well be reflected in the technical problems associated with land use. Each site represents a distinctive type of environment and at each of its occurrences it will provide a similar range of habitats for man, plants and animals. Thus the site is a unit of landscape, or in other words a taxonomic land type, which for most practical purposes can be regarded as having similar possibilities and similar problems for land use wherever it recurs. The individuality of sites needs to be taken into account in subsequent technical studies, but larger areas of land represented by land units or land systems will form the basis of development planning.

Land unit

In practice, the land unit is usually a group of related sites which has a particular land form within the land system and wherever it occurs again it would have the same association of sites. The simplicity or complexity of the land form accepted as the unit of study is in part determined by changes in genetic factors (e.g., parent material, internal drainage) that are reflected in changes in soil and/or vegetation but not in the land form itself. The land units are the broader components of the land system that are illustrated and described in the land-system descriptions

Land system

Land units are geomorphologically and geographically associated to form patterns which recur in the landscape. The boundary of the pattern generally coincides with that of some discernible geological or geomorphogenetic

282

feature or process. Within the pattern it is the same land units which recur. Where a different assemblage of land units begins, it is a different land system. Thus by subdividing a region into its land systems (patterns) and describing each in terms of its component land units and sites, it is possible to reduce the complexity of the region to a reasonable number of types which can be comprehended without having to resort to the time-consuming practice of actually mapping each occurrence of each land unit or site. The areas of the land systems can be directly calculated from the maps, and the proportionate areas of the land units within each approximately estimated. Where land units can be identified on aerial photographs their proportions can be measured by appropriate sampling techniques. Experience has shown that this way of 'looking at country' is quickly picked up by others not directly concerned with this kind of work and that it assists them substantially in their own activities and appreciations.

The usual scales of working are described in the section on methods. At this point it is necessary to mention only that the concept can be used at a variety of scales. In the early stages of its development it was found desirable to introduce three terms: simple land system, complex land system and compound land system (Christian & Stewart, 1953). A simple land system is one composed of clearly definable land units which recur in association to form a simple recurring pattern such as an old peneplain with remnants of old drainage depressions, restricted to a climatic range which does not involve major vegetation or soil transitions due to the climate factor. A complex land system would be the combination of two or more such systems, for example, a similar peneplain uplifted and dissected at intervals by parallel streams, with two distinctive patterns, the peneplain pattern and the pattern of scree and colluvial dissection slopes, lower slopes, floodplains, stream levees and stream channels. The two simple land systems are geomorphogenetically related and this is the main distinction between the complex and compound land systems. The latter would be represented by a number of isolated igneous intrusions into a sedimentary area. The intrusions may represent outliers of a more extensive igneous land system elsewhere but are isolated within the sedimentary land systems with which they are less closely related geomorphogenetically than the components of a complex land system.

However, at a more detailed level of working some land units of the foregoing examples, such as the floodplains of the river systems, would be regarded as simple land systems themselves. Their gently sloping or undulating topography could be subdivided into a number of smaller, less diverse, land units with slight differences in altitude but probably important differences in depth of flooding, and in consequence, recognizable differences in vegetation and, most likely, soils also.

Ideally this sort of fine sub-division would be made irrespective of the smallness of size of any units but in reality the surveys have a practical task to achieve, namely the provision of the essential preliminary information for resource planning and development as quickly as possible. Where sites are very small in size they can well be disregarded or their presence merely noted, thereby saving a good deal of effort. They may need to be more closely examined in a subsequent survey.

The practical value of land systems
The usefulness of a land system is determined by the land units that comprise it. On the basis of these units a land system can be discarded from further immediate consideration or given specialist attention by agriculturists, foresters, pedologists, geologists, hydrological engineers or others. Thus a land-system survey is a valuable approach to the channelling of further effort into the areas where it can be most rewarding and of different kinds of technical personnel to those areas where their special skills, or combinations of skills, are most needed. Examples are shown in Tables 6.3 and 6.4, two ecosystems in the Vanimo Area, Papua New Guinea (CSIRO, 1972).

The assessment of the land-use value of land types is discussed in other parts of this article. At this point, it should be stressed that the second stage of investigation will differ for different sites, land units and land systems. In one case, a more thorough mineral-resource study may be indicated for selected areas, or a detailed forest assessment may be the appropriate investigation; in another it may be agronomic experiments to determine the possibilities of arable agriculture or to correct faults in an existing agriculture or to introduce a major change in agricultural land use; another area with water resources may require the attention of the hydrological engineer before the agriculturist or the forester is brought in, or it may be an area in which the road engineer is needed to make further examination of land-surface characteristics in relation to road-transport development.

Apart from its contribution to the planning of investigations in the next stage in an orderly and selective way there is another importance in having a permanent framework of land types such as the land-system map provides. The framework has equal value as a guide to where and how widely the results obtained from further investigations at one location or local experience may be expected to apply. Thus if agricultural experimentation is conducted on a sample of a site, or successful land use has already been achieved, the results can be expected to apply to other occurrences of that site. However, different sites, even though apparently similar in many respects, must be suspected of responding differently until proved by trial to do otherwise. This will require testing but once it is established that the results do apply to a sample of a second site then likewise they can be expected to apply to all other occurrences of that site.

A third value of the land-system framework is in its use as a common basis of sampling for subsequent studies. Where data are to be collected for statistical, economic, education, health, biological or other equally divergent purposes, there is an advantage if the geographic unit used for sampling is common to

284

each. This makes the data so much more useful because of the possibilities of direct correlation and collation. The land-system approach provides a logical common basis of sampling because it identifies the inherent environments which differ according to the basic features of land rather than according to any arbitrarily selected criteria of a more temporary or superficial significance.

For these three reasons, however, it is important that the land-system framework should be a truly basic and permanent one which future surveys will not alter although they may add more detail.

The nature of present land use is information of an essential, supplementary kind which must be recorded in the course of survey but it does not provide satisfactory criteria for distinguishing basic land types. Present land use may not be a true indication of future land use possibilities. It is only when land use has been very intensive for a long time and has become itself an inherent feature of land, closely correlated with other inherent features, that it is a useful distinguishing feature for basic surveys.

The Australian Land Research approach distinguishes conceptually between the two aspects, resource inventory and resource assessment. Assessment will vary according to the technical knowledge available and this will change with time. Although survey and preliminary assessment may go on concurrently, the latter should not be allowed to influence the selection of the basis units which are mapped.

In a number of its surveys, therefore, the CSIRO has produced a land-use group map as well as a land-system map. The land-use group map shows the present interpretation of possible land use, based on local experience and other information but it is regarded as a temporary map that will be modified as soon as more information on how best to use land is available. In contrast, the land-system map, the map of basic land types, should persist as the permanent basic reference for any purpose.

(End of quote from Christian & Stewart, 1968)

Table 6.3. Papua New Guinea: Papul Land System — 61 sq. miles = 158 sq. km. (CSIRO, 1972).

Land Forms (Plates 13, 14) — Flood-plains and river terraces up to 1 mile wide developed along larger rivers. Flood-plains consisting of meandering channels, scroll plains, and flood-plain terraces are generally 900–1500 ft wide and bounded by well-developed terraces. In most areas only one terrace observed rising 6–15 ft above flood-plain. Where lowest terrace is extensive it is also referred to as alluvial plain. In places a second terrace 30–45 ft above flood-plain occurs. Terrace surfaces have irregular micro-relief up to 1.5 ft and are crossed by shallow channels.
Terrain Parameters. — Altitude: 120–500 ft. Relief: nil (1.5 ft). Characteristic slope: very gentle (0.5°).
Streams and Drainage. — Flood-plains regularly flooded during rainy season but terraces are above high-water level. Flood-plains and terraces drained by shallow creeks. Edges of terraces dissected by steep-sided gullies.
Geology. — Recent alluvium.

Vegetation. – Differs in three major occurrences. In the east along the Piore River, tall forest with rather open canopy (Fo), and on scroll plains, seral stages developing from cane-grass vegetation (Gt/Fmo).

In the central part along Sereri and Bilia Creeks, tall forest with open canopy with scattered light-toned crowns (Fod) is the major forest type, with smaller-crowned forests in upper reaches (Fosd) and on some ill-drained terraces of Sereri Creek (Fos), and mid-height forest with rather open canopy (Fmo) in upper reaches of Bilia Creek.

In the west along the Pulan River, tall forest with rather closed canopy (F) covers most of land system; tall forest with rather open canopy with light-toned crowns (Fod) and seral stages (Gt/Fmo) occurs further downstream.

Soils (8 obs.) – On scroll plains and flood-plain terraces probably similar to those of Pual. Soils on lower terraces alkaline, can be calcareous; on upper terraces usually neutral to weakly acid or weakly acid to acid. This trend of increasing acidity is due to increased leaching, first of carbonates, then of exchangeable cations, with increasing age. Nearly all soils are undeveloped.

Soils are moderately deep to very deep, moderately to slowly permeable, and generally imperfectly drained. Few soils are poorly drained, while youngest soils tend to be well drained. Soil texture generally fine, but some medium-textured soils and coarse-textured subsoils also occur. Consistently fine-textured soils on wider, plain-like terraces.

Nitrogen contents low to moderate; phosphate contents vary greatly, but appear to be low in acid soils, moderate in neutral to weakly acid soils, very variable in alkaline soils; potash moderate to very high in weakly acid to alkaline soils, low in acid soils.

Population and Land Use. – Population 737 distributed over 7 villages. Land use negligible.

Forest Resources. – High. Very high stocking rate forests (F and Fo) cover 34% (24% and 10% respectively), high stocking rate forests (Fod, Fos, and Fosd) cover 43% (30%, 9%, and 4%), and low stocking rate forest (Fmo) covers 4%. Distribution of forest described under Vegetation. Access generally good but may be hampered on flood-plains by flooding and in minor areas by poor drainage.

Agricultural Assessment. – High capability for improved pastures, moderate for arable crops and irrigated rice, low for tree crops. Flooding renders narrow stretches along major streams almost useless for development and drainage deficiencies are major limitation. Rather high to very high soil pH is common additional hazard for tree crops. Gullies and surface unevenness on terraces reduce capability for arable crops and irrigated rice.

Engineering Assessment. – Only slight topographical limitations for road construction and favourably located for regional road connection inland between Vanimo and Aitape and/or Lumi. Bridge construction would be a major cost item but cheaper low-level flood-way crossings appear suitable. Soil materials generally of low suitability for road construction. Only limited amounts of sand and hard gravel in river beds.

Soils CH, MH, CL, with some ML and SM in subsoils and near streams: nearly always very deep.

Land system mapping has not been found to be a direct road to resource evaluation in Melanesia, nor, by itself, has soil mapping (Brookfield & Hart, 1971). Such mapping covers the islands with a bewildering mosaic of small areas; since the detail of ground survey is not equal to the detail of the maps, the significance of such fine areal differentiation is very doubtful. There are two alternatives, or complementary ways in which to proceed. The complex measurement of attributes and their tabulation suggest that finer results might be achieved by abandoning the general reconnaissance based on the

Table 6.4. Papua New Guinea: Isi Land System – 229 sq. miles = 594 sq. km. (CSIRO, 1972).

Land Forms (Plates 6, 8, 9, 11). – Very low and in places ultra-low hills and ridges mostly rising from alluvial plains. Ridges generally form irregularly branching pattern, slopes moderately steep to steep and frequently slumped causing locally great differences in slope steepness. Slumps are mainly on lower and middle slopes; some also on upper slopes causing knife-edged steep-sided ridge crests. Slumps have commonly 15–24 ft high back walls and 18–24 ft wide slump benches. Upper slopes are generally convex, middle and lower slopes concave. Ridge crests are very narrow, locally narrow, some flat topped. Crestal slopes are uneven but summit level of ridge crests and hill tops is rather even. Gullies cut back into middle slopes and are 6–18 ft wide, 3–18 ft deep, 75–90 ft apart. Size of gullies increases downslope.

Terrain Parameters. – Altitude: 15–300 ft. Relief: ultra-low and very low (30–120 ft). Characteristic slope: moderately steep (14–17°). Grain: very fine (360 ft).

Streams and Drainage. – Dense irregular pattern of small streams mostly of first and second order. Larger streams are mapped out separately as Basu (7). Most stream gradients are 2–3°.

Geology. – Pliocene marl and mudstone.

Vegetation. – Tall forest with a rather open irregular canopy with scattered light-toned crowns (Foid) covers about three-quarters; on the remainder is less open forest (Fid).

Soils (14 obs.). – Very fine to fine-textured and nearly always slowly to very slowly permeable soils. Most are moderately shallow, locally moderately deep. Shallow soils are confined to some very steep slopes. Soils are generally well drained, locally imperfectly drained, and on slump benches poorly drained. Soil reaction varies with weathering from alkaline to strongly acid. More acid, moderately to strongly developed soils tend to occur on crests, upper slopes, and slumped slopes. More alkaline, slightly to moderately developed soils tend to dominate on lower and steepest slopes.

Low to moderate soil nitrogen contents that tend to be highest on moderate slopes; phosphate is low to very low; potash is high to very high and tends to be highest on lower and steeper slopes, lowest on upper slopes.

Population and Land Use. – Population of 285 distributed over 3 villages. These are situated near Ossima on narrow ridges near sago-supplying valley floors of Kabuk. About 3% of land system used for gardening.

Forest Resources. – High stocking rate forests (Foid, Fid) cover 92% (77%, 15%), remainder is stocked by small areas of various forest types including secondary forest. Access is moderate, main limitations being moderately steep slopes and minor areas of imperfect to poor drainage on slump floors.

Agricultural Assessment. – Very low capability for arable crops and moderate capability for improved pastures, topography being main limitation. Pasture establishment may be difficult on very clayey soils, but would probably provide best possible protection against slumping and soil creep. Control of gullying requires special attention. Slow permeability and local drainage deficiencies and soil alkalinity reduce capability for tree crops to low.

Engineering Assessment. – Instability of slopes, particularly when cut by roads along contour and absence of suitable road-building materials are major problems in road construction. Much culverting needed but rarely substantial bridges.

Soils are mainly CH, rarely MH; generally moderately deep to shallow, with deep to very deep soils only on slump and valley floors.

land system, and by proceeding in one or both of two ways. Areal sampling, within strata determined perhaps by visible features such as terrain, altitude, vegetation type, could yield a scatter of small areas within which data on all possible measurable attributes might be collected. Once the most significant

287

ecological factors have been established, it is then time to proceed to regional extrapolation, and to the more detailed study of areas that are either important at the present time because they are still occupied, or potentially important because they seem to have a high capability for development. There seems little need to cover a whole country at an equal and intermediate level of detail.

The alternative approach is reconnaissance at a level of very much greater simplicity than the land system or the soil type. Heyligers' map of major environments of Bougainville is useful, although they certainly demand subdivision, as a first approximation in the analysis of distribution of population, something quite beyond its primary purpose of explaining the plant cover. While Melanesia could be divided into many dozens of land systems, the gain to understanding, according to Brookfield & Hart (op. cit.), might be greater if the region were instead to be divided into twenty or thirty major environments, capable of local subdivision as required.

The British surveys of the Land Resources Division offer interesting suggestions for achieving tolerable results at far lower costs than the Australian surveys (Brookfield & Hart, op. cit.). In the Pacific territories, small Land Resources Division teams have proceeded to a land unit/land system definition by methods similar to those of CSIRO, except that the basis of the definition is more ecological and less genetic; land systems are then grouped into land regions, of which there would be some six or eight on a large island in the Solomons group. Within land systems, sample areas are selected only for detailed investigation of soils, plants, hydrology, topography and geology. The use of sample areas represents a great saving in cost; it permits identification of both areas and problems worthy of closer investigation, and provides a methodology through which specific investigation of selected sites for development purposes can be integrated into the survey systems as a whole. Grouping into land regions approaches a classification of the land surveys into major environmental classes.

6.3.2 Definition and survey of land suitability/capability

The best known system is that of the U.S. Department of Agriculture. This is an interpretative system which uses a soil survey map as the basis and which brings the individual soil map units into groups with similar management requirements (Moormann, 1973). The capability of grouping is designed to help users of land to interpret soil maps, and to make broad generalizations based on potentials of soils and limitations in their use and management. The system is primarily concerned with the risk of erosion and to a lesser extent with other hazards such as wetness, shallowness of soil, salinity or alkalinity. Eight classes are recognized, with subclasses at the secondary level at which four kinds of limitation are recognized — risk of erosion, wetness, drainage or overflow, and limitations in respect of rooting zone and climate. At the tertiary

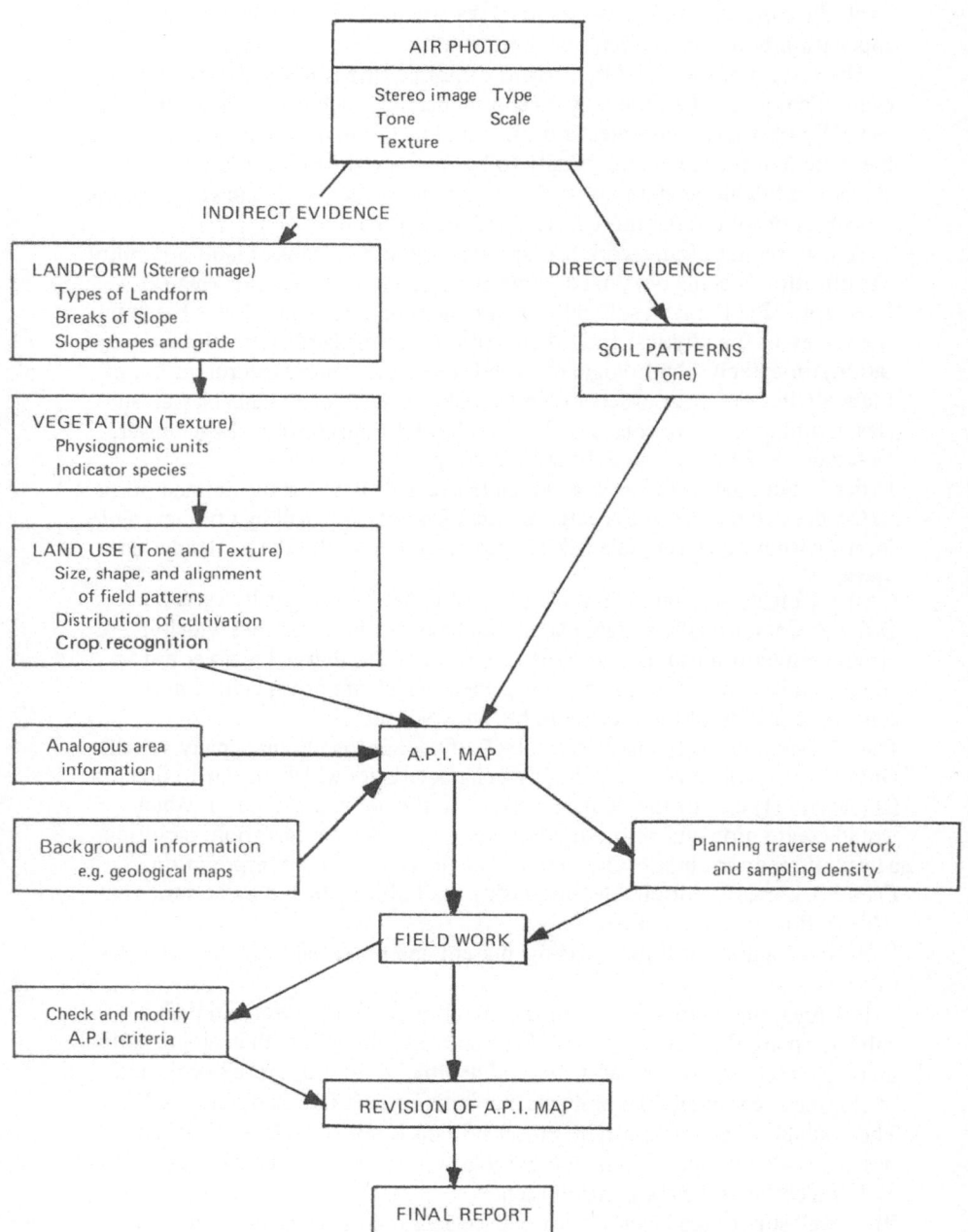

Fig. 6.3. Airphoto interpretation in soil survey (Stobbs, Lecture Notes). The type and scale of air photography determines the amount of detail which can be identified and interpreted by means of the stereoscopic image, tone and texture of the photo patterns. The diagram refers to the use of panchromatic black and white photography (see pp. 338–342).

level, the capability unit provides more specific and detailed information for application to individual fields on a farm.

The system of the U.S. Department of Agriculture is not generally applicable over much of Asia because it assumes a moderately high level of management, generally associated with mechanization and high rate of inputs. The use of the system in such areas has usually to be based on the interpretation of soil characteristics alone, without sufficient agronomic data or a knowledge of the dynamics of soils under more intensified management.

A new standard framework for land evaluation by means of land suitability classification is being developed under the auspices of FAO (Brinkman & Smyth, 1973). The land suitability component of land evaluation is based on the survey of the physical and other attributes of the land (soils, climate, vegetation, topography, hydrology, etc.) and therefore requires interpretation of these attributes. Three orders of land suitability, each embracing classes, subclasses and units, were considered by the Expert Consultation, Wageningen, October, 1972 (Brinkman & Smyth, 1973, pp. 75–7).

Order 1: Suitable land; Land on which (sustained*) use for the defined purpose in the defined manner is expected to yield benefits that will justify recurrent inputs without unacceptable risk to land resources on the site or in adjacent areas:

Class 1.1 highly suitable; 1.2 moderately suitable; 1.3 marginally suitable.

Order 2: Conditionally suitable land; Land having characteristics which in general render it unsuitable for (sustained) use in the defined manner but which, subject to conditions of management which are not specified in the general definition of the use, could be rendered suitable.

Class 2.1 conditionally highly suitable; 2.2/3 conditionally marginally suitable.

Order 3: Unsuitable land; Land having characteristics which appear to preclude its (sustained) use for the defined purpose in the defined manner or which would create problems of production, upkeep and/or conservation, requiring a level of recurrent inputs unacceptable at the time of the interpretation.

Class 3.1 presently unsuitable, but having limitations which may be surmountable in time; 3.2 unsuitable.

(Figs. 6.4 a and b for flow charts for present and potential suitability classifications).

In Korea, the factors determining suitability of land for wetland padi cultivation are the characteristics of the land and the soil, such as slope, natural drainage, texture, erosion, available soil depths, content of gravels or degree of stoniness, conductivities and acid sulphatic layer (Chun Soo Shin, 1971). The availability of water and irrigation systems is not considered. Each soil has been rated in one of four suitability groups as below, with six sub-groups in each indicating soil classes within each:

P1   well-suited: land suitable for padi rice without a need for special practices of development or management; no special limitations or hazards

* the desirability of qualifying use as *sustained* is under active debate

290

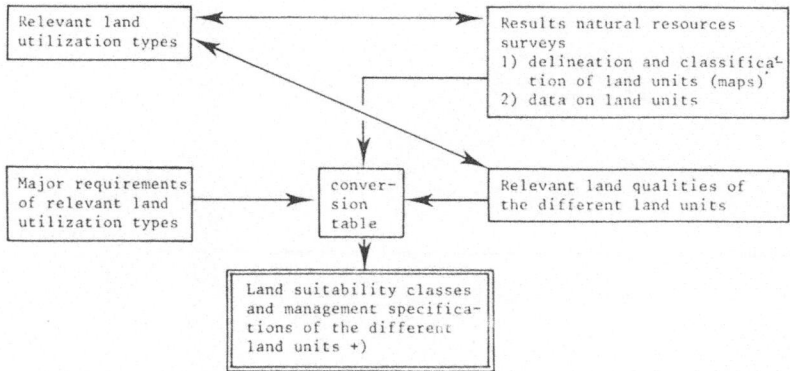

Fig. 6.4a. Expert Consultation: Flow chart for present suitability classification (Brinkman & Smyth, 1973).

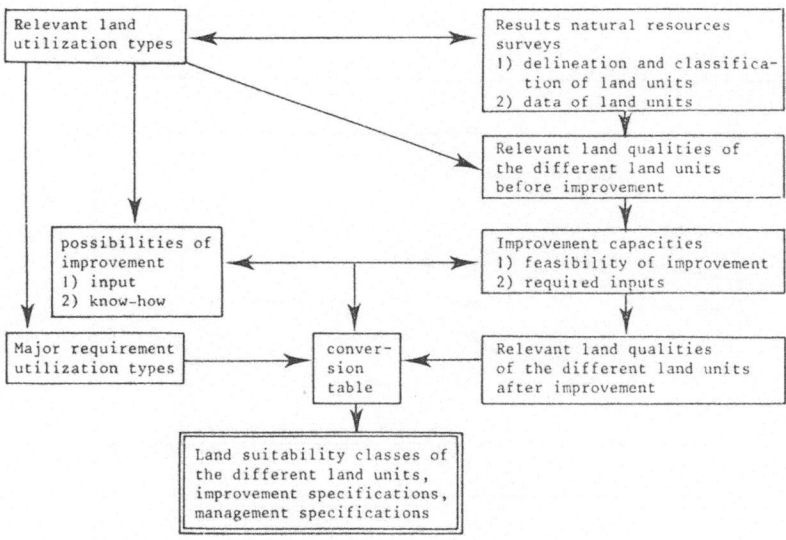

Fig. 6.4b. Expert Consultation: Flow chart for potential suitability classification.

P2   moderately suited: land suitable for padi rice with simple special practices; moderate hazards and limitations

P3   poorly suited: land suitable for padi rice with the application of difficult special practices; severe hazards and limitations

P4   very poorly suited: land of limited or questionable suitability for padi rice because of very severe hazards and limitations, and the need for very difficult special management practices.

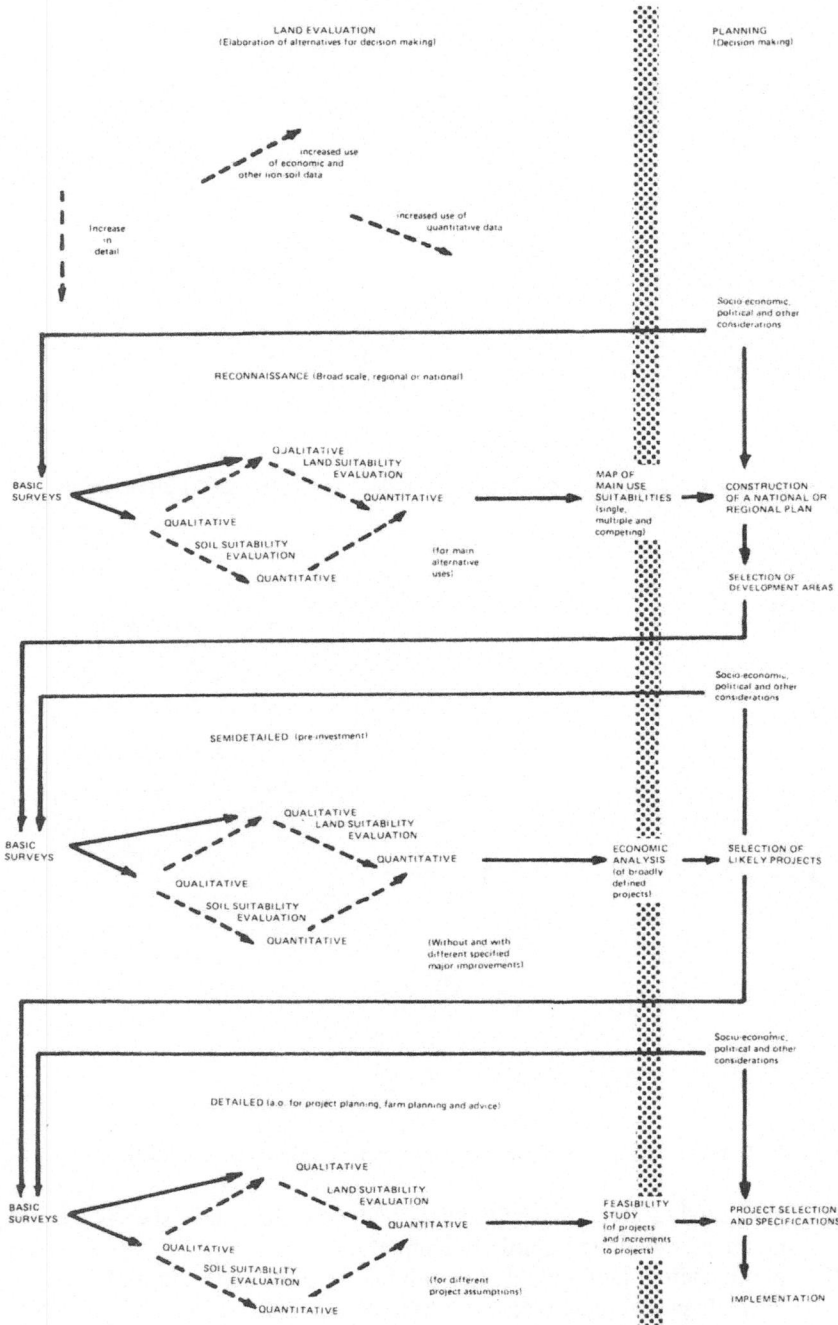

Fig. 6.5. Expert Consultation: Chart of land evaluation activities (Brinkman & Smyth, 1973).

292

Table 6.5 shows the rating criteria of suitability groups adopted for padi land in Korea. Rating criteria are shown in other Tables of the original, for upland (dryland) crops, orchards and mulberry plantations, woodland, pasture and hayland; these are summarized in Table 6.6.

Table 6.5. South Korea: Rating criteria of suitability groups for padi land (Chun Soo Shin, 1971).

Soil-related items	P1	P2	P3	P4
Natural drainage	Somewhat poor	Mod. good poor	Good very poor	
Soil texture	Fine clayey Fine loamy Fine silty	Fine clayey Fine loamy Fine silty	Coarse loamy Coarse silty	Sandy
Available soil depth (cm)				
hard rock hard pan	> 100	100–50	50–20	50–20
gravels cobbles sandy	> 100	100–50	50–20	20–10
Gravels, cobbles on the surface	< 3%	3–15	15–50	50 <
Rocks, bouldery stoniness	None	None	None	None
Salt cond. (mm hos/cm at 25°C)	< 4	4–8	8–16	16 <
Depth to acid sulphatic layer (cm)	> 100	> 100	100–50	50–20
Soil erosion	None Slightly	None Slightly	Eroded	Eroded
Slope percent				
P1 P2	2	2–7	7–15	15–30
P3	–	–	< 7	7–15
P4	–	–	–	< 7

Table 6.6. South Korea: Composite guide for land reclamation (Chun Soo Shin, 1971).

Soil-related items	Land use Rec. P (padi)	U (upland)	O (orchard and mulberry)	Pasture and Hay G1 (intensive)	Pasture and Hay G2 (extensive)	F (woodland)	W (Recreation and wild life)
Slope per cent	<15	<15	<30	30	30–60	30<	–
Natural drainage	Somewhat poorly Mod. well (Fine loamy or clayey soils)	Well Mod. well	Well Somewhat Excessively	Welll Mod. well	Somewhat Excessively Excessively	Well	Any
Soil texture	Fine loamy Fine clayey	Fine loamy Fine clayey	Fine loamy Coarse loamy	Fine clayey Fine loamy Coarse loamy	Coarse loamy Sandy	Fine loamy Coarse loamy	River wash Sandy or cobbly
Available soil depth (cm) hard rock hard pan	>50	>50	>100	>50	>10	>25	–
cobbles gravels sandy	>25	>25	>10	>10	–	>10	–
Stoniness, rockiness or bouldery on the surface	None	None	None stony	None stony	Bouldery rocky	None rocky	Rock out crops
Soil erosion	None eroded	None eroded	None severe	None eroded	Severe	None severe	Gullied
Salt cond. (mmhos/cm at 25°C)	<16	–	–	–	–	–	30 <

In 1968, the (British) Land Resources Division commenced a study of the Northern and Luapula Provinces of Zambia. The Report (*Land Resources Study no.* 18) is to be published in five volumes, totalling 1,000 pages, for the use of administrative and technical personnel. In addition to the Report in the usual format, a computerized data bank of soil and site information has been set up, which will enable reinvestigation of land capability whenever modifications of the system adopted suggest themselves. It also makes it possible to review the effects of different inputs.

In the course of this work, it was agreed that a revised land capability classification was required for the area of open field (including mechanized) agriculture in Northern Zambia (Lang, 1973). For this purpose, a *pro forma* was introduced which can be used to classify the land capability of auger boring sites fairly quickly. The *pro forma* is divided into two parts. The upper part comprises a series of boxes in which soil and site observations at an auger site can be recorded. The lower part contains instructions whereby the information needed to classify the capability of the site can be looked up in tables or calculated.

It is understood that land capability surveys are in progress or completed in the Philippines. For Indonesia, see Chapter 7.

6.3.3 Land quality

The Expert Consultation at Wageningen defined this type of characteristic: 'A single land quality is a complex attribute of land which, when used as a diagnostic criterion, acts in a manner clearly distinct from the actions of most other land qualities in its influence on the suitability of land for a specific kind of land use. The expression of each land quality is determined by a set of interacting single (or compound) land characteristics having different weights in different environments depending on the values of all characteristics in the set.'

A major land quality may be used as a diagnostic criterion reflecting limitations to land suitability. It may be rated and quantified and suitability class limits specified using the rated criteria. The Expert Consultation adopted a list of major land qualities for rural land use (from Beek & Bennema, 1972), as a first approximation to give an idea of their intended nature. Only those major land qualities are shown which relate to agricultural use. Many similar and some quite different major land qualities would become relevant to use for wildlife and recreation, village areas, fishing, waste disposal, etc.

1. Major land qualities related to plant growth

    – availability of water
    – availability of nutrients
    – availability of oxygen for root growth

- availability of foothold for roots
- conditions for germination (seed bed etc.)
- salinization and/or alkalinization
- soil toxicity or extreme acidity
- pests and diseases related to the land
- flooding hazard
- temperature regime (including incidence of frosts)
- radiation energy and photoperiod
- wind and storm as affecting plant growth
- hail and snow as affecting plant growth
- air humidity as affecting plant growth
- drying periods for ripening of crops and at harvest time

2. Major land qualities specifically related to animal growth

- hardships due to climate
- endemic pests and diseases
- nutritive value of grazing land
- toxicity of grazing land
- resistance to degradation of vegetation
- resistance to soil erosion under grazing conditions
- availability of drinking water
- accessibility of the terrain

3. Major land qualities related to natural product extraction

- presence of valuable wood species
- presence of medicinal plants and/or other vegetation extraction products
- presence of fruits
- presence of game for meat and/or hides
- accessibility of the terrain

4. Major land qualities related to practices in plant production, in animal production or in extractions

- possibilities of mechanization
- resistance to erosion
- freedom in the layout of a farm plan or a development scheme, including the freedom to select the shape and the size of fields
- trafficability from farm to land
- vegetation cover in terms of favourable or unfavourable effects for cropping

6.4 The application of aerial photography and space technology

6.4.1 Aerial photography and interpretation.

It might have been preferable if this section had been placed immediately after 6.1 to stress its association with the basic inventory, rather than with the interpretative stages of land evaluation. This would seem appropriate in that analysis of air photographs is usually an opening move in the game (A. J. Smyth, FAO, personal communication, 1974).

This approach is confirmed by ITC/Enschede, which proposes the following procedure in land survey, in ten steps, each consisting of separate activities which nevertheless belong together and influence each other. When it is not possible to do step 2, step 1 is followed directly by 3: 1. photo perusal, photo analysis, preparation of preliminary legend; 2. preliminary field inspection; 3. preliminary photo-interpretation and generalization; 4. tracing and colouring of preliminary maps; 5. field work, correlation and classification; 6. final photo-interpretation and mapping; 7. drawing final land and land-attribute maps; 8. evaluation; 9 and 10. reporting and reproduction.

Another standard operational procedure, using stereoscopic pairs of aerial photographs in conjunction with air photo-mosaic reproductions for terrain investigations at all levels is: 1. study of all available maps, geologic and agronomic references; 2. initial stereoscopic study of aerial photographs; 3. selective field reconnaissance after consultations with local staff on the spot; 4. selective field sampling and provisional control of air photo boundaries; 5. laboratory testing and summary of data; 6. more detailed study of aerial photographs and mapping; 7. selective field checking and detailed control of air photo boundaries; 8. transfer of controlled air photo boundaries to base maps; 9. preparation of reports and tables and reproduction of maps (Jones, 1969).

The time taken by a CSIRO Land Research team for a survey in one season in the field is generally 15 to 18 months, as follows: 1. pre-field work — collection of existing information, preliminary interpretation of aerial photographs, selection of sampling site and planning of field itinerary — three months; 2. field work — three months; 3. final interpretation of aerial photographs — three months; 4. specialist evaluation of field data and co-ordination of descriptions of land systems — three to four months; 5. preparation of report — three to five months (Christian & Stewart, 1968).

Aerial photographs have been employed by the Land Resources Division of the British Directorate of Overseas Surveys at all levels of intensity, from exploratory and broad reconnaissance investigations to detailed surveys and development plans for specific enterprises (Table 6.1). Generally, it is at the least intensive early stages of survey, when time and money available for ground control are minimal, that the greatest benefits are to be gained from stereoscopic examination and interpretation of aerial photographs. Air photo-interpretation enables a comprehensive picture of the landscape and the distribution on it of vegetation and land use to be obtained far more rapidly than from ground observation alone. Without aerial photographs, the data resulting from exploratory and broad reconnaissance surveys would take far longer to acquire, and the need for them would often be overtaken by events before they were presented (M. G. Bawden, I. D. Hill, G. Murdock & P. Tuley, mimeographed note, 1971).

Land systems are patterns of landscape recognized on aerial photographs and mapped to define areas of similar terrain with a similar environment. Breaks of slope often form the major boundaries both between and within land systems, and can be usefully mapped from aerial photographs, even if their

297

significance can be deduced only after ground control. There is, however, always a danger of deducing too much from aerial photographs at the preliminary stage of a reconnaissance and of thereby introducing errors which could subsequently lead to incorrect conclusions. It is therefore essential at this stage to concentrate on analysis or photo-recognition, and to use for this analysis and its interpretation such terminology and symbolization as clearly distinguishes between identification of objects and deduction of their significance. Mapping breaks of slope or vegetation boundaries on aerial photographs thus provides an objective analytical framework which can subsequently be interpreted. The boundaries between provisional land systems and the characteristics of each of the land facets can be recognized (see Beckett et al. under 6.6). Many land system boundaries coincide with breaks of slope, which are also landform boundaries. Breaks of slope cannot be recognized with certainty on print laydowns, and thus in many areas of complex terrain the landform pattern cannot be appreciated without a detailed stereoscopic analysis (Bawden, 1967).

The aerial photograph is a pictorial representation of the ground, on which all detail is shown in its correct relative position. If there is failure to obtain the basic terrain information required by study of stereoscopic air photos, then it is not so much the fault of the photographs but of the person using them, who has not progressed sufficiently or completed enough research or selective field checking to obtain the maximum amount of data which are reasonably extractable from the stereo image (Jones, 1969).

The application of the technique to southern Asia is discussed by Verstappen (1969); in addition to the usual survey of geomorphology, geology, soils and water resources, reference is made to calculations of the relative period in areas under shifting cultivation and to the special features of central Rajasthan; widespread lime concretions at depths of 30 to 120 cm. prevent addition of rainfall to groundwater, necessitating the construction of tanks for storage of surface-water; their siting has a profound influence on the sociology of the region. In North-West Frontier Province (Pakistan) aerial survey contributed to an evaluation of the dry farming or *sailaba* fields. *Sailaba* cultivation is based entirely on surface water, particularly (imperfect) sheet flood: water is collected on the field where it infiltrates; silt particles washed down by the water are a natural fertilizer. The relation between the generally sloping and slightly convex surface of the fan, the gradual spreading of the surface water, and the location of the *sailaba* fields (as also of irrigated fields) are shown clearly on aerial photos, and agricultural potential can thus be evaluated (Verstappen, op. cit.).

Aerial photographs have been invaluable in the study of deltas in particular, in Java and New Guinea, because they reveal, if combined with essential field checking, a variety of features of terrain which cannot, or can only with great difficulty be obtained by other means. There are three aspects: a. delineation of individual parts and elements of the delta; history of delta; interpretation of dissected older parts; mapping of vegetation, soil, hydrology, land use, and mangrove coasts; b. dynamic interpretation of deltas and delta coasts; changes in river courses; variations in frequency or force of onshore winds, coastal movements; and photogrammetry; mapping of shorelines; underwater features, determination of depth, etc. (Verstappen, 1966).

In surveys in Africa and other countries by the (British) Land Resources Division, stereoscopic examination of pairs of photographs facilitates the recognition of differences in land form, soil and vegetation, and boundaries may be marked on photographs. For example, in the Land Use Survey of Malawi, it has been possible to apply statistical sampling to air photo coverage, to provide adequate information economically, about the distribution of types of cultivated, uncultivated and uncultivable land.

New techniques involving colour and infra-red colour make it possible for a person with normal vision to distinguish a large number of colours compared with a few grey tones (more grey tones can be distinguished with transmitted than reflected light). The Land Resources Division finds that results of analysing vegetation, soil and land use patterns on true and infra-red colour air photographs are only marginally better than results from black and white prints. The use of infra-red colour facilitates remote sensing in the determination of the grass cover and/or the quantity of herbage present; it is also of value in range inventory. The recent development of high-altitude photography and the use of multispectral sensors to identify natural formations is unlikely to supersede conventional photography for some years. Most workers will continue to use films recording the visible and near infra-red bands of the spectrum exposed at medium altitudes in fixed-wing aircraft (Rains, 1970).

Grassland, as defined by Pratt, Greenway & Gwynne (1966) can be recognized on aerial photographs by its smooth grey textureless tone and its location (seasonally flooded plains and sites with impeded drainage, apart from high altitudes). Although grassland can be recognized and the density of tall shrubs and trees determined on 1:60,000 scale photography, it is most difficult to assess the density of woody plants in low shrub savanna at this scale. The problem was resolved for resource surveys of Botswana by analysing the distribution and type of the burning patterns. The characteristic burning pattern may appear as black or shades of grey. A pattern from fires over a period of seven years is an indication of good forage production. Patterns from deliberate burning suggest that deterioration is occurring. One must, however, be careful in interpreting fire patterns until the nature of the forage and density of the shrub can be checked (Rains, 1970).

The International Institute for Aerial Survey and Earth Sciences, Enschede, has reported on vegetation patterns in a savanna region of Northern Nigeria, during the UNDP/FAO survey of the soil and water resources of the Sokoto Valley. The interpretation of the relation between these and other patterns, clearly visible on aerial photographs, to the edaphic conditions played an important part in speeding up the reconnaissance survey. Both the main patterns, the pseudo-dune and gully-pattern, and the black and white dot pattern could be subdivided into sub-patterns in which various pattern elements were distinguished (Zonneveld, de Leeuw & Sombroek, 1971).

The Dutch workers would agree with British workers that the ordinary, old-fashioned black and white panchromatic film will remain the cheapest and still the most valuable tool far into the future. Thermal sensing (far infra-red) will never be useful for general grassland surveys. Radar will certainly be interesting for small-scale survey, but so far there is little experience with it. The more conventional remote sensing like near infra-red (preferably combined with visible radiation in false colour) is of particular value on vegetation types that have much open space, as in arid or semi-arid zones. The contrast between bare soil and patches covered with vegetation is sharp, and enables a good study of patterns to be made (I. S. Zonneveld, personal communication, 1972).

6.4.2 Space technology and observations

FAO held a technical consultation on the application of remote sensing to the management of food and agricultural resources in Rome, September, 1971 (FAO, 1971). Contributions included a review of FAO's activities, and documents, grouped under the heads of techniques, programmes and applications, included statements from the United States of America (A. B. Park, R. H. Miller), Brazil (F. de Mendonça), Mexico (S. Padilla Guzmán & F. García Simon) and India (P. R. Pisharoty of the Indian Space Research Organization, Ahmedabad, Gujarat), application to developing countries (G. A. Long, France, L. Sayn-Wittgenstein – Canada); and application to survey of special resources (geology, W. A. Fischer; agriculture – M. G. Baumgardner; soil resources – A. R. Mack; forest and range – R. N. Colwell; fisheries – W. H. Stevenson and E. J. Pastula Jnr.).

Three years later, it is noted that FAO is rapidly expanding its activities in remote sensing (J. A. Howard, personal communication, 1974) by the establishment of a Remote Sensing Unit at Headquarters and the purchase of an additive viewer for providing tailored colour imagery for field projects and for project formulation; equipment is also being obtained for the analysis of ERTS tapes (see below).

UNESCO reviewed the role of remote sensing in the programme on Man and the Biosphere at the first session of the International Coordinating Council in Paris, November, 1971. A progress report appears in *Nature and Resources*, January/March, 1974. It seems unlikely that UNESCO can contribute usefully either to the design of satellites or satellite-borne remote-sensing equipment, or to the direct collection and analysis of the data they obtain. This requires a sophisticated technology and will continue to be done in and by launching countries. The role of UNESCO may be more with regard to the interpretation of satellite data and their use for practical purposes. There is a need to work out such methods of interpretation of data on natural resources of the earth,

300

collected from outer space, which can be considered as standard routine methods for general acceptance, similar to those which exist for aerophoto-interpretation or the use of satellite data for meteorological purposes.

Photography from space by NASA Earth Resources Technology Satellite (ERTS) can make a major contribution to the survey, interpretation and mapping of land resources (Wilford, 1974). From an orbit of 570 miles, the ERTS makes fourteen complete orbits a day, crossing the polar regions; it covers the same ground once every eighteen days. The satellite is equipped with cameras and remote-sensing instruments for mineral prospecting, mapping, crop inventories and the monitoring of pollution.

The photographs so produced show forests as red, urban areas as blue; they record conditions invisible to the human eye. The photographs are created from three scans, one visible blues and greens, one visible red, and one invisible infrared, which is strongly reflected by healthy vegetation. As electronic data are received on the ground from each scan, they are transferred into black and white images with the aid of a computer. The three scans are then combined into a single colour composite, usually by projecting them on an appropriate colour filter and superimposing the three on a sheet of colour film. The selection of the colour filters is governed by the need to make features recorded by infrared visible. Blue-green tones are printed in blue, red in green, and infrared in red, as has been routine practice in aerial photography. The photographs obtained from the ERTS scans can therefore be interpreted by the methods to which investigators are already accustomed.

Signals from satellite scanners are received simultaneously as long as the ERTS is within receiving range of ground stations in California, Maryland, Alaska and Saskatchewan; scans of more distant areas are stored in videotape for night transmission to receiving stations when they come within range. In this way most of the earth can be kept under almost constant surveillance, and subtle changes, whether natural or man-made, may be detected from one scan to the next.

According to a NASA report, many of the major crops can be identified sufficiently well for inventory from space; these are accurate for 90 per cent of the time, and are produced at one-twentieth of the cost of similar statistics from conventional ground or aerial surveys. Forest fires and damage by floods, even in the remotest areas, can be assessed economically and rapidly. In the United States, early warning of flooding and of summer water deficits may be obtained by constant monitoring of winter precipitation and snow accumulation in mountain ranges. Snow surveys can be made with sufficient precision to assist in the control of hydraulic power from dams. Satellite photographs have shown how protection and good management of the Ekrafane Ranch in the Sahel makes it stand out green with grass in the midst of the prevailing overgrazed desert (2.5). The data obtained from satellite photographs can be automatically transformed into usable map products, including land use maps; new geological features may be found, even in well-mapped areas.

Infrared maps from ERTS have been used to chart the underground water supplies of the United States. They have also been adopted to detect the distinctive heat signatures of different crops such as wheat, maize and cotton, and to sense the difference in the heat emitted by healthy and blighted crops. This technique is still in the experimental stages, but it is hoped to prove the ability of spacecraft to provide an accurate census of agricultural crops and to give early warning of infestation. 'Out of these experiments could come the basis for a forest and rangeland inventory of Iran; a crop inventory in Brazil; methods of spotting crop infestation; geological maps of Spain, Australia, Africa and the western United States; volcano surveys of Central America, and detection of biologically-rich fishing areas off the coast of Chile' (Wilford, op. cit.). One of the most desolate areas of central Asia, Sinkiang, was formerly one of the most unknown regions of the earth. ERTS has shown, by the almost total absence of red in the photographs, bare mountains and arid deserts where no vegetation can survive. The dune pattern criss-crossing Takla Makan (west of the Gobi Desert) can be used in the study of the prevailing winds. Geologists can recognize older sandstone formations which, in other desert areas of the world, often contain rich reserves of water or petroleum.

Howard (1973) considers that it is sometimes useful to regard remote sensing as confined to the aerial collection of data which can be presented in image form, or to the collection of non-image data directly associated with image sensing. The interests of the FAO Remote Sensing Unit cover the full ambit of remote sensing – aerial photo-interpretation, aerial photogrammetry, thermal scanning and side-looking radar (SLAR), the last being used in a land-use study in Indonesia where there is virtually continuous cloud cover.

FAO has used or is using ERTS imagery in projects in Bolivia (general thematic mapping), Botswana (vegetation mapping), Colombia (land-use, forest types), Ecuador (land-use, forest types), Ethiopia (hydrology), Indonesia (general thematic mapping), Morocco (geomorphology), Philippines (land-use) and the Sudan (soils, geomorphology, vegetation). Skylab imagery is being examined in Ecuador, the Philippines and the Sudan. Howard (op. cit.) discusses under six heads the applications and relative merits of ERTS imagery. The comparatively limited spatial resolution and the absence of stereoscopy in these images are partly offset by their spectral information content (Howard & de Kock, 1974). The systems regarded as most effective in extracting relevant information from the signal are electronics-based. Howard & de Kock (op. cit.) justify the use of the sub-optimal and comparatively cheap additive viewer as applied to examples from developing countries.

Barrett (1971) – pre-ERTS – examines how the meteorological satellite photographs and nephanalyses of the American Environmental Sciences Services Administration (ESSA) might be employed in the study of the complex climatological patterns in the tropical Far East. Paying attention especially to those climatic elements and factors that cannot be mapped satisfactorily because of the irregular scatter of stations on the ground, Barrett attempts evaluations of a. mean monthly cloud cover, through the direct evidence of nephanalyses, b. monthly rainfall distribution, estimated indirectly from nephanalyses and using a method involving data from satellites and conventional weather records, and c. the distributions of the chief weather-producing arrangements of cloudiness, plotted from the direct evidence of satellite photographs and nephanalyses.

6.5 Cartography and presentation of results

French Institute, Pondicherry
Gaussen (1949, 1959) has proposed maps of vegetation together with environmental conditions at small (1:1,000,000) or medium (1:200,000) scales. The main points may be summarized. The vegetation, both the natural and that created by man's activities (agriculture, plantations), is presented on the main map. Six insets are placed around the main map. Four of these show administrative divisions and hypsometry, geology and lithology, soil types and bioclimates, to permit correlations with the vegetation. The remaining two show the '*plésioclimax*' (potential vegetation) and agricultural regions having the same crops and same cultivation practices.

All the agricultural areas are left in white on the main map, as opposed to the coloured areas that represent the natural vegetation. The symbols of cultivated crops, often carrying a statistical value, are placed on the white background. Horizontal lines indicate human action in the landscape, such as plantations, shifting cultivation or irrigation canals. Vertical bands indicate proportions, such as crops of major, intermediate and minor importance in the inset on agriculture. Land use of every district is also presented.

Every physiognomic stage of the natural vegetation is represented by a definite pattern. The grass savannas are indicated by a fine network of small dots; if some shrubs are scattered amidst the grasses, they are represented by large dots. Thickets are shown by crosses, woodlands by strokes, and the forest type in full colour. Thus the lower the stage of degradation, the less the intensity of colour.

The choice of colour given to a series of vegetation takes into account the ecological conditions, for example, blue for high rainfall, yellow for poor rainfall, and red for high temperature. Between these extremes, other classes of rainfall and temperature are represented by intermediate colours of the spectrum. Mean temperature of the coldest month is indicated in red if more than 20°C., in orange if between 15 and 20°C., and in yellow if between 10 and 15°C. By combining the colours of the corresponding classes of rainfall and temperature, the resultant colour of the climatic complex is obtained. The soil factor may also intervene to modify the tint given by the climate.

The colour of the ecological complex is used in full tone to represent the

303

plésioclimax stage of the series, if the *plésioclimax* happens to be a dense forest; the other degraded physiognomic stages of the series are shown by the conventional patterns mentioned above, but in the colour of the series. The crops of dry ecology are in red and orange, those of wetter ecology are in blue or violet.

The climatic implication on the vegetation is through colour, that of the soil either by colour or special graphical representation. The biotic factor cannot be represented directly on the map. However, a comparison between the main map showing the present state of vegetation and the inset of *plésioclimax* showing the potential vegetation type reveals the degree of degradation brought about by man in an area.

As the same principles of cartography are used throughout the world (climatic analogies by colours and physiognomic representation by definite patterns), it is simple to compare any two regions. This in turn assists foresters and agriculturists in the interchange of economic species from one region to another (P. Legris, personal communication, 1974).

World Land Use Survey

In 1949, the World Land Use Survey Commission proposed a classification of land use; this was an attempt to get a standardized method for recording data in the field and for its presentation in map form, so that there might be an international common language for the ways in which land is used or occupied whatever the climate or nature of the terrain. The ultimate concern of the Commission was agriculture or agricultural potential. The classification of categories of land use recommended to be recognized and mapped were: 1. settlements and associated non-agricultural lands (dark and light red); 2. horticulture (deep purple); 3. tree and other perennial crops (light purple); 4. croplands; a. continuous and rotation cropping (dark brown); b. land rotation (light brown); 5. improved permanent pasture, managed or enclosed (light green); 6. unimproved grazing land; a. used (orange); b. not used (yellow); 7. woodlands; a. dense (dark green); b. open (medium green); c. scrub (olive green); d. swamp forests (blue green); e. cut-over or burnt-over forest areas (green stipple); f. forest with subsidiary cultivation (green with brown dots); 8. swamps and marshes — fresh and saltwater, non-forested (blue); 9. unproductive land (grey).
A strong body of opinion would like to have a tenth category added for information about areas of water.

The Commission believes that this classification appears to be the nearest approach to an internationally acceptable system of notation for land use and vegetation cover. Further subdivisions are to be expected to meet local conditions and larger scales of map. For example, Malaysia has suggested two slight changes in notation: 'grassland' for 'grazing land', because it is not always possible to decide whether areas of low vegetation cover are actually being used for grazing; 'unused land' for 'unproductive land', so as not to prejudge the value of land under hitherto unknown practices.

The above notes come from the discussion at an international symposium on techniques for land use and related surveys held in London, April, 1970 (World Land Use Survey, 1970), following the paper: J. Kostrowicki, Poland – data requirements for land use survey maps.

Other relevant contributions were: E. Csáti, Hungary – graphic methods of data evaluation for land use maps; D. P. Bickmore & I. S. Evans, London – some recent advances in automatic cartography (including some aspects of special interest to developing countries).

CSIRO Land Research

Base-maps are required in two phases of resource surveys, for working plans and for the final mapping.

Work plans need not be accurate. Uncontrolled mosaics are useful for the compilation of existing information and the planning of field work. They allow rapid regional assessment of features that are distinct on aerial photographs. In some cases aerial photograph patterns are not very distinct or their relations to ground characteristics are complicated by man-made features. In such cases mosaics are of less value. By transferring the essential information to simple planimetric maps and adding distinctive colours, regional relations are more easily interpreted and planning of field work is facilitated.

Maps showing the final mapping of land resources may be either mosaics or planimetric maps. Mosaic maps are more useful for personnel who are concerned with further field use of the information for either planning or more intensive surveys or for planning development. It is standard practice in Australian reconnaissance surveys to provide mosaics marked with mapping boundaries to authorities responsible for development.

In base-maps for mapping resources, absolute accuracy (in geodetic terms) is less important than accuracy of local detail of topography and cultural features, to which the land resource sub-division boundaries are relative. It is only if existing maps of the appropriate scale are less accurate (as assessed from aerial photographs) than the accuracy required for land resource boundaries that the preparation of new base-maps from the aerial photographs is warranted. Most countries have a set of standard sheet sizes, map scales, projections, and ground control which should be taken into account in planning new mapping for resource surveys.

The post-war programme of medium- and small-scale mapping in Australia is an example of the integration of a national mapping programme with the requirements of resource survey. In the absence of a national geodetic survey and a need for rapid mapping for reconnaissance resource studies, a plan was evolved: a. aerial photography at a scale of 1:50,000 initially, but recently at 1:84,000 with a superwide-angle lens; b. preparation of uncontrolled photomaps for use in reconnaissance studies of geology and land resources; c. estab-

305

lishment of ground control by astrofix, and spot heights by barometric (in some cases airborne) traverses; d. preparation of planimetric maps at 1:250, 000 by slotted template assemblies fitted to that of ground control. Detail is filled in with the aid of 'sketchmasters', but in areas of great relief, e.g. New Guinea, the stereotope is used for planimetric plotting. The planimetric maps have either been used directly for resource base-maps or have been compiled at reduced scales of 1:500,000 or 1:1,000,000. A national geodetic survey of horizontal control by first-order tellurometer traverses controlled by Laplace azimuths has also been conducted. Supplementary control will be obtained with airborne equipment for measuring distance (Aerodist) and by second-order tellurometer traverses based on helicopter transportation. Supplementary vertical control within the third-order levelling is obtained by various combinations of fourth-order levelling, helicopter-borne barometric heighting and radar altimetric profiles. This work allows the rapid expansion of the present small amount of fully contoured maps at 1:250,000.

The preparation of large-scale maps from aerial photographs by photogrammetric means is the major use of aerial photographs. It is, however, most important for personnel concerned with the study or development of resources to consult the aerial photographer and the photogrammetrist from the commencement of planning of the project, in order that the scale of the aerial photographs and scale and accuracy of the map will be appropriate (Christian & Stewart, 1968).

6.6 Land inventory and data bank

Canada Land Inventory
This comprehensive survey of land capability and use was designed as a basis for land use and resource planning for agriculture, forestry, recreation, wildlife and sport fish. The Federal Government approved the programme under the Agriculture and Rural Development Act in 1963. National classification systems were drawn up through the co-operative efforts of provincial and federal departments responsible for resource development. The area covered embraces approximately one million square miles, including all of the Atlantic Provinces, the settled portions of Ontario and Quebec, and the Western Provinces. Each province has been responsible for the inventory within its own borders, with financial and technical assistance and central co-ordination from the Federal Government. Now that the work of the Canada Land Inventory is essentially completed, the data gathered are providing a basis for more detailed studies in regional areas.

Mapping is carried out at two scales. Land capability maps for agriculture, recreation, ungulates, waterfowl and sport fish, and maps of present land use have been compiled in manuscript form at a scale of 1:50,000. Excepting sport fish, the land capability maps are published in a multi-colour format for public distribution (50 Canadian cents each, from Map Distribution Office, Department of Energy, Mines and Resources, Ottawa). A central computerized data bank has been set up to store all map data and related information. Retrieval of data within or between sectors will be possible.

The agricultural maps indicate classes and subclasses according to the Soil Capability Classification for Agriculture. The mineral soils are grouped into 7 classes and 13 subclasses according to the potential of each soil for the production of field crops. The classes indicate the degree of limitation imposed by the soil in its use for mechanized agriculture. The subclasses indicate the kinds of limitations that individually or in combination affect agricultural use (adverse climate, poor soil structure, erosion danger, low fertility, inundation, low water-holding capacity, salinity, stoniness, shallowness to bedrock, undesirable soil characteristics, adverse topography, excess water).

Capability for forestry is rated according to 7 classes depending upon the capability of the land to grow commercial timber in areas stocked with the optimal number and species of trees (Class 1 best, Class 7 cannot yield timber in commercial quantities). This rating considers land in its natural state, without improvements such as fertilization and drainage. With improved forest management, productivity may change, limitations may be overcome and class changes may become necessary. The capability mapping is accomplished through interpretation of air photographs and field surveys. The assignment of land units to a capability class is made on the basis of all known or inferred information about the unit, including subsoil, soil profile, depth, moisture, fertility, land form, climate and vegetation.

Canada Land Inventory Publications

Report No.
1 Objectives, Scope and Organization 66 pp. Revised 1970 Reprinted 1972.
2 Soil Capability Classification for Agriculture 16 pp. Reprinted 1972.
3 The Climates of Canada for Agriculture 24 pp. 19 maps 1966
4 Land Capability Classification for Forestry 2nd Edition 36 pp. Revised 1970 Reprinted 1972.
5 The Economics of Plantation Forestry in Southern Ontario. D. V. Love & J. R. M. Williams 46 pp. 1968
6 Land Capability for Recreation 70 Photographs 2 map examples 110 pp. 1970
7 Land Capability for Wildlife Half-tone, stereo and colour, illustrations 29 pp. 1970.
8 Soil Capability for Agriculture in Nova Scotia Maps and tables 45 pp. 1970
9 Landowners and Land Use in the Tantramar Region of New Brunswick 195 pp. 1968.

Other Documents

An introduction to the Canada Geographic Information System
A Technical Introduction to the Canada Geographic Information System
Land Use in Canada – Reprint from *Canadian Geographical Magazine*. D. F. Symington, 15pp. 1968.
Towards Integrated Resource Management, Report of the Sub-Committee on Multiple Use, National Committee on Forest Land 67 pp. 1970
A Bibliography of Social and Economic Research Pertaining to Rural British Columbia. Dean S. Goard & Gary Dickinson 2nd Edition 43 pp. 1971

A Guide for Resource Planning – The Canada Land Inventory Bilingual explanatory folder with colour illustrations 8 pp. 1970.
The Productive Capacity of the Natural Resources of Manitoulin Island
A Working Document (10 maps) David R. Cressman. 195 pp. 1968.
Outline of Canada Land Capability Classification for Sport Fish 5 pp. 1966
A Guide to the Classification of Land Use for the Canada Land Inventory 19 pp. 1968.

United States of America
New York State: 'The land use and natural resource inventory of New York State has mapped land use and collected the data in a computer system; soil, geological and agricultural characteristics have been added to the computer system, and other physical, economic and social data will be added later. The inventory was made from recent aerial photographs, with ground control, at a scale of 1:24,000, and transferred to topographic base maps. After mapping and production of overlays, cells of one square kilometre were identified on the maps as a geographic referencing system for data storage and retrieval. The Universal Transverse Mercator grid system was used, with 140,000 cells covering the State. Data were then summarized by cell: percentages of a cell in a certain area land use; numbers of items; numerical types (presence or absence); or mileages of various point count items. These data were then key-punched on special coding forms, stored on direct access discs (IBM 2316), and became available for quantitative analysis and display.

'The Center for Aerial Photographic Studies at Cornell University, which did the work under contract to the Office of Planning Co-ordination of New York State, with assistance from the Laboratory of Computer Graphics at the Graduate School of Design at Harvard University, has developed two computer programmes that anyone can use, even without experience with computers or computer programming. These programmes are: DATALIST, which produces inexpensive direct listings and summaries capable of arithmetic and logical manipulation; and PLANMAP, which produces simple or weighted computer graphic displays of area or point data or combinations of the two. PLANMAP, a computer graphic programme, may also be used to identify and display grid cells with certain very specific qualities or combinations of qualities involving any of the presently coded 130 land use characteristics.

'Agricultural land use in New York State is classified first as active (in commercial use) or inactive (fairly recently removed from agriculture). Active areas are delineated according to use by major enterprises – orchards; vineyards; horticulture; cropland intensively used for cash crops; and land used more extensively for crops related to dairy and poultry, pasture and speciality farms. Inactive agriculture classifications include land fairly recently removed from active agriculture but not yet committed to forest regeneration, and also land waiting to be developed or under construction for urban uses' (from FAO *Soils Bulletin* 22, being part of the background documentation for the Expert Consultation, Wageningen, 1972 – Brinkman & Smyth, 1973).

CSIRO Division of Soils
A pragmatic approach has been adopted, working with small data bases where
it is fairly certain that some help may be given with a specific problem raised
by colleagues (A. W. Moore, personal communication, 1974; draft of paper
for International Soils Congress, Moscow, on 'Computer storage of soils data:
use of a generalized file management system' – A. W. Moore, W. T. Ward &
C. H. Thompson).

Three file management systems developed for scientific use, the Belgian,
the British Columbian and the CSIRO soil micromorphology file, are all
specific, each being written to do one particular job. As an alternative
approach, the Division of Soils has experimented with a generalized file management system designed for commerical use. This has the advantage that the
cost of developing the system can be amortized over a large number of users.
The Division has now established eight files dealing with data of soils and vegetation.

The main facilities required in a generalized file management system are
those of data editing, validation, interrogation based on sophisticated logic,
hard-copy report generation and interfacing with a common high-level proce-
dural language, e.g. Fortran. The Division has used one system, Infol, written
in Compass for the Control Data Corporation (Palo Alto, Cal.) 3600 computer.
By means of simple control statements, it is possible to interrogate files using
sophisticated logic, produce indexes by inversion of a whole file about specified
attributes, sort data, and put them out as reports, or on magnetic tape or other
storage media. There are some problems and comments relating to software,
nature of data collected, compatability of data from different sources, file
size, back-up and costs. Experience shows that, even with the constraints imposed by the non-procedural language of Infol, it can be useful in processing
soil data.

CSIRO Division of Land Use Research
Land resource surveys made in Australia and elsewhere make use of the patterns
of landform and vegetation seen on air photos to predict the attributes of
land. Numerous attributes such as slope directly affect most forms of land use;
attributes are also important for extrapolating soil properties from a very few
observations (Speight, 1971). A flexible scheme is to list all those attributes of
land form and vegetation that are known or thought to have predictive value,
and to ensure that mapped boundaries enclose areas that are internally homogeneous, and that differ from each other in terms of these attributes. Every
mapped region is considered to be unique. A decision whether to declare it to
be similar to another region in terms that are significant for various kinds of
land use can be made (and subsequently changed) in the light of an accumula-
ting body of data on the relations between the descriptive attributes and the
attributes to be predicted. The methodical recording of attributes of individual
mapped regions implies an ability to manipulate the resulting large volume of

data. A system is being developed around a data bank that contains both descriptions of mapped regions and graphic information including regional boundaries and topographic data.

Computer systems for the storage and display of information for regional planning commonly use a matrix of regular grid cells as the areal units to which descriptive attributes are attached. This Division has adopted a system which does not use grid cells, but which allows data to be stored against points, lines and regions representing the true position of, for example, cadastral, topographic or biophysical boundaries (Cook & Johnson, 1973). Lines and region boundaries may be as complex as desired. The attribute data are stored in a structure designed for the application. Although more expensive to establish, this approach is said to have a number of advantages over more commonly used systems. This computer system provides for the storage of lines and points (Fig. 6.6a to e). Lines may represent linear features or delineate region boundaries. Attributes may be attached to points, lines and regions.

'There is essentially no limit to the extent of the stored map, within a co-ordinate range specified during establishment. Although the system does in fact sub-divide the map into parts similar in some respects to conventional map sheets, the user may think of the map as one continuous coverage. The system is designed to deal with one map, usually an attribute complex map; it is not designed to compare or combine numerous maps of the same locality. In other words, all the maps of a locality which it is desired to access are combined into one complex map before input to the system. This approach has been adopted to avoid the logical and computational difficulties caused by the proliferation of very small regions when maps are overlaid.

'The descriptive data associated with any point, line or region may be as extensive as desired and may be given a structure convenient for the application. Furthermore, links may be established across descriptors on the basis of similarity in any defined respect. However, both the structure of the data and any associative links must be specified during establishment.

'The computer system is in two parts, developed independently. One part handles the map base, the positional and topological information representing the points, lines and regions stored. The other handles the attribute base, containing the descriptive data. Links are provided between a map element in the map base and its associated descriptive data in the attribute base.

'User access to the map base is by keyword command (such as INPUT, EDIT, DISPLAY) for standard operations, and by specially written Fortran program for non-standard operations. Access to the attribute base is by Fortran program.' (Cook & Johnson, 1973).

Great Britain
In a paper to the Photogrammetric Society in February, 1967, J. N. R. Jeffers of the (British) Nature Conservancy asked the question: 'what are the implications of the existence of electronic digital computers to the planning

and conduct of land-use surveys based on the analysis of aerial photographs?' His own reply is as follows:

'In considering this application of electronic digital computers, it is necessary to record three basic decisions: 1. all calculations to be made during the course of the survey will be carried out by the computer; 2. the basic data from the survey will be recorded in such a way that they may be input to the computer without any manual intervention, e.g. no key-punching of handwritten data; and 3. the final tables and maps, regarded as the end-product of the survey, will be produced by the computer in a form suitable for direct reproduction. It is important to stress that these decisions should be made before the start of the survey. An attempt to introduce electronic computers after the survey has been commenced may result in some increases in efficiency, but the full gain from their introduction will not be achieved.

'The object of a land-use survey of a developing country is to determine, for certain administrative units and natural land types, the pattern of land-use within a number of defined categories. The survey is to be based on photo-interpretation of sample points superimposed on recent photography of the specified areas. One of a set of numbered templets is chosen at random for each photograph, placed on the photograph, and the land-use category on which each of the numbered dots on the templet falls recorded. The desired results from the survey are assumed to be as follows: 1. estimates of the proportions of the land surface occupied by each land-use category, together with standard errors of the estimates, for defined administrative and natural units; 2. estimates of the area occupied by each land-use category, together with standard errors of the estimates, for defined administrative and natural units; and 3. maps of the land-use within broad categories at several different scales, for planning and illustrative purposes.

'The flow diagram for such a survey, assuming that the three basic decisions are taken, is given in Fig. 6.7 in the form of a network planning diagram. The diagram assumes that, while the aerial photography is being done, the forms on which basic data are to be recorded (assumed in this case to be cards) are printed, that the intensity of the sampling is calculated, and that the necessary templets are prepared. The first stage at which the implications of the use of a computer become apparent is in the actual recording of the land-use categories observed on each photograph. The object is to create, at the point of recording, data which can be passed direct to the computer without any further manual intervention' (Jeffers, 1967).

An organized store of terrain data in which every item is conserved requires an index based on the kind of terrain it represents, and the latter requires a terrain classification, or a list of the different kinds of terrain, such that: 1. every site belongs to a terrain class; 2. the user can identify the terrain class at any site, and 3. all sites within one terrain class are sufficiently similar, so that it is reasonable to predict terrain conditions at all from observations at a few (Beckett, Webster, McNeil & Mitchell, 1972).

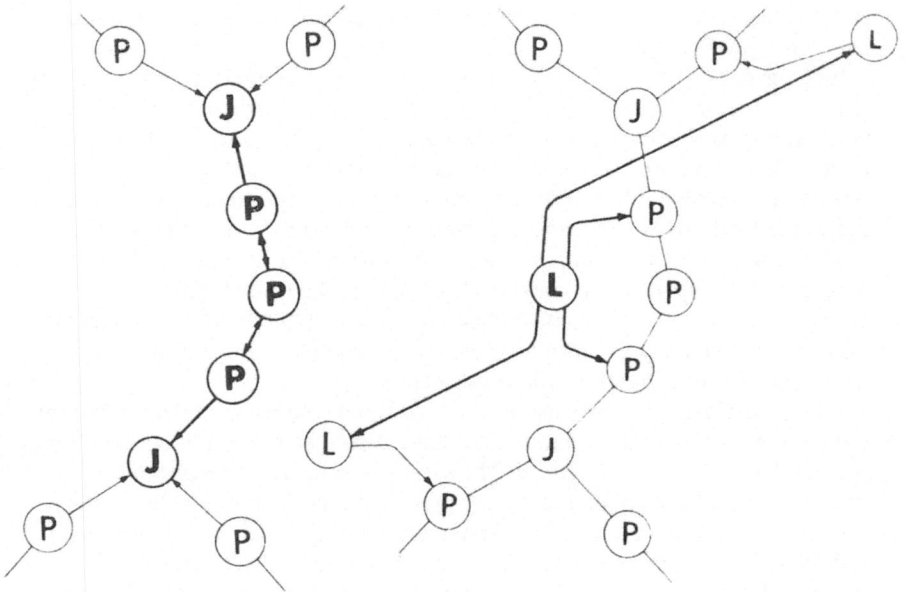

Fig. 6.6a. CSIRO, Data storage system (Cook & Johnson, 1973); Map structure components, points and junctions.
Fig. 6.6b. Map structure components, lines.

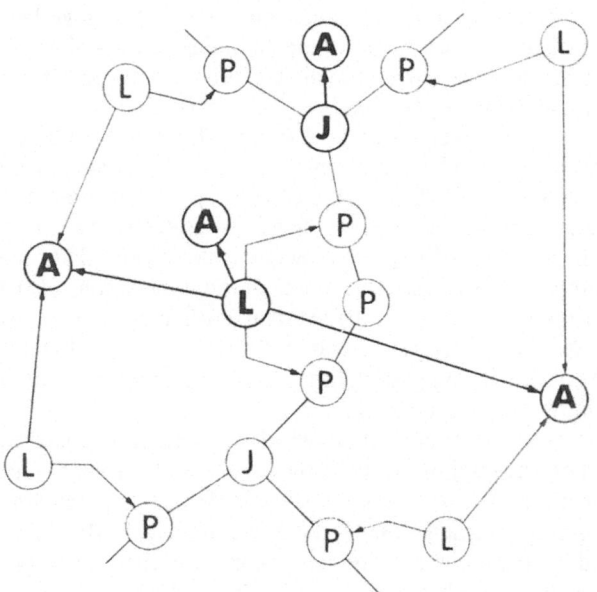

Fig. 6.6c. Map structure, components, attributes.

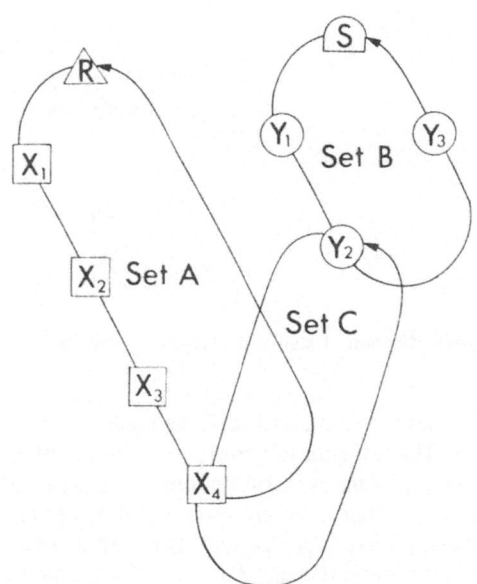

Fig. 6.6d. Attribute base, a simple data structure.

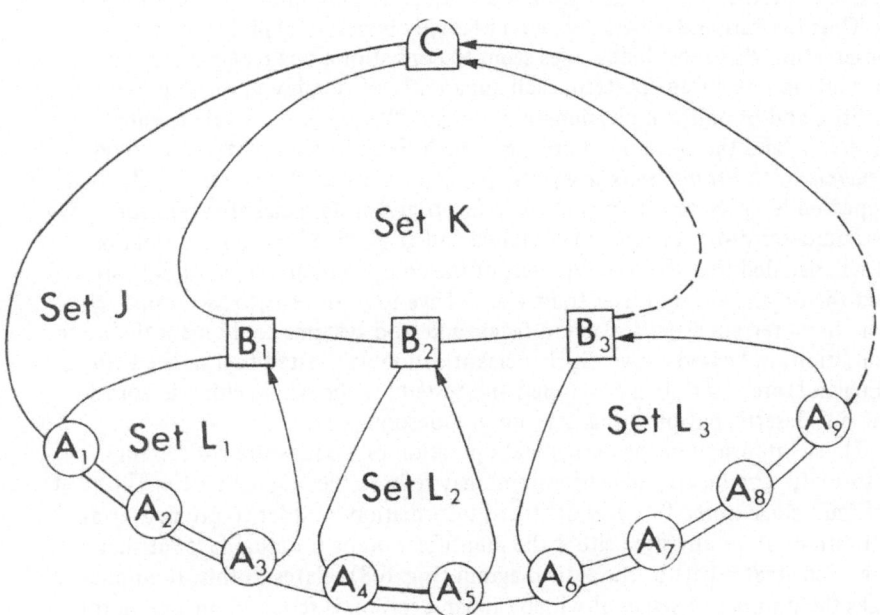

Fig. 6.6e. Attribute base, a complex data structure.

313

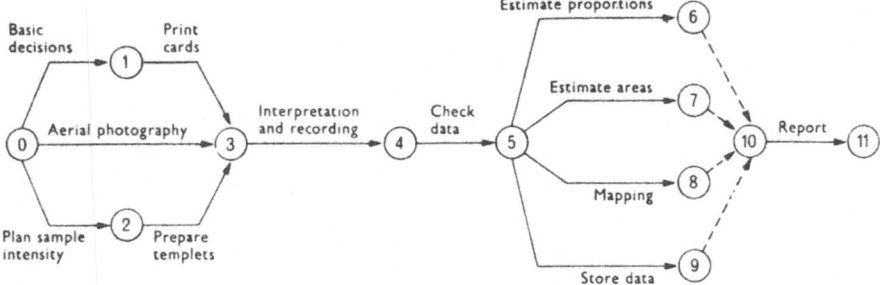

Fig. 6.7. Network diagram of stages in land-use survey (Jeffers, 1967).

These workers eliminate terrain classifications made by specialists in a number of disciplines. The information content of geographical regions is too small; the world cover of soil maps is too limited; the legends of most geological maps are too palaeontological; geomorphological mapping creates more problems of interpretation than it solves, and its benefits are unproved; morphological mapping is laborious and its claims are unproved; when terrain is classified in terms of measurable properties shown to define its suitability for practical purposes, most such parameters can be determined only with considerable expertise and detailed ground access, and rarely in inaccessible areas.

Thus Beckett and others (op. cit.) became obliged to exploit the empirical observation that most landscapes seem to consist of a few recognizable components in a recurrent pattern, each apparently of roughly uniform potentialities and of uniform physiographic origin. 'We called the smaller units *facets* . . . and the compound units of which these formed parts were *homogeneous* or *recurrent landscape patterns*, and finally *land systems* . . . ' It then appeared justified to attempt to use a facet/land system classification for terrain descriptors for indexing detailed and generalized terrain information. It was decided that the development of the equipment and logic of data store and the organization of user trials would have to await tests to ascertain whether a terrain classification in facets and land systems could meet the three conditions indicated above. Such a classification was tested first in the Oxford (England) area, and then developed and tested for the whole climatic zone of the hot deserts, but omitting any site in monsoon Asia.

Those interested in the design and operation of a data store for terrain information on facets and land systems may refer to the original article. The most difficult stage in the retrieval of stored information in order to predict terrain conditions at an unvisited site is the identification of the local facet at that site. The greater part of the flow diagram (Fig. 6.8) relates to this. It emphasizes the number of stages at which human decision is required on a compromise between less than perfect alternatives (Beckett and others, op. cit.). See also Jeffers (1970) on modern statistical techniques which may be adopted in land

314

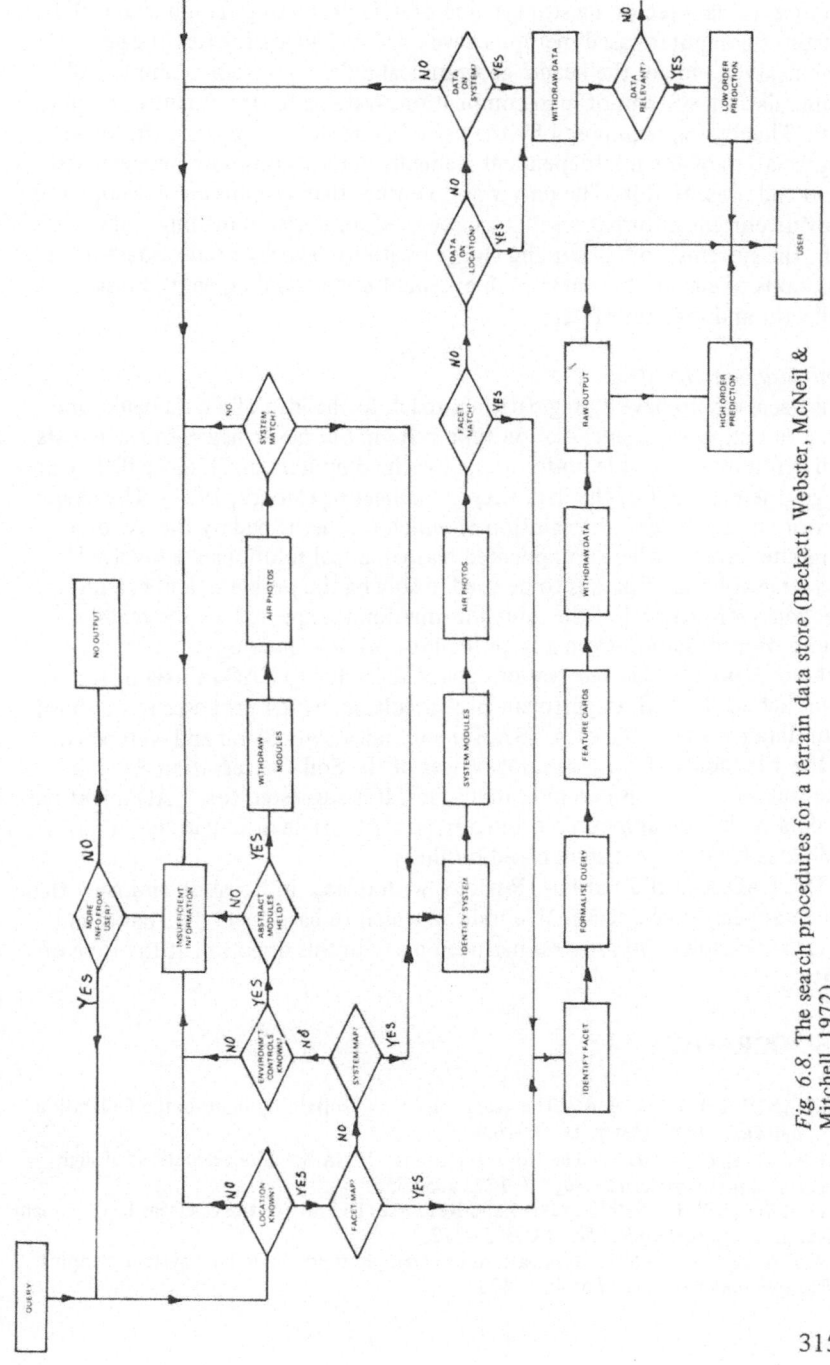

Fig. 6.8. The search procedures for a terrain data store (Beckett, Webster, McNeil & Mitchell, 1972).

315

use surveys to provide material suitable for the improved methods of data capture, of data recording storage, and of data processing. Tomlinson (1970) discusses computer-based methods developed in Canada for treating geographical data under the heads: geographical indexing systems, simple grid manipulation, systems of map compilation, systems for treatment of graphic data. The discipline imposed by the use of a computer requires a thorough appreciation of the interdependent elements in the continuum between acquisition and presentation. The only way to ensure that systems are developed to provide current information that can be used for decision-making and to prevent the systems from becoming clogged with irrelevant or out-of-date information is to ensure that there is an excellent understanding between the collector and the user of data.

Soil data bank for Asia
It is necessary to have a pragmatic approach to the idea of a data bank; one must not expect a science-fiction type system, but nevertheless soil scientists and agronomists must learn to interact with computers much more if they are to get the most out of the data they are collecting (Moore, 1971). The characteristics of a problem, the solution of which can be helped by the use of a computer, are: a. where complicated mathematical solutions are involved; b. a large volume of data is to be used in solving the problem, and c. a high frequency of retrieval of the same information is expected, or the ready retrieval of information from a large file is of prime concern.

Moore (op. cit.) has reservations about the value to FAO in Asia of the approach of the U.S. Department of Agriculture, which proposes to establish four data banks (systems) in USA, for soil, range, woodland and watershed, in five files controlled by a senior officer of the Soil Conservation Service. The soil series name is common to all files. It is suggested that FAO might tag all data with geographical co-ordinates, either latitude or longitude, or coordinates from some system based on them.

The FAO Asia and Far East Regions, with offices in Bangkok and New Delhi, would appear to be a reasonable unit on which to base a soil data bank, but so far as is known, no progress has been made in this direction at the time of writing.

BIBLIOGRAPHY

ALEXANDER, J. B., – 1964 – The evolution of land suitability maps in the Federation of Malaya. *J. Trop. Geogr.* 18: 1–16.
BARRETT, E. C., – 1971 – The tropical Far East: ESSA Satellite evaluations of high season climatic patterns. *Geogr. J.* 137: 535–555.
BAULKWILL, W. J., – 1972 – The Land Resources Division of the Overseas Development Administration. *Tropical Sci.* 14: 305–322.
BAWDEN, M. G., – 1967 – Applications of aerial photography in land system mapping. *Photogrammetric Record* 5: 461–473.

BAWDEN, M. G., I. D. HILL, G. MURDOCK & P. TULEY, – 1971 – Applications of Aerial Photography to Integrated Soil and Land Survey by the Land Resources Division in Nigeria. Land Resources Division, Tolworth Tower. mimeo. 13 pp.

BECKETT, P. H. T., R. WEBSTER, G. M. McNEIL & C. W. MITCHELL, – 1972 – Terrain evaluation by means of a data bank. *Geogr. J.* 138: 430–456.

BEEK, K. J. & J. BENNEMA, – 1972 – Land Evaluation for Agricultural Land Use Planning. An Ecological Methodology. Dept. Soil Science and Geology, University of Agriculture, Wageningen. 61 pp. mimeo.

BRINKMAN, R. & A. J. SMYTH (eds.), – 1973 – Land Evaluation for Rural Purposes. Summary of an Expert Consultation, Wageningen, The Netherlands, 6–12 October, 1972. International Institute for Land Reclamation and Improvement, Wageningen. 116 pp.

BROOKFIELD, H. C. & D. HART. – 1971 – Melanesia: A Geographical Interpretation of an Island World. Methuen, London. 464 pp.

CANADA LAND INVENTORY – various dates – Reports nos. 1–9 (see pp. 000–000).

CHRISTIAN, C. S., – 1958 – The concept of land units and land systems. Proceedings Ninth Pacific Science Congress, 1957, vol. 20: 74–81.

CHRISTIAN, C. S., J. N. JENNINGS & C. R. TWIDALE, – 1957 – Geomorphology. In 'Guide Book to Research Data for Arid Zone Development,' pp. 51–65. Arid Zone Research 8. UNESCO, Paris.

CHRISTIAN, C. S. & G. A. STEWART, – 1953 – General Report on Survey of Katherine-Darwin Region, 1946. Australian Land Res. Ser. no. 1. CSIRO.

CHRISTIAN, C. S. & G. A. STEWART, – 1968 – Methodology of integrated surveys. In 'Aerial Surveys and Integrated Studies. Proceedings of the Toulouse Conference,' pp. 233–280. UNESCO, Paris.

CHUN Soo-Shin, – 1971 – Land suitability classification in Korea. In 'Soil Survey and Soil Fertility Research in Asia and the Far East,' pp. 215–225. World Soil Resources Reports no. 41. FAO, Rome.

COMMONWEALTH SCIENTIFIC AND INDUSTRIAL RESEARCH ORGANIZATION, – 1972 – Land Resources of the Vanimo Area, Papua New Guinea. Land Research Series no. 31. CSIRO, Canberra. 126 pp.

CONKLIN, H. C., – 1967 – Some aspects of ethnographic research in Ifugao. *Trans. New York Acad. Sci. ser. II*, 30: 99–121.

CONKLIN, H. C., – 1972 – Philippines: North Central Ifugao Land Use Series (set of 8 maps). American Geographical Society, New York. Scale 1:5,000, 29 inches × 34 inches.

COOK, B. G. & B. V. JOHNSON, – 1973 – A computer data bank for regional planning. Paper to First Australian Conference on Urban and Regional Information Systems. Hunter Valley Research Foundation, Newcastle, 15–16 November, 1973. 10 pp.

DENT, F. J., – 1969 – General land suitability for crop diversification in peninsular Thailand. *Rep. Rubber Res. Centre, Thailand* 7: 1–50.

FOOD AND AGRICULTURE ORGANIZATION OF THE UNITED NATIONS, – 1971 – Aircraft and satellite remote sensing for developing nations. FAO, Rome. 36 pp.

FAO, – 1973 – Syllabus on Agricultural Sector Analysis, Part 3. Alternative Approaches and Models for Sector Analysis. FAO, Rome. 122 pp.

FAO, – 1974 – Approaches to Land Classification. Soils Bulletin 22. FAO, Rome. 120 pp.

GAUSSEN, H., – 1949 – Projets pour diverses cartes du monde a 1:1,000,000: la carte écologique du tapis végétal. *Ann. Agron.* 19. 78 pp.

GAUSSEN, H., – 1959 – The vegetation maps. *Inst. Fr. Pondichéry, Trav. Sect. Sci. Tech.* 1(4): 155–179.

GHAFFAR, Abdul bin Hamid, – 1972 – The impact of land clearance for agriculture on forestry. *Malayan For.* 35: 47–53.

HOWARD, J. A. – 1970 – Aerial Photo-Ecology. Faber and Faber, London. 325 pp.

HOWARD, J. A., – 1973 – Recent application of remote sensing to FAO activities. Second Conference of the IUFRO Working Group on Remote Sensing, Freiburg, Germany (also issued as FAO document AGD(RS) 1/73, mimeo, 7 pp.).

HOWARD, J. A. & R. B. de KOCK, – 1974 – Additive viewing as an interpretative technique. pp. 189–192. Atti Ufficiali del XIV Convegno Internazionale Tecnico-Scientifico sullo Spazio. Rome.

JEFFERS, J. N. R., – 1967 – The use of electronic computers in land-use surveys based on photo-interpretation. *Photogrammetric Record* 5: 465–469.

JEFFERS, J. N. R., – 1970 – Modern statistical techniques in land use surveys. In 'World Land Use Survey Occasional Paper no. 9. New Possibilities and Techniques for Land Use and Related Surveys,' pp. 65–72. Geographical Publications, Berkhamsted.

JOHNSON, P. L. (ed.), – 1969 – Remote Sensing in Ecology. University of Georgia Press, Athens. 254 pp.

JONES, R. G. B., – 1969 – Some applications of air photo-interpretation to agricultural development planning. Paper to Institution of Agricultural Engineers, Silsoe, Bedford. 10 pp.

KING, R. B., – 1970 – A parametric approach to land system classification. *Geoderma* 4: 37–46.

LANG, D. M., – 1973 – Handbook of the Land Capability Classification Method for Open-Field (Including Mechanized) Agriculture in Northern Zambia. Supplementary Report 9, Land Resources Division, Tolworth Tower. 32 pp.

McCORMACK, R. J., – 1971 – The Canada land use inventory: a basis for land use planning. *J. Soil and Water Conservation* 26: 141–146.

MALAYSIA, STATE OF SABAH, – 1973 – Land Capability Classification. Technical Sub-Committee on Land Capability Classification of the State Development Planning Committee. Technical Monograph 1. Kota Kinabalu. 22 pp. + 15 plates.

MELLOR, J. W., – 1973 – Accelerated growth in agricultural production and the insectoral transfer of resources. *Econ. Develop. and Cultural Change* 22: 1–16.

MONMONIER, M. S., – 1971 – Digitized map measurement and correlation applied to an example in crop ecology. *Geogr. Rev.* 61: 51–71.

MOORE, A. W., – 1971 – Regional Soil Data Bank for future evaluation and projection. In 'Soil Survey and Soil Fertility Research in Asia and the Far East,' pp. 185–198. World Soil Resources Reports no. 41. FAO, Rome.

MOORE, A. W., W. T. WARD & C. H. THOMPSON, – 1974 – Computer storage of soils data: use of a generalized file management system. Paper to International Soils Congress, Moscow.

MOORMANN, F. R., – 1973 – Classification of land for its use capability and conservation requirements. Paper to FAO/SIDA Regional Seminar on Shifting Cultivation and Soil Conservation in Africa, Ibadan, Nigeria, 2–21 July, 1973. Mimeo. FAO, Rome. 11 pp.

PANTON, W. P., – 1966 – Land capability classification in the States of Malaya, Malaysia. Proceedings Second Malaysian Soil Conference, Ministry of Agriculture and Co-operatives, Kuala Lumpur.

PANTON, W. P., – 1970 – Application of land use and natural resource surveys to national planning: the Malaysian experience. In 'World Land Use Survey Occasional Paper no. 9, New Possibilities and Techniques for Land Use and Related Surveys,' pp. 129–136. Geographical Publications, Berkhamsted.

PRATT, D. J., P. J. GREENWAY & M. D. GWYNNE, – 1966 – A classification of East African rangeland. *J. appl. Ecol.* 3: 369–382.

RAINS, A. B., – 1970 – The evaluation of African rangeland using aerial photographs. In 'Proceedings Eleventh International Grassland Congress,' pp. 99–101. University of Queensland Press. St. Lucia.

RAINS, A. B. & M. A. BRUNT, – 1971 – An evaluation of air photography for land

318

resource surveys in the tropics. In 'Proceedings Seventh International Symposium on Remote Sensing of Environment,' pp. 2319–2327. University of Michigan, Ann Arbor.

RIQUIER, J., – 1974 – A study of parametric methods of soil and land evaluation. In 'Approaches to Land Classification,' pp. 47–53. Soils Bulletin no. 22. FAO, Rome.

ROBERTSON, V. C., – 1970 – Land and water resource planning in developing countries. *Outlook Agric.* 6: 148–157.

SCOTT, I. M., – 1974 – Soils of the central Sarawak lowlands. Sarawak Department of Agriculture, Soil Survey Memoir no. 2.

SCOTT, R. M. & J. G. SPEIGHT, – 1971 – Soil–landform–vegetation relationships on a small catchment. Paper to Section 21 – Geographical Sciences. Australian and New Zealand Association for the Advancement of Science. 43rd Congress. mimeo. 12 pp.

SPEIGHT, J. G., – 1971 – Landform description from air photos. Paper to Section 21. Australian and New Zealand Association for the Advancement of Science. 43rd Congress. mimeo. 10 pp. + appendices.

SPEIGHT, J. G., P. C. HEYLIGERS & R. M. SCOTT, – 1973 – Extrapolation in integrated surveys. Paper to Symposium on Methodology of Integrated Survey and Mapping (Section 21: Geographical Sciences). Australian and New Zealand Association for the Advancement of Science. 45th Congress. mimeo. 9 pp.

STEWART, G. A., – 1962 – Agricultural development in monsoonal Northern Australia. *J. Econ. Bot.* 16: 161–170.

STEWART, G. A. (ed.), – 1968 – Land Evaluation. CSIRO/UNESCO Symposium. Macmillan, London. 392 pp.

STOBBS, A. R., – 1967 – Recent trends in land resource investigation. Paper no. J.7 to Commonwealth Survey Officers' Conference. mimeo. 10 pp. + appendices.

STOBBS, A. R., – 1968 – Some problems of measuring land use in underdeveloped countries: the land use survey of Malawi. *Cartographic J.* (December): 107–110.

STOBBS, A. R., – 1970 – Soil survey procedures for development purposes. In 'World Land Use Survey Occasional Paper no. 9. New Possibilities and Techniques for Land Use and Related Surveys,' pp. 41–63. Geographical Publications, Berkhamsted.

SYMINGTON, D. F., – 1968 – Land use in Canada. *Canadian Geogr. Mag.* reprint. 15 pp.

TAKAHATA, S. & Y. HAYAKAWA, – 1971 – Grassland vegetation analysis using remote sensing techniques. 1. Aerial photographs taken from balloons and identification of the forage. *Res. Bull. Hokkaido Nat. Agric. Exper. Station* no. 99: 124–129.

TOMLINSON, R. F., – 1970 – Computer based geographical data handling methods. In 'World Land Use Survey Occasional Paper no. 9. New Possibilities and Techniques for Land Use and Related Surveys,' pp. 105–120. Geographical Publications, Berkhamsted.

UNESCO, – 1971 – Seminar on Integrated Surveys: Range Ecology and Management, Jodhpur, India, 9–27 November, 1970. UNESCO Field Science Office for South Asia, New Delhi. 204 pp.

VERSTAPPEN, H. T., – 1966 – The use of aerial photographs in delta studies. In 'Scientific Problems of the Humid Tropical Zone Deltas and their Implications. Proceedings of the Dacca Symposium,' pp 29–33. UNESCO, Paris.

VERSTAPPEN, H. T., – 1969 – Aerial survey and rural development in south Asia. In 'Problems of Land Use in South Asia. Yearbook of South Asia Institute. Heidelberg University,' ed. U. Schweinfurth & M. Domrös, pp. 1–6. Otto Harrassowitz, Wiesbaden.

WILFORD, J. N., – 1974 – From space, a rounded view of the earth. *New York Times,* 10 February.

WONG, I. F. T., – 1971 – The Present Land Use of West Malaysia (1966). Ministry of Agriculture and Lands, Kuala Lumpur. 61 pp.

319

WORLD LAND USE SURVEY, – 1970 – New Possibilities and Techniques for Land
Use and Related Surveys. Occasional Papers no. 9. Geographical Publications,
Berkhamsted. 138 pp.
WRIGHT, R. L., – 1972a – Principles in geomorphological approach to land classification.
Z. Geomorph. N. F. 16: 351–373.
WRIGHT, R. L., – 1972b – Some perspectives in environmental research for agricul-
tural land-use planning in developing countries. *Geoforum* 10: 15–33.
WYATT-SMITH, J., – 1964 – A preliminary vegetation map of Malaya with descriptions
of the vegetation types. *J. Trop. Geogr.* 18: 200–213.
YATES, F. & H. R. SIMPSON, – 1960 – A general program for the analysis of surveys.
Computer J. 3: 136–140.
ZONNEVELD, I. S., – 1972 – ITC Textbook of Photo-Interpretation. Vol. VII.
Use of Aerial Photographs in Geography and Geomorphology. Chapter VII. 4:
Land Evaluation and Land(scape) Science. ITC, Enschede. 106 pp.
ZONNEVELD, I. S., P. N. de LEEUW & W. G. SOMBROEK, – 1971 – An Ecological
Interpretation of Aerial Photographs in a Savanna Region in Northern Nigeria. ITC
Publication Series B. no. 63. Enschede. 41 pp.

7 SOME SURVEYS, COMPLETED OR PLANNED

This Chapter contains brief particulars regarding a number of surveys for which information was available and also the location of certain projects which were based on field surveys. It is not claimed that the list is by any means complete; other surveys are being conducted by international agencies, bilateral aid organizations and research foundations, but details were not available at time of going to press.

Pakistan: Indus Special Study. International Bank for Reconstruction and Development/Hunting Technical Services (HTS), in association with Sir Alexander Gibb and Partners, London, and International Land Development Consultants N.V. of the Netherlands.

Potential development of land and water resources for irrigated agriculture in the Indus Basin, designed to assist Government in overall planning of the Basin, to formulate priority projects up to the year 1975, and to identify essential inputs – appraisal of the economic viability of the Tarbela Dam – HTS responsibility largely devoted to setting up and operating a series of studies of water courses and farm studies throughout the northern and central parts of the plains, to assess the potential and constraints at farmer level, as an extension of the earlier studies on the Lower Indus.

Pakistan: IBRD. An IBRD economic mission to Pakistan undertook a critical and detailed review of the agricultural sector. HTS provided an agricultural economist to review previous studies, terms of trade between industry and agriculture, pricing and the best use of available resources.

Pakistan: FAO Land and Water Development Division, Soil Resource Development and Conservation Service. Strengthening of the national Soil Survey Unit; reconnaissance soil survey and land evaluation; semi-detailed survey of agricultural areas.

Pakistan: Lower Indus Project: Water and Power Development Authority, Lahore/study by HTS in association with Sir M. MacDonald and Partners. Rehabilitation and improvement of the southern part of the Indus Plains – nearly 5 million ha. – first an investigation of waterlogging and salinity in irrigated land in order to design measures for control and reclamation – second, complete development planning for the region, with breakdown into individual projects for implementation in a specified order of priority over a period of twenty to twenty-five years.

Pakistan: Advisory Planning Group, Overseas Development Administration, London/Water and Power Development Authority, Lahore. Study by HTS in association with Sir M. MacDonald and Partners.

Preparation of project planning reports for development within context of regional plan for lower Indus — one tubewell irrigation and drainage project prepared to pre-investment stage, a second begun — a major new Indus River barrage and canal complex planning study included.

Pakistan: Khairpur Operational Advisory Group. Land and Water Development Board. Lahore. Study by HTS in association with Sir M. MacDonald and Partners. Operational Advisory Groups consisting of agricultural adviser, adviser on land reclamation and specialist in farm planning and extension to assist project director in establishing crop intensification and land reclamation in relation to available water supplies.

Pakistan: Mangla Watershed Management Study. Water and Power Development Authority, Lahore/HTS.

General study of the River Jhelum catchment above dam site at Mangla — mapping land use and land capability, ecological studies for forest and pasture management, design of soil conservation measures, geological studies and socio-economic appraisal of present land use techniques and of cost/benefit ratios.

India: Central Arid Zone Research Institute.
Integrated surveys in the States of Rajasthan, Haryana, Gujarat and Karnataka (see 5.3.3).

India: Government of India.
All-India soil survey. Headquarters Indian Agricultural Research Institute, New Delhi, with four regional centres located at Delhi, Nagpur, Bangalore and Calcutta, representing four major soil regions.

India: IBRD
Agricultural markets project appraisal mission — a credit specialist assigned by HTS was primarily concerned with the evaluation of lending channels and existing credit facilities, and the preparation of proposals for future improvements.

India: Government of India/United Nations Development Program/FAO Pre-investment survey of forest resources.

India: FAO, Land and Water Development Division, Soil Resource Development and Conservation Service.

Soil survey and soil and water management research and demonstration in the Rajasthan Canal area. Survey of soil resources in this semi-arid to arid area for irrigation development, establishment of two experimental stations for development of suitable soil management techniques.

India: French Institute, Pondicherry. Survey and mapping of vegetation, covering entire country.

India: Saurashtra University, Department of Biosciences, Rajkot, Gujarat.
Integrated Survey of River Narmada Upper Catchment, Madhya Pradesh.
This project, sponsored by the Council of Scientific and Industrial Research, is being executed in four phases: ecological survey of forest vegetation, standing biomass of present communities and some dominant tree species, hydrological cycles, and grasslands. Prescriptions may then be given regarding

measures for flood and erosion control (*J. Ind. bot. Soc.* 51: 356–73, 1972).
India: International Coffee Organization/HTS.
Collaboration in mission with terms of reference to review the status of the
coffee industry in eastern India, and to identify projects suitable for financing,
and to formulate plans for organization and operation of the projects.

Nepal: FAO, Land and Water Development Division, Soil Resources Develop-
ment and Conservation Service.
Feasibility study of irrigation development in the Terai.
Nepal: Kathmandu Valley. ODA/HTS
From data supplied by Department of Housing and Physical Planning,
Kathmandu, maps were prepared. depicting land use, administrative units,
and additional data compiled from air photo-interpretation for the forest areas.
Nepal: Ministry of Health/Thomas A. Dooley Foundation.
National health survey to supply baseline quantitative data to assist in planning
future health work. Items surveyed included environment, topography,
climate, rainfall, crops, livestock, sanitation and housing – rural Nepal represented one of few areas in which a traditional pre-industrial ecological balance
between man and his environment could be studied, greatly modified by
rapid modernization during the 1960's. (See R. M. Worth & N. K. Shah –
1969 – Nepal Health Survey, 1965–1966. University of Hawaii Press,
Honolulu.)
Nepal: Asian Development Bank Special Funds.
Kanki irrigation, approved 14 December, 1971; Chitwan Valley development,
approved 14 December, 1972.

Bangladesh: IBRD/International Development Association.
Factors influencing land use and possibilities for agricultural development.
Bangladesh: IBRD
Brahmaputra Right Bank Flood Embankment Project. HTS assigned two
specialists to participate in a mission to review the general effectiveness of
this project and potential of that portion of the IDA-financed tubewell project
falling within the projected area.
Bangladesh: Sylhet and Chittagong Tea Irrigation Project. HTS in association
with Sir M. MacDonald and Partners.
Periodic droughts in tea-growing areas a serious concern – assessment of sur-
face water supplies and of groundwater resources – agronomic and soil studies
to assess irrigation needs – economic analysis.
Bangladesh: FAO, Land and Water Development Division, Soil Resource
Development and Conservation Service.
Reconnaissance soil survey and land evaluation; semi-detailed survey of agricul-
tural areas.
Bangladesh: Dacca north, Irrigation and Drainage Project. HTS in association
with Sir M. MacDonald and Partners, and Pakistan Techno-Consult Ltd.
Irrigation and drainage feasibility study over an area of 80,000 ha. north of

Dacca, subject to annual flooding – consideration of flood control measures, irrigation by canals and tubewells, and provision for river transport systems – necessary to plan for nine development components later integrated into a comprehensive project capable of phased development.

Bangladesh: Tubewell Development Project: Agricultural Development Corporation/HTS and Sir M. MacDonald and Partners, Assistance in planning and execution of programme for the sinking of 3,000 tubewells to serve a cropping area of 70,000 ha. in seven districts in the northwest – establishment of three field trial farms and assistance in extension and training, water management and plant protection.

Sri Lanka: FAO/UNDP.
Soil survey of selected areas in connection with the Mahaweli Ganga irrigation and hydro power survey, and with agricultural diversification of tea and rubber growing areas.

Sri Lanka: Asian Development Bank.
Mission undertaken to review finance of the Uda Walawe Feasibility Report (HTS collaborated).

Burma: FAO, Land and Water Development Division.
Soil survey of the Settang River Valley.

Malaysia: Prime Minister's Department, Economic Planning Unit; Survey of West Malaysia for preparation of maps of land capability (6.2.4).

Malaysia, West: Jengka Project. Federal Land Development Authority. HTS Study in association with Tippetts, Abbett, McCarthy and Stratton, New York. Preparation of master plan for settlement and development of an area of about 200,000 ha. in what was largely primary rain forest – detailed soil survey, forest inventory, agricultural surveys, land capability survey and proposals for specific crops for each type of land (see 6.2.4).

Malaysia, West: Tanjong Penggerang and Johor Tengah. Government of Malaysia and ODA, London. Study by HTS in association with Binnie and Partners, London, the Overseas Development Group of the University of East Anglia and Shankland, Cox and Associates.
Preparation of comprehensive master plan for the economic and social development of some 300,000 ha. in the State of Johor over the development period 1971–1990 – exploitation of timber resources, including setting up a timber complex for processing, and recommendations for future management of forest reserves – agricultural development to permit implementation by national land development authorities, development of smallholdings and some participation by the private sector – special emphasis on possible crop diversification and integration of livestock enterprises.

Malaysia: Government of State of Sabah: Survey of land resources, land capability (6.2.4).

324

Malaysia: Government of State of Sarawak: Soil survey of the central Sarawak lowlands (3.3.2).

Malaysia: Government of State of Sarawak/HTS: Miri-Bintulu Regional Planning Study over two years to evaluate total resources, physical, human and financial, of an area of 1.4 million ha — preparation of regional plan for utilization in a programme of development — first stage to consist of land-use zonation plan in outline, and detailed regional planning of selected areas within the whole. For information regarding the work and publications of H. S. Morris, C. A. Sather & Hatta Solhee on the sociological aspects of the Study, see *Borneo Research Bulletin*, Vol. 6, no. 2, p. 53, August, 1974.

Malaysia: Government of State of Sarawak: Classification of land (1948) into five classes: reserved land, mixed-zone land, native area land, native customary land, and interior area land.

Malaysia: Asian Development Bank, Capital Resources and Special Funds Besut agricultural development, approved 22 September, 1970; Sabah land development, approved 1 August, 1974.

Brunei: Land capability survey. Commissioner for Development, Government of Brunei/HTS: Survey as basis for planned development of infrastructure, roads and other services — extensive use of air photographs, mobile field teams and transport by helicopter in difficult, densely forested terrain — further objective was identification of areas for land development.

Indonesia: Agro-Economic Survey (A.E.S.)

This inter-ministerial body founded in 1965 for policy-oriented research started a rice intensification study (funded principally by Ford Foundation), which was initially conceived as an attempt to determine the impact of the introduction of high-yielding varieties on income of rice farmers, use of labour, need for credit, and decisions about allocation of resources. The R.I.S. evolved later to encompass the entire farm operation, labourers, middlemen and village leaders, and ultimately fundamental problems of rural change, income and opportunities for employment, the necessary data being collected by in-depth interviewing of sample farmers over five years.

Assessment of agricultural resources and condition of rural society in 14 initial projects — specialists from Departments of Agriculture, Plantations, Public Works and Home Affairs, and from universities, working in ad hoc teams as suggested by Professor of Rural Sociology, Bogor (see de Vries, *Bull. Indon. Econ. Studies* 5: 73–77, 1969). See also general economic surveys including data on land use, crop areas, crop production, population, rural incomes and holdings, and food consumption: examples — West Java (Daroesman, *Bull. Indon. Econ. Studies* 8: 29–54, 1972), Bali (Daroesman, *Bull. Indon. Econ. Studies* 9: 28–61, 1973), East Nusatenggara (Makaliwe &

Partadireja, *Bull. Indon. Econ. Studies* 10: 33–54, 1974), and north Sulawesi
(Boediono, *Bull. Indon. Econ. Studies* 8: 66–92, 1972).

The A.E.S. now proposes (October, 1974) to follow up the R.I.S. with a
broader, integrated, five-year study of rural dynamics (R.D.S.), designed to
a. describe in detail the patterns and trends over time of income and employment for all major groups in selected villages, b. analyze the interplay of
forces between people, capital, technology, government programmes and
services from the private sector, and c. provide base data for planners at all
levels. These data will relate to systems of agriculture, distribution and use of
resources, allocation of labour and sources of income, household expenditures,
rural institutions, migration and population, impact of new technology and
access to services, in villages selected by purposive rather than random
sampling, primarily in defined ecological regions in West Java in the first
instance.

Such an investigation requires the use of multiple levels of analysis and
multiple research methods. For example, the human units of analysis whose
economic behaviour is to be observed, interpreted and compared must comprise at least the following: individuals; households; small groups of neighbours,
kin or others among whom economic interactions are frequent; socio-economic
and occupational strata; whole hamlets; whole villages; and larger administrative
units up to the level of the kabupaten. At each of these levels, different
research methods may be required.

Indonesia: Soil Resources Institute, Bogor/UNDP/FAO, Land and Water Development Division: Land capability appraisal project.

Special emphasis placed on production of crops for food, fibre or industrial
purposes; population very unbalanced, with Java, Madura and Bali constituting
8 per cent of the land area, but having up to 75 per cent of the population
(see Chapter 4.3).

Report (with supporting maps) made at a map scale of 1:2,500,000,
appraisal being primarily based on 'desk studies' using data at present
available – definition of broad areas promising or not promising for more
detailed studies.

Criteria include the design of Land Development Units based upon
delineated geographic areas similar in climate, soils, physiography, relief,
land use, water resources and management, soil resources and management,
and crop production. The Development Units are then grouped into five Land
Capability Appraisal classes, based upon limitations to use by certain Land
Utilization Types and specified levels of management.

Areas of the country are suitable for development for specific kinds of
crops or land use. In Phase II of the Project (commenced March, 1974), new
data will be collected on soils, land, farming systems, etc. for use in refining
Land Capability Appraisals.

Indonesia, Java: Ministry of Public Works and Power/Directorate-General of
Water Resources Development/Asian Development Bank: Teluk Lada Irriga-

326

tion Project. Preparation of viable project for integrated development in the Teluk Lada Plain — survey of irrigation, drainage, flood protection, rural infrastructure — ADB to provide services of eight experts for nine months.

Indonesia: Asian Development Bank, Special Funds: Tadjum irrigation, approved 17 June, 1969; Gambarsari-Pesanggrahan irrigation rehabilitation, approved 23 December, 1970; Sempor Dam and irrigation, approved 2 December, 1971; Wampu River flood control and development, approved 4 April, 1972.

Technical Assistance: Foodgrain production, rural credit survey, advice to Ministry of Agriculture, feasibility study on Sempor Dam construction, and Wampu River flood control.

Indonesia: Soil Research Institute, Bogor/FAO, Land and Water Development Division Survey of Upper Way Sekampung catchment, Sumatra.

Indonesia: Soil Research Institute, Bogor/Directorate of Transmigration/FAO, Study of land-use dynamics in two sample areas in south Sumatra subject to both official and spontaneous transmigration.

Indonesia: Soil Research Institute, Bogor, Soil survey for agricultural development in south Sulawesi. Preparation of soil map. Definition of land capability, mapping of present and recommended land use (M. Soepraptohardjo in FAO World Soil Resources Report no. 41, 151–155).

Indonesia: Government of Indonesia/IBRD/FAO, Survey of potential areas for major livestock development project and selection of areas in Sulawesi and other islands.

Indonesia: Directorate of Land Use (E. Made Sandy), Survey of successive stages of land use in Java since the first arrival of small, primitive communities with little lasting effect on vegetation; settlement at 25 m. above sea level, reasonably flat for shifting cultivation and safe from flooding; subsequently settled cultivation — initial clearance of land for rice fields watered from streams to avoid seasonal droughts, and construction of villages; use of some fields to store water for irrigation of rice padis, some fields changed into mixed fruit and vegetable gardens; gradual increase in pressure of human population and further deforestation — good forest covers capable of conserving water and preventing erosion become progressively less in area; further progressive damage to land resources, drying up of marshes around lakes, seepage of sea water to rice padis increasing their salinity, salinization of drinking water.

Indonesia, Java: /Kali Progo/Ojo Basin/ODA/HTS/Sir M. MacDonald and Partners, To establish a basic hydrological network, to assess availability of water from all sources, to determine existing water requirements, and to prepare a plan for future development in the Basin — to examine need for remodelling and rehabilitation of existing irrigation schemes — survey supported by geological, pedological and agricultural surveys, and submitted to critical economic analysis.

Indonesia, Java: Ngandjuk-Kediri (Middle Brantas) Project.

ODA, HTS/Sir M. MacDonald and Partners, To assess availability of water

supplies for irrigation from both surface and underground sources — to plan
an increase in supply for the expansion of irrigated agriculture.

Indonesia, north Sumatra: Smallholder Rubber Survey: Government of Indo-
nesia/Commonwealth Development Corporation/HTS, Survey of smallholder
rubber production and processing facilities to select areas for detailed examination with the aim of improving overall rubber production.

Indonesia, Madura: Government of Indonesia/ODA/HTS/Sir M. MacDonald
and Partners, Two-stage study to identify opportunities for development of
water resources, particularly groundwater, for the improvement of water
supplies for agricultural and urban use.

Thailand: Land and Water Development Division, FAO, Assessment of soil
fertility conditions; large field experimental programme on soil management
and fertilizer use; best rates and application of fertilizers for main crops;
assisting Government in development of fertilizer policy.

Thailand: Land and Water Development Division, FAO, Strengthening soil
survey and land classification — establishment of a national soil survey and
land classification unit in the Land Development Department — reconnaissance
survey and evaluation of soil resources — soil survey of selected areas.

Thailand: Land and Water Development Division, FAO, Experimental and
demonstration farm for irrigated agriculture — development of suitable
crop rotation and irrigation techniques adapted to the difficult environmental
conditions of northeast Thailand.

Thailand: Royal Thai Forestry Department/FAO
Mae Sa Integrated Watershed and Land Use Project
Catchment area considered from ridgetop to ridgetop as one socio-ecological
unit; total resource survey conducted; best land allocated to intensive
agriculture/terraced padi, remainder becoming orchards and commercial
forests, with tribal settlers made responsible for prevention of fire and for
protection from intruders who might try to clear land for the growing of
opium poppy.

Thailand: University of Kyoto, Center for Southeast Asian Studies, Interdisciplinary study of the potential for rice production in the Chao Phraya
Basin by specialists in geology, irrigation technology, soil science and crop
science (3.5).

Thailand: Oil Palm Development: Ministry of Agriculture/HTS, Preliminary
appraisal of the potential for developing commercial oil palm production in
southern Thailand, leading to a specific project.

Thailand: Yom Basin Study: Government of Thailand/ODA/HTS/Howard
Humphreys and Sons, Full development of the water resources of the Mae
Yom Basin and its complementary land resources.

Thailand: Yom Basin Soil Survey: Rural Irrigation Department, Government
of Thailand/HTS/Howard Humphreys and Sons, Semi-detailed and detailed
soil survey of an area 3000 sq. km., for the preparation of maps of soils and

328

land classification to facilitate planning of canals and supporting structures.
Thailand: Chiengmai University Farm. Faculty of Agriculture, University of Chiengmai/ODA/HTS, Detailed soil survey and land capability classification with recommendations for future developments.

Thailand: Water Resources Investigations. Government of Thailand/ODA/HTS. To prepare recommendations for the phased development of the water resources of the tributaries of the River Yom in Phrae and Sukhotai Provinces in northern Thailand, to be integrated with groundwater development and the main river projects.

Thailand: Yom Tributaries Study. Rural Irrigation Department/HTS/Howard Humphreys and Sons/Sir M. MacDonald and Partners, This study follows that of the Yom Basin (see above) — to identify and prove the surface and groundwater resources of the tributary areas — preparing a plan for their utilization in agricultural development.

Thailand: Southern Lands Settlement Study. Department of Public Welfare/ODA/HTS, in association with members of REDECON. Survey of physical, human and economic resources of five large settlements, totaling 300,000 ha — physical resources classified according to characteristics and qualities of soil and terrain to identify land for further development — second stage preparation and evaluation of a phased programme.

Thailand: Regional Planning Study of South Thailand. Government of Thailand/ODA/HTS/REDECON/Shankland Cox Partnership and Cooper Brothers, Examination of total resources of the region, identification of key projects and survey of opportunities for industrial growth.

Thailand: Asian Development Bank Technical Assistance
Accelerated rural development programme, approved 23 September 1969; agricultural development programming in the Nong Wai pioneer irrigated agriculture project, approved 9 October, 1969.

Indochina: Mekong Basin Development Programme, The most ambitious and largest development programme ever attempted — involves a series of forty to fifty mainstream and tributary dams, with the major objectives of power production, flood control and irrigation — overall co-ordination provided by Mekong Committee, established by the Economic Commission for Asia and the Far East of the United Nations* — technical assistance and other support provided or pledged by nearly thirty nations and a dozen international organizations — three main groups of factors arising from the development projects will have ecological consequences: 1. the downstream consequences of altered river flow when dams are constructed; 2. the consequences of the creation of lakes when dams are constructed, and 3. consequences of resettlement, irrigation and other changes in land use. See also SEADAG Paper no. 34, Speculations on the Results of the Possible Edaphic Changes resulting from

* now ESCAP

329

the proposed Mekong Basin Development Projects', Mary McNeil, New York, April, 1968; Report prepared for USAID by L. M. Talbot, 1970, 'Effect of Mekong Development on Biotic Factors, particularly Wildlife, Parks and Reserves'; L. A. Cohen, in *Focus* (Amer. Geogr. Soc.), 'International Co-operation for Development: the Mekong Project', June, 1972; Report prepared by G. Conway & J. Romm for the Ford Foundation Office for Southeast Asia, Bangkok, 1973, 'Ecology and Resource Development in Southeast Asia', pp. 45–47.

Khmer Republic: Pilot Station for Irrigated Agriculture, Battambang. FAO, Land and Water Development Division. Survey of soil and other environmental conditions of the area, in irrigation and drainage network of the Prek Thnot river, a few km. south of Phnom Penh.

Laos: Asian Development Bank, Special Funds, Tha Ngon agricultural development, approved 10 March, 1970. Technical Assistance, Integrated agricultural development programme for Vientiane Plain, approved 16 October, 1968; Casier Sud pioneer agriculture, approved 28 March, 1973.

South Vietnam: Asian Development Bank, Special Funds, Binh-Dinh irrigation, approved 16 December, 1971; Go Cong pioneer agricultural project, approved 17 December, 1973.
Technical Assistance, Tan An irrigation, approved 4 December, 1973.

Philippines: FAO, Land and Water Development Division, Assessment of soil fertility conditions; experimental programme on soil management and fertilizer use; best rates and application of fertilizers for principal crops; assistance to Government in development of fertilizer policy. Extensive project for appraisal of soil resources to commence in 1975.
Philippines: Department of Anthropology, Yale University. Applications of integrated survey and air photo-interpretation in an anthropological study of north-central Ifugao (see 6.2.5).
Philippines: Government of Philippines/University of the Philippines/UNICEF/WHO. Assisted Rural Health Demonstration and Training Centre. Project at Novaliches — observations on environmental sanitation, maternal and child health and vital statistics, control of community diseases, health and nutrition education surveyed ten years after inception of project in 1950. Incidental observations on geographical, topographical background, climate, total and seasonal rainfall, range of temperatures, soils, erosion, agricultural systems; social structure, credit, improved farming economy following introduction of fertilizers and pesticides (T. V. Tiglao — 1968 — *Health Practices in a Rural Community*. University of the Philippines, Quezon City).
Philippines: Asian Development Bank, Capital Resources, Davao del Norte irrigation, approved 22 November, 1973, Special Funds, Cotabato irrigation,

approved 18 November 1969, Technical Assistance, Laguna de Bay water resources development study, approved 8 August, 1972; Agusan del Sur irrigation, approved 23 May, 1974; Pulangui irrigation, approved 9 July, 1974.

South Korea: FAO, Land and Water Development Division, Assessment of soil fertility conditions; large field experimental programme on soil management and fertilizer use; best rates and application of fertilizers for principal crops; assistance to Government in development of fertilizer policy.
South Korea; FAO, Land and Water Development Division, Strengthening soil survey unit of the Plant Environment Department — reconnaissance survey of the country — semi-detailed and detailed survey of selected areas.
South Korea: FAO, Land and Water Development Division, Pre-investment survey of Naktong River Basin — survey of soil and other environmental conditions and establishment of pilot areas for integrated agricultural development.
South Korea: FAO, Land and Water Development Division, Experimental and demonstration work on soil management and soil conservation techniques in three representative watersheds and recommendations for a national action programme on soil conservation.
South Korea: Asian Development Bank, Capital Reserves, Andong Dam multipurpose development approved 6 October, 1970. Technical Assistance, Namgang—Imjin area development, approved 17 October 1973.

International Institute for Aerial Survey and Earth Sciences (ITC)
Following a recommendation at the UNESCO Conference in Toulouse in 1964
on principles and methods of integrating aerial survey studies of natural
resources for potential development, the ITC–UNESCO Institute for Integrated
Surveys was established by agreement with the Government of the Netherlands,
first at Delft in 1965, and later transferred to Enschede.

The main objectives of the Institute are: a. research and instruction concerning the application of techniques for the preparation, execution and evaluation of integrated surveys, in particular regarding developing countries, and b.
promotion of the application of integrated surveys through co-operation with
organizations actively engaged in development programmes at the international
level.

Study and instruction for postgraduate participants are therefore focused
on:
— the operating ability in multidisciplinary teams
— the adaptation of one's own specialization to the requirements of the project
— formulation of procedure for data collection as required for the objectives
  of the survey
— study of responsibilities and interrelations between disciplines in a survey
— composition of team reports with appropriate conclusions and recommendations, and
— review of individual case studies by the participants.
Instruction is provided at two levels: a. a course of seven to eight weeks for
team leaders in development projects, project makers, senior executives and
high-level employees of governmental and private organizations who are or
will be responsible for the integration of research, and b. a course of nine
months' duration promoted by UNESCO, intended for specialists qualified
in different disciplines of earth sciences, social sciences and technology involved in investigations of natural resources, who receive lectures in the land
sciences and photogrammetry, climatology, sociology, economics, engineering,
geology and medical ecology; selected examples of typical projects are
studied; practical exercises are given in aerial photo-interpretation, thematic
cartography, and the planning of projects.

Having devoted the initial years to the establishment of the Institute, it is
proposed to promote courses for participants from specific regions or countries.
For the purpose of practical training, the Institute sees the need for a series
of areas for field work, of different types and in different environments,
where staff and students may participate in the actual execution of projects.

The programme of the subdepartment of integrated surveys at ITC has developed more in the direction of the *management* of natural resource survey (I. S. Zonneveld, personal communication, 1974). The teaching programme is mainly directed to the question of how to manage a survey, namely, what type, when and where is it to be carried out, and how are survey activities related with each other and with the stage of planning and execution of the project for which the survey is being made.

Other courses at ITC provide answers to the questions: how does one conduct the various types of survey involved, how are the disciplines interrelated, and what are the techniques to be applied. Several subdepartments of the Natural Resources Department organize courses on natural resources surveys such as geological surveys, soil surveys, geomorphological surveys, forestry surveys. In order to meet the need for integration between these various surveys, a new interdepartmental course has been established which is operated by the various subdepartments simultaneously. This course is called 'Rural Surveys'. The former courses on rural geography, vegetation and agriculture have been combined in this course, because it has been judged better to integrate these related subjects (see I. S. Zonneveld, lecture notes on plant ecology below). This is done for two reasons: the techniques used are rather closely related, and individual attention to them would involve duplication – for example in survey techniques of cultural (land use) and natural vegetation and land units, etc., combined with rural geography and with a background of soils, geomorphology, geology, climatology statistics. Above all, these disciplines have so many points of contact that combination enhances their overall value in a truly integrated rather than cumulative sense.

Unlike the still independent ITC/UNESCO Integrated Survey Course, this Rural Survey course deals not with 'total project survey', but is limited to some important aspects of any survey project for development planning, especially land evaluation itself. It deals with the techniques of survey, including photointerpretation, the techniques of classifying land and its attributes, and the process of evaluation of the basic data into values to be used by the planners and executors, that is (*pragmatic*) land evaluation. Such a course might also be called a kind of 'integrated survey' according to the current use of the term outside ITC. It is hoped that in the future this Rural Survey Course will progressively lead to integration of the ITC work itself.

Land Resources Division, Overseas Development Administration
This Division organizes a number of individual training courses overseas for the staff of government services engaged in the planning of farm and ranch development and in soil and water survey.

Staff of the Division assists with training courses held at the National College of Agricultural Engineering at Silsoe, Bedfordshire, on the use of aerial photographs in environmental studies, with special emphasis on the evaluation of land resources in developing countries. By attending these cour-

333

ses, the participant becomes familiar with the geometry of the aerial photo-graph and its stereoscopic examination, the stereoscopic extraction and re-cording of terrain data, and the appreciation and possible uses and limitations of aerial photography (see lecture notes of A. R. Stobbs on aerial photo-interpretation below).

National College of Agricultural Engineering, Silsoe, Bedford, England
The curriculum of the third year B.Sc. degree contains the following subjects among some thirty options:
Systems methodology and quantitative methods
Resource survey methods
Agricultural development systems
Postgraduate syllabuses include:
Resource survey methods — principles of integrated planning, aerial photography, photogrammetry, photo-interpretation, remote sensing, techniques for collection of ecological and socio-economic facts, preparation of physical plans
Work system design — physical, physiological and psychological characteristics of man in relation to work.
Water resource development
Land resource planning — methods in different parts of the world, techniques of classifying natural resources, satellite photography, engineering principles in planning.
Surveying and land forming
Operational research — mathematical models, linear programming, integer programming, simulation, network analysis.

Two short courses were offered in 1974:
1. Aerial photographic interpretation for land resource appraisal, 15 to 19 July, a. basic uses, types and geometry of air photographs; b. land resources appraisal; c. interpretation of vegetation; d. statistical sampling surveys and interpretation of land use; e. interpretation for soil survey for dryland and irrigated land; f. interpretation for land classification.
2. The use of statistics in resource analysis, 23 to 27 September, with special reference to the development of regional and subregional structure plans in Great Britain.

University of Reading
A postgraduate course in integrated land resources survey commenced in October, 1974, designed to cover two years, and to include a period of field work, probably in the tropics or subtropics. The syllabus will include: a. relevant aspects of the earth and physical sciences, principles of meteorology and climate; principles of geology, mineralogy, petrology and petrogeography; geomorphology, pedology, principles of hydroiogy; ecology. b. techniques of

334

land resource survey, including principles and methods of integrated terrain analysis; integrated survey field methods, preparation and practice, project planning and report compilation. c. land use planning and plan implementation. This course includes physical planning and its relation to social and economic planning, watershed management and conservation practice; irrigation, drainage and land reclamation; agricultural planning and development; engineering planning and development.

A. RELEVANT ASPECTS OF THE EARTH AND PHYSICAL SCIENCES

Unit 1 *Principles of Meteorology and Climate*

Introduction to weather and climate. Regional climatology; climate classification; climatic change. Micro-climatology, especially in relation to agriculture and plant (crop) production. Climate as a resource.

Unit 2 *Principles of Geology, Mineralogy, Petrology and Petrography*

The physical, chemical and optical properties of the more common rock forming minerals and rock types, together with their nature and mode of occurrence; a detailed consideration of the rock cycle of weathering, erosion, transport, deposition, lithification, uplift and orogeny; concepts and principles underlying historical geology.

Unit 3 *Geomorphology I*

Systematic and dynamic geomorphology relating fluvial, glacial, coastal and aeolian processes of erosion and deposition to landforms, paying particular attention to tropical regions. The treatment is thematic and includes a consideration of new concepts and theories, experimental and quantitative techniques, and models.

Unit 4 *Geomorphology II*

Spatial and chronologic geomorphology. This includes consideration of the different theories of origins, processes and history of the world's regional landform patterns, especially in the Quaternary. It leads on to a consideration of different types of geomorphological regionalization, the dynamic interrelationships of such regions and the characteristics of their constituent sites.

Unit 5 *Pedology I*

The distinction between soils and soil material; physical and mechanical properties of soils including soil texture, soil structure, soil pore characteristics and thermal properties of soil; clay minerals; iron and aluminium oxides; soil water and soil solutions, including saturated and unsaturated flow; soil acidity; interactions of clays, sesquioxides and humus; organic matter; soil fauna and humus forms.

Unit 6 *Pedology II*

The nature, morphology, genesis, classification, areal pattern of distribution and applied aspects of 3-dimensional bodies of soil. This includes the distinction between rock weathering and soil formation; the physical, chemical and biological agents and mechanisms involved in soil formation; quantitative techniques for establishing soil morphology; soil site history; soil use and capability; soil erosion, conservation and improvement.

Unit 7 *Principles of Hydrology*

Hydrometeorology; the atmosphere, its heat balance and meteorological elements. Precipitation; frequency, intensity and duration. Hydrological cycle; evaporation, run-off and infiltration; hydrographs. Floods; estimation and routing; statistical methods.

335

Reservoir storage and yield. Geohydrology; ground water flow and storage; chemistry of ground water. Water balance; studies of catchment areas, estimations and estimations from insufficient data.

Unit 8 *Ecology I*

Concept. The ecosystem; biogeochemical cycles, energy flow, primary productivity, world production, secondary and tertiary productivity. The plant in the ecosystem; the environment of the plant above and below the ground, plants and their environment. The making of the plant community; tolerance range of species, adaptation, competition, ecological tolerance range, initiators of succession, the processes of succession, the 'climax' vegetation. Vegetation today; stability, zonation, history of British vegetation. Study of local habitats.

Unit 9 *Ecology II*

Uniformity and size of communities, primitive classifications. Aims of classifications of communities. Analysis of communities for classification; reconnaissance and intensive survey, aerial survey, transects, quadrats. Physiognomy and analysis of structure and scale. Methods of classification for various purposes; physiognomic and floristic classification, methods of general survey and intensive survey. Mapping; aim of map, boundary definitions, cartography.

B. TECHNIQUES OF LAND RESOURCES SURVEY

Unit 10 *Principles and Methods of Integrated Terrain Analysis*

Concepts and methods of integration among the different scientific and practical disciplines concerned with land evaluation, laying emphasis on aerial photographic interpretation, and the use of remote sensing and associated semi-automatic analyses and display techniques. It also considers the co-ordination of the acquisition, storage, retrieval and evaluation of natural resources information from aircraft, balloon, rocket and satellite platforms as well as from ground survey. Some existing integrated surveys are studied in terms of the techniques used and the success achieved.

Unit 11 *Cartography, Survey and Elementary Photogrammetry*

The evaluation, compilation and drafting of topographical, atlas and thematic maps: the principles of surveying with practical field work, computation and adjustment of results: and simple stereoscopy, including height and slope measurements, graphical construction of cross sections, radial triangulation and the use of simple plotting instruments.

Unit 12 *Integrated Survey Field Methods: I Preparation*

This course is essentially the preparation for the field work associated with the projects and dissertation. It involves an introduction to field methods used in geological, soil, hydrological, geomorphological and ecological surveys.

Unit 13 *Integrated Survey Field Methods: II Practice*

Quantitative methods for identifying and describing soils, rocks, vegetation and weather in the field, with particular stress on the establishment of grades and classes for single parameter and land capability mapping; field notebook notation; sampling techniques; the definition and interpolation of field mapping units; sampling techniques; plotting of field data in the form of maps; construction of legends; transcription of mapping units to classification units; and the interpretation of laboratory data leading to the interpolation of land capability units.

336

Unit 14 *Project Planning and Report Compilation*

The staffing of planning teams and the integration of specialists; critical path analysis applied to the organization of team work; archive work, check lists of equipment; procedure for establishing base and forward camps; communications; support staff and liaison work. Critical path analysis applied to report preparation; report layout and design; user requirements of reports.

C. LAND USE PLANNING AND PLAN IMPLEMENTATION

Unit 15 *Physical Basis of Planning*

This is a seminar course in five parts: 1. Physical planning and its relation to social and economic planning; 2. Watershed management and conservation practice; 3. Irrigation, drainage and land reclamation; 4. Agricultural planning and development; 5. Engineering planning and development.

France

Up till 1966, the Centre d'Etudes Supérieures de Géologie et Géophysique à l'Ecole Nationale Supérieure de Pétrole et des Moteurs de l'Institut français du Pétrole (IFP) provided a certain number of grants to overseas engineers who wished to improve their knowledge of the field of photo-interpretation. From 1966, the course took on a new orientation. It then dealt chiefly with agricultural, pastoral and forest development. The Société Géotechnip (affiliated to the IFP) was given responsibility of organizing this new course, in association with other research organizations in France, especially those at Toulouse, Montpellier, the research laboratories of Kodak, the Laboratory of Applied Geomorphology, Strasbourg, the National Forest Inventory, etc. The overall scope and programme of this course have been given by Leneuf & Rossetti (1967). It is presumed that, since Géotechnip is no longer active in the field of aerial photography and the specialists on the staff have dispersed, the course itself has now been discontinued (see 3.6.1).

Thailand

From 1967 to 1973, the Government authorities in Thailand collaborated with FAO, acting as executing agency for UNDP, in a project entitled: Strengthening soil survey and land classification (see Chapter 7). Because of the importance of aerial photography to the project, the Head of the Soil Survey Division of the Department of Land Development supported the FAO specialist (D. C. Schwaar) in a training programme designed: a. to establish an air-photo section and to train its personnel in the filing of airphotos and the preparation of base maps using topography, photo-mosaics and transfer techniques, and b. to introduce soil surveyors to modern techniques of photo-interpretation adapted to local conditions.

It was recommended that a photographic atlas of Thailand should be compiled, showing soil series, soil families and their corresponding photo-images, and also the characteristic land-use units and patterns in the country.

337

Lecture notes: Plant ecology – vegetation science – I. S. Zonneveld, ITC, Enschede, Netherlands
Headings indicating aspects covered:

I   Introduction – related sciences, important concepts, role of vegetation science in integrated surveys.

II   Basic properties of environment influencing plants – social relation between plants, ecological laws regarding plants, vegetation and environment.

III   Synmorphology – constituents, life form system, symbols, Raunkiaer, Ellenberg, Iversen & Zonneveld, Dansereau; three-dimensional form, structure, size and shape, vertical and horizontal aspects.

IV   Classification – guiding principles, sources of confusion between requirements of separation and of integration of characteristics (properties) and guiding principles (forming factors), continuum concept in relation to classification, ordering of basic data.

V   Study of succession (syndynamics) – concepts of climax and succession.

VI   Survey of vegetation – aims of survey and character of map, difference between a legend and a general classification, kind of maps, possibilities of various sensing methods, landscape and vegetation mapping, interrelation of scale, accuracy and purpose; sampling, relation between accuracy of sampling and mapping scale, distribution of sample spots, mesological/ecological sampling, relation and difference between sampling and mapping; recognition, analysis and interpretation from aerial photographs, evaluation of vegetation maps, vegetation mapping using aerial photo-interpretation.

VII Synecology and application

Figures: Land-forming factors and attributes and their interrelation; Spheres of environment of plant life; Mosaic-like patterns, zonation, alternation; Scheme of a continuum with two gradients; Levels of synthesis in vegetation survey; Scheme of parametric land classification or resource evaluation; Scheme of surveying and mapping vegetation using aerial photo-interpretation.

Lecture notes: Air photo interpretation of soils. – A. R. Stobbs, Land Resources Division, Overseas Development Administration, Great Britain.

1. Part of the work of assessing land quality for human use comprises identification, description, analysis and classification of soil profiles. It also requires that the soil units so identified shall be displayed visually in map form to demonstrate their extent and distribution. The role of air photo interpretation in soil survey has become increasingly important as photo coverage spread until few areas now lack coverage by at least one set of photography and many are now covered by a succession of photo contracts of different ages, scales and types. The advantages to be gained in terms of time, detail and precision have been set out many times.

2. The following notes refer to the use of medium-scale panchromatic black and white photography (1/20 000 – 1/80 000), except where otherwise specified.
3. Soil is a three dimensional object possessing both extent and depth. At any given moment in time, the soil profile at any selected point is the result of the continuous interaction of climate upon rock, continuously modified by the effect of vegetation growing upon it, and the use made of it by humans. It is clear that features concerned with the description of the soil profile cannot be seen on the air photo (the depth of the soil; arrangement of horizons; their colour, texture, structure, and consistency; inclusions). We are therefore concerned only with those aspects visible to us through stereoscopic inspection of the photograph. These primarily refer to the soil site in relation to the landscape, the patterns imposed on it by human use and vegetative growth, its origin and form in response to geology and climate. With experience it is then possible to make deductions concerning the type of soil, from the visual data.
4. Three qualities of the air photograph are utilized: 1. Stereoscopic image (derived from overlapping photo pairs); 2. Tone (the various shades of grey on the photo); 3. Texture (the stipple pattern produced by vegetation)
They are utilized in two ways:
1. Photo identification (of the patterns formed on the photo by the interaction of the above three qualities. Such patterns are delineated by drawing boundaries). 2. Photo interpretation (assessment of the significance of those patterns in soil terms, leading to the preparation of a photosoil classification table and a description of the units in that classification).
5. The more difficult the terrain, and the more inaccessible the area, the more important the photograph becomes as a tool for the soil surveyor, though it can never completely replace field work for the reasons specified in paragraph 3. As a tool, the air photo is used in 3 successive stages: 1. In pre-field work, to delineate all those photo patterns which may reflect changing soil characteristics and thus to produce a hypothetical soil map. 2. In field work, to plan a network of traverses and inspection points designed to examine the hypothetical soil map thoroughly, and to facilitate extrapolation and inter-polation. Also (in the absence of adequate maps) as a navigational aid; and, usually in the form of mosaics, to provide a base map for the draughting of the soil boundaries when confirmed. 3. In post-field work, using revised interpretative criteria derived from field work, to modify and elaborate the hypothetical soil map by interpolation between the field traverses, which will have been carefully marked on the photos.
6. In both pre- and post-field work two classes of evidence are observed on the photos
1. *Direct* evidence; 2. *Indirect* evidence.
7. Direct evidence is the extent to which the surface of the soil shows in the photograph as differing tones. In assessing tones it must be borne in mind that between adjoining photographs, and more particularly between areas or strips of photographs the tones may differ though the soil is the same, and vice versa. This is because: 1. The reflectance properties of the soil may change with varying wetness or dryness at different times, or the presence, absence, or growth of grass upon it. 2. The reflectance properties may also change the tonal values due to a varying angle of light between the source, the object and the camera (changing time of day; aircraft's course). 3. The tonal value for a soil surface will also change as a result of cultivation.
With black and white photography it is often difficult to decide whether smooth or very faint, fine, stipples are the direct result of soil reflectance, or reflect a pattern of grass.
In general it may be taken that the more arid the area the greater the likelihood of the soil showing up directly on the photo as a result of the absence of vegetation.
8. Indirect evidence is much more frequently used than direct evidence. This requires identifying and assessing the pedological significance of patterns primarily reflecting other aspects of the physical environment.
9.1 Geomorphological evidence is commonly the most important source of useful infor-mation relating to soil patterns. The following geomorphological factors are involved.

1. Identification of the *landforms* present $\begin{cases} \text{piedmont} \\ \text{escarpment} \\ \text{etc} \end{cases}$

2. Description of the *slope shape* $\begin{cases} \text{convex} \\ \text{concave} \\ \text{combined} \\ \text{complex} \end{cases}$

3. Measurement of the *angle of the slope*

4. Identification and marking of *breaks-of-slope*

The basis of interpreting landforms into soil qualities rests on the premise that, 'given the same lithological rock type, under the same type of climate, the same landform will give rise to the same, or a closely similar, soil type'. This has to be modified under the wettest tropical climates due to the fact that, despite differing rock types giving rise to differing soil parent materials on different landforms, the dominant soil forming factor is often the hot wet climate. This can give rise to closely similar soil qualities due to such factors as intensity of weathering and leaching, despite the influence of other factors.

9.2 Breaks-of-slope are of great importance because they are invariably soil boundaries, although soil boundaries are not necessarily always breaks-of-slope.

9.3 Slope shape is important because it may be associated with catenary distribution of the soil.

9.4 Slope grade is significant in indicating the likelihood (in association with drainage pattern) of permanent or seasonal waterlogging; or at the other extreme, of slopes so steep as to create danger of erosion, or even the lack of any nature soil profile.

9.5 Landforms can provide useful information in two ways: 1. They can be broadly classified as 'erosional' or 'depositional' with important pedological implications as a result. Erosional landforms include hills, escarpments, gorges and plains that are either dissected or with moderate to steep undulations. Depositional landforms include those associated with floodplains and outwash plains. Examples are river terraces, river levees, sloughs, old river channels, alluvial fans and piedmonts. 2. They can be indicative, through their shape, per se, and through the regional geological structure, showing as a photo pattern, of the lithological rock type, from which it may be possible to draw useful conclusions concerning the likely mechanical constitution (texture and particle size) of the soil and, by way of deductions relating to the minerals released by chemical weathering of the rock, of the possible nutrient status and chemical composition of the soil. In doing this it is necessary to take account of climate as influencing the weathering process (deep weathering, rapid or slow weathering, leaching or non-leaching etc). Useful information on rock type may also be deduced by identification of the type of drainage pattern (parallel, sub-parallel, trellis, reticulate, etc) and by its density (closeness of the primary, secondary, etc. streams).

9.6 The interpretative procedure for geomorphological evidence is therefore: 1. Identify the landforms; i.e. a. Depositional or erosional; b. Piedmont, escarpment, alluvial fan, etc.; c. nearly level, 0 − 2°, gently undulating, 2 − 5°, moderately undulating, > 5° etc. 2. Mark all breaks-of-slope to delineate these units and where necessary, their subdivisions.

(e.g. alluvial fan $\begin{cases} \text{fan head trench} \\ \text{outwash tail} \end{cases}$

3. Classify and if desired evaluate all slopes. The smooth, regular slopes of plains and plateaux with few marked breaks-of-slope may have to be arbitrarily subdivided.

10.1 Vegetation may also be a useful source of indirect evidence. It may be used in two ways: 1. By classifying the type of vegetation, usually in physiognomic terms (e.g. scattered large trees 5 per acre); or frequent large emergents among close small trees; in such terms as open savanna woodland, grassland with scattered trees, close tall forest, etc. 2. By identification of indicator species. The possibility of doing this will depend

340

upon the scale of the photography, a highly characteristic photo appearance of the species (e.g. palms) and the knowledge that the species in question is strongly associated with certain soil characteristics.

10.2 Within a given topo-climatic zone, vegetation types are usually associated with: 1. drainage classes (commonly in flat or gently undulating terrain); 2. markedly different soil types derived from widely different parent materials. This may be at the Great Soil Group level, rather than the soil series level.

10.3 Vegetation patterns not only obscure the tonal differences of the soil but, particularly with dense woodland to tall forest types, they tend to obscure the geomorphological evidence by 'flattening' the topography, or even causing an apparent relief inversion.

10.4 Texture is the main photo characteristic by which vegetation patterns are identified. Where indicator species are being observed, stereo image is useful in helping to identify by means of the crown size and shape; tone may also be significant due to the different light reflectance properties of different types of leaf (e.g. pine needles, hard glossy evergreen leaves, dull, soft deciduous leaves, palm fronds, etc.).

10.5 Vegetational boundaries are often tentative since the changing nature of the vegetation communities can reflect qualities other than soil. The commonest is, of course, human interference. Human interference can take the forms of: 1. Clearing (for cultivation). Identified by residual field patterns enclosing regrowth of different sizes. 2. Cutting over woodland for firewood and building supplies (note: this may be identified by the selective nature of the cutting on some occasions, which in turn may provide useful information on an indicator species' distribution). 3. Burning. This can usually be identified by very dark tones; a point or short front of origin upwind, from which the fire will spread, often in a roughly delta shape in a downwind direction, until intercepted and halted by a natural or man made break (cleared land, river, swamp, road). The dark tone of the burnt area may also have a series of very fine lighter toned lines across it from the subsequent passage of people and animals.

11.1 Land Use is the third of the important sources of indirect evidence. It reflects the following soil qualities: 1. Slope (may be modified by aspect) N-S-E-W in relation to direction and angle of sun. 2. Fertility; 3. Drainage and water supply; 4. Mechanical composition (case of tillage).

11.2 The characteristics of land use observed for pedological purposes are: 1. Density of cultivation (this may be quantified as so many acres/hectares per unit area and accordingly classified as 'sparse', 'moderate', 'dense'). 2. The characteristic size, shape and alignment of fields. It must be remembered however that local factors such as tribal practice can significantly change size and shape without any soil change.

11.3 A general principle applies, namely that the more primitive the agricultural system, generally the closer the adaptation to soil type. Capitalised agriculture, with fertilisers and soil ameliorants, highly mechanised and with techniques for watering and draining, can frequently choose sites for other reasons in addition to soil. It may lay out its field pattern to facilitate the use of machinery and in so doing ignore soil boundaries.

11.4 Crop patterns can generally only be identified if the scale of the photography is suitable, and may then only be useful if it is in some way an indicator (e.g. rice on floodable ground). In general the range of soil characteristics appropriate to a given crop is usually so wide as to vary primarily with topo-climatic zone, though with some, limiting factors such as pH, or preferred conditions such as heavy or light texture, acid or alkaline conditions, may give general indications (e.g. tea in Malawi on Ferrisols). However, their distribution is too patchy as a rule to allow any definitive soil boundaries to be drawn. To a lesser extent this last point also applies to the other observable qualities of land use.

12. Zootic factors are the final and least important source of indirect evidence. The only pattern commonly encountered is that resulting from termite activity. It is common to find termite patterns reinforcing evidence from other sources, but the size and density of the termite mounds may at times provide independent information about the nature of

341

the soil, provided the patterns can be related to known termite species in the area under inspection.

13. Types of photography. Each type of photography has its own particular significance for the soil interpreter. 1. True colour. This aids him by resolving the need to interpret shades of grey. It is usually at once apparent whether the observer is looking at the soil or at a fine cover of grass. Where the soil is visible, the colour differences enable boundaries to be accurately drawn. This is particularly useful in relatively smooth topography without marked breaks-of-slope, where the soils form a catenary association related to that topography. It may also assist the observer through permitting a clearer division of vegetation types based on foliage colour. 2. Infra-red. This may help through its value for vegetation interpretation. 3. False colour infra-red has the valuable attribute of sensitivity to degrees of wetness, hence assisting in defining drainage classes and their limits. It can also assist through allowing a better distinction of vegetation types.

14. Photo scales affect the degree of detail which can be observed.

14.1 Photography can also be divided as follows: A. 1. Vertical – by far the most common common: a narrow zone astride the path of the aircraft; 2. Oblique – useful in vegetation counts and assessment of landform shapes and slopes. Asymmetrically distributed astride the path of the aircraft; 3. Low angle oblique – analogous to photographs from mountain tops or tall buildings, i.e. panoramic. May have similar advantages to 2 but should only be used to complement normal vertical air photography.

B. 1. Satellite and other small scale: advantageous in delineating major regional patterns. In this respect takes the place of the mosaic in pre-field work by drawing attention to boundaries, often primary ones, which might not be noticed on single prints, because of being so wide and diffuse as to encompass a whole print or more. 2. Medium scales: have been discussed, as the majority of available photography is of this class. 3. Large scales (> 1/10,000) are associated with land planning, soil conservation, engineering (soil mechanics) and project execution generally.

15. There is a general principle that, the more detailed and large scale the survey, the relatively less important air photographs become as a tool in soil mapping and description.

16. The general principle for pre-field A.P.I. is to plot as many potential boundaries as possible, and to build up a table classifying the patterns the boundaries enclose. Provided the photographic difference of the pattern is proven, the boundary is a justifiable one, even should it subsequently be deleted as insignificant. It is a maxim that any justifiable photo boundary should in fact reflect some changes in soil characteristics, even should that change subsequently be identified as too insignificant to retain.

17. Simultaneously with the pre-field A.P.I. background information (e.g. geological maps and reports of the area) is sought and used 1. to assist in establishing A.P.I. criteria; 2. to assist in describing the photo patterns pedologically. Likewise analogous areas are sought which have already been examined and which can throw additional light on the area under observation as 1. and 2.

BIBLIOGRAPHY

LENEUF, B. & C. ROSSETTI, – 1967 – A propos d'un programme de stage sur l'application de méthodes photographiques aériennes à l'étude de certains facteurs de la mise en valeur agricole, pastorale et forestière. Actes du IIe Symposium International de Photo-Interpretation. VII 21–25. Editions Technip, Paris.

INDEX: SUBJECTS

Malaria   218, 219, 220, 227
Mammal(s)   56, 160, 168, 225
Man, as component   8, 9, 10, 13, 33, 36, 43–60, 61, 64, 66, 157, 162, 165, 224, 230
– as economic constituent   212, 214– 221, 224, 231, 266
– geography   6, 7, 165, 168, 173
– influence on resources   18, 29, 31, 33, 34, 52, 61, 63, 72, 153, 154, 162, 164, 170, 173, 175, 222, 235, 241, 260
Mangrove   140, 275
Manila hemp   77/78
Mansonia   159
Mathematics, livestock planning   199– 205, 228
Mediterraneity, of climate   105, 168
Melaleuca leucodendron   81, 82
Meteorology   31–35, 175, 176
Micrometeorology   112/113, 115–119, 172, 176
Micronutrient(s)   170
Migration, of peoples   74
Milk production, planning   199–205, 229, 232
Millet   180/181, 226, 255
Mineral resources   274, 277–279
Models   8, 12, 21–25, 26, 35, 40, 42, 46, 60, 62, 63, 64, 103, 131, 156, 158, 164, 174, 218, 232, 271, 317
Monsoon climate   8, 9, 34, 88–94, 95, 120, 163, 164, 170, 171, 233, 235
Mosquito ecology   226
Mountain(s) (see altitude, high)

Natural selection   19
Nematode(s), sampling   187, 189, 227
Niche   21, 36, 62, 160
Nitosols   82
Nomads   9, 54/55, 73, 247
Nutrition, animal   200, 210, 222, 246, 264
– human   9, 43/44, 46, 48, 49, 65, 175, 199, 212, 219–221, 222, 224, 225, 228
– plant   8

Oestrus   208
Oil palm   77/78, 153, 216
Ombrothermic diagram   90, 103/104, 124
Oncomelania   159
Onion   254
Orang utan   153

Orange   254
Orogeny   10
Oryza sativa (see also rice)   191
Ova storage   208

Pagan Gaddang   260
Paleoecology   104, 140
Palynology   140
Panicum   41
Pantidotale   155
P. maximum   157
Parameters, use in analysis   1, 14, 23, 26, 31, 32, 39, 92, 102, 107, 111, 122, 156, 173, 261/262, 265, 269, 318, 319
Pennisetum benthamii   19
P. purpureum   157
Pennisetum (hybrid Napier)   157
P. typhoides   246, 249
Phaseolus vulgaris   181
Physiology, photosynthesis   38, 61, 64, 102, 110, 111, 229
Phyto-ecology   167
Pig   199, 202–205, 207, 208, 209, 213, 225, 226
Pigeon pea   249
Pigweed   192–194
Pineapple   98, 216, 217, 226, 243
Pinus merkusii   81
P. roxburghii   151
Planosols   82
Plant sociology   161, 165, 167
Plésioclimax   141
Plusia   190
Pluviothermic quotient   168
Podzols   81
Polarization, livestock production   205, 228
Population, definition   19
Population density   9, 10, 212, 214, 215
Potato (Irish)   109, 196, 222, 255
– sweet   202–205, 255
Poultry   199, 203, 205, 206, 207, 223, 224–227, 229
Probability, climate   113/114, 119–123
Production in agriculture and ecology   26
– productivity   19–21, 26–28, 40, 62, 63, 64, 102, 140, 225
Prosopis cineraria   41
Protein(s), holo- and hetero-   19
– quality   181, 183/4, 222, 223, 224, 226, 227, 229

347

INDEX: GEOGRAPHICAL NAMES